CALIFORNIA NATURAL HISTORY GUIDES

MAMMALS OF CALIFORNIA

California Natural History Guides

Phyllis M. Faber and Bruce M. Pavlik, General Editors

MAMMALS
of California

REVISED EDITION

E. W. Jameson, Jr.
and Hans J. Peeters

Illustrated by Hans J. Peeters
Skull Drawings by E. W. Jameson, Jr.

UNIVERSITY OF CALIFORNIA PRESS

To the memory of A. Starker Leopold, 1913–1983

California Natural History Guides No. 66

University of California Press
Berkeley and Los Angeles, California

© 2004 by the Regents of the University of California

Library of Congress Cataloging-in-Publication Data

Jameson, E. W. (Everett Williams), 1921–.
   Mammals of California : E. W. Jameson, Jr. and Hans J. Peeters; illustrated by
Hans J. Peeters ; skulls drawings by E. W. Jameson, Jr.—Rev. ed.
      p.   cm. — (California natural history guides ; 66)
   Rev. ed. of: California mammals. c1988.
   Includes index.
   ISBN 0–520–23581–9(case).—ISBN 0–520–23582-7 (pbk.)
   1. Mammals—California.   I. Peeters, Hans J.   II. Jameson, E. W. (Everett
Williams), 1921– California Mammals.   III. Title.   IV. Series.

QL719.C2J35   2004

599′.09794—dc22—dc21   2003061288

27   26   25   24   23   22
10   9   8   7   6   5   4

The paper used in this publication meets the minimum requirements of
ANSI/NISO Z39.48–1992 (R 1997) (*Permanence of Paper*). ♾

Cover: Golden-mantled Ground Squirrel *(Spermophilus literalis),* by Hans J.
Peeters.

The publisher gratefully acknowledges the generous
contributions to this book provided by

the Gordon and Betty Moore Fund
in Environmental Studies
and
the General Endowment Fund of the
University of California Press Associates

CONTENTS

Plate section follows page 202

ACKNOWLEDGMENTS

It is a pleasure to indicate the many friends who have provided invaluable help. They have assisted us in numerous ways, and to them we tender our most profound thanks. These include S. Bunnell, H.L. Cogswell, R. Cole, J. DiDonato, A. Engilis Jr., G. Gould, D.F. Hoffmeister, W.J. Houck, G.S. Jeffers, D.A. Kelt, T. Koopmann, H. Leach, W.Z. Lidicker Jr., R.D. Mallette, F. Marino, L.G. Marshall, R. Mohr, B. Nelson, R.L. Rudd, B. Schulenberg, V. Simpson, C. Stowers, P.Q. Tomich, and D. Updike. J.A. Junge and R.S. Hoffman have generously allowed us to reproduce their drawings of shrew skulls. In addition, we should like to express our appreciation for the editorial assistance of the late O.P. Pearson, the late A.C. Smith, and the late A.S. Leopold. All of these friends have given generously of their time and have made unique contributions to this book.

The second edition was reviewed by Professor James Patton, to whom we are deeply indebted. His meticulous reading and many comments on our first draft have been a tremendous help in the final preparation of this edition. Despite Dr. Patton's assistance, any errors belong to the authors. We owe a deep debt of thanks to Professor R.S. Lane and Professor Jerold Theis for their very valuable help with the section on wild mammals as carriers of pathogens.

We are also grateful to our editors, Laura Cerruti and Scott Norton, for their dedication to the second edition. Their enthusiasm and sound editorial advice were a constant assist throughout the preparation of this edition, which also greatly benefited from Lynn Stewart's expertise.

Map 1. California Counties

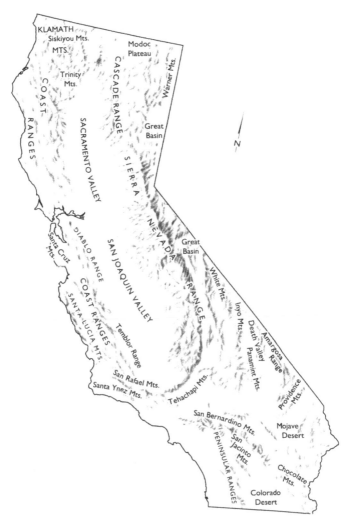

Map 2. California Topography

MAMMAL ECOLOGY

INTRODUCTION

The Pacific Coast has a diverse mammal fauna reflecting its wide variety of climates and habitats, as well as its vast area and elevational range. Much of the climatic diversity is the result of mountain building over the last 3 to 4 million years and concurrent drastic fluctuations in weather. The high elevations of the Klamath Mountains, the Cascade Range, and the Sierra Nevada trap sufficient rain to support extensive coniferous forests on the western slopes and create rain-shadow deserts to the east. Along the coast, cool winds from the north account for persistent remnants of redwood forests and associated flora and fauna.

Differences in elevation alone provide for a rich fauna, as illustrated by the occurrence, only a few kilometers apart, of lowland forms such as skunks *(Mephitis* and *Spilogale)* and alpine mammals such as the Pika *(Ochotona princeps)* and Wolverine *(Gulo gulo)*. The many rivers are the habitat of Beaver *(Castor canadensis)*, River Otter *(Lutra canadensis)*, and Mink *(Mustela vison)*. Formerly, California's Central Valley was populated by herds of Pronghorn *(Antilocapra americana)* and Tule Elk *(Cervus elaphus nannodes)*, in addition to an abundance of Grizzly Bears *(Ursus arctos)*. The populations of these large mammals are now greatly reduced or nonexistent in the coastal states.

This book gives geographic distributions of species in general terms; topographic features (shown in Map 2 at the beginning of the book) aid in associating range with habitat. The range given for each species of land mammal is an approximation based on known occurrences. Although we intend the range maps to be accurate, you may fail to find some species where they are reported to occur and encounter others where they have not previously been known. Geographic distributions are in a constant state of flux; some changes are temporary, others long lasting.

STUDYING MAMMALS

Some Hints

Some large mammalian groups are bewildering in their complexity. Chipmunks (*Neotamias* spp.), pocket mice (*Chaetodipus* and *Perognathus* spp.), kangaroo rats (*Dipodomys* spp.), pocket gophers (*Thomomys* spp.), and others present major difficulties in identification. We hope that this volume will assist in the recognition of all Pacific Coast mammals. The keys, illustrations, and descriptions, when used together, will distinguish all but a few of the most similar species.

The world of mammals is essentially one of darkness and smells. Because most of the abundant species are entirely or mostly nocturnal, they are unfamiliar to all but the few people who seek them out. Many species move from daytime resting places only after dark and return before dawn. They communicate largely through odors, most of which go undetected by the human mammal, and sounds, which are frequently high pitched and of low volume. We are, therefore, usually aware only of those few species that regularly move abroad in daylight: chipmunks, ground squirrels, tree squirrels, and a few others. Occasionally we glimpse a fox or a Virginia Opossum *(Didelphis virginiana)* as it is briefly illuminated by automobile headlights, but few people have observed kangaroo rats, flying squirrels (*Glaucomys* spp.), or the Water Shrew *(Sorex palustris).*

To learn about wild mammals you must learn to recognize indications of their presence and, when possible, see them move about naturally. Generally, you cannot watch wild mammals as readily as wild birds. Direct observation in daylight is feasible for only a few species, such as chipmunks, ground squirrels, and tree squirrels. Armed with binoculars and a wealth of patience, however, you can begin to understand the behavior of ground squirrels and other diurnal mammals.

To observe nocturnal species, you may need special equipment (see the list of suppliers at the end of this section). For example, you can obtain infrared light sources and binoculars that convert infrared to visible wavelengths. Because the mammalian eye does not detect infrared light, wild mammals are unaware that they have become illuminated. With such equipment you can observe a fox, Raccoon *(Procyon lotor)*, or Northern Flying Squirrel *(Glaucomys sabrinus)* under natural conditions. A spotlight covered with translucent red plastic is sometimes almost as satisfactory as an infrared light. Also, many species are relatively undisturbed by illumination from a handheld flashlight, provided that the beam is not directed at them. Some observers have found that creatures such as kangaroo rats and pocket mice pursue their activities without inhibition in dim artificial light.

Small mammals that remain in the open can be observed at night with the aid of luminescent capsules. Translucent capsules containing Cyalume (which produces a cold light) can be temporarily attached to fur with rubber cement. They produce light for up to six hours and fall off after a day or less. Automatic cameras tripped by touching a wire or breaking a beam of light are also available; these can be set along trails, by waterholes, or at bait stations. Camera traps are typically set for midsized to large mammals. The U.S. Department of Agriculture, Pacific Southwest Research Station, provides an excellent manual (General Technical Report PSW-GTR-157) that details the use of camera stations, track plates, snow tracks, and so forth for carnivorans; survey sheets are provided with this publication to encourage the public to provide needed information.

In contrast to bird songs, which have been studied in detail, the vocalizations of mammals have not been extensively explored. Most mammals utter brief sounds, usually "call notes," that may be superficially similar among several species. Careful studies have, however, demonstrated a variety of utterances produced under different circumstances (see the section "Senses: Vocalization"). Vocalization is an area in which talented amateurs may make original contributions to the understanding of mammalian behavior.

Sophisticated hunters have long known that battery-powered hearing aids make audible many low-volume sounds made by both birds and mammals. A hearing aid is of greatest benefit when there are no competing sounds from freeways and urban areas. Hearing aids are very useful at night, especially in open areas with little vegetation to disrupt sound transmission. Bat detectors are also available.

Portable tape recorders are now of excellent quality and operate over a broad range of wavelengths. Recorded vocalizations can be converted to visual signals on an oscilloscope. Their details can then be compared, and specific sounds can be correlated with specific behaviors of both free-ranging and captive mammals.

Small mammals can be captured in a variety of homemade or commercially available traps. To make a simple live trap, wire an ordinary mouse trap to a small tin can, and tie a piece of hardware cloth to the loop of the trap so that the can is closed off by the wire mesh when the trap is sprung (fig. 1). More refined live traps in many sizes can capture mammals ranging in size from a shrew to a fox.

Steel jaw traps are now illegal in California except when used for conservation. A jaw trap that is properly modified by cushioning the jaws with burlap wrapped with tape and by hammering the spring down slightly to weaken it will not injure the animal. Steel traps, when so modified, are a valuable conservation tool. They enable the specific removal of predators—the Red Fox *(Vulpes vulpes)*, for example, where it is a threat to the endangered Clapper Rail *(Rallus longirostris)*.

Many students of small mammals use ordinary snap traps. They are inexpensive and readily available but kill the animals they catch. Nevertheless, they may be preferable to live traps if the goal is to study internal anatomy. Snap traps may also be useful for species that frequently enter empty cabins and can cause extensive damage to blankets, pillows, and food. Although deer mice (*Peromyscus* spp.) and the House Mouse *(Mus musculus)* are the most common intruders, wood rats (*Neotoma* spp.) and the Northern Flying Squirrel *(Glaucomys sabrinus)* frequently take up residence in attics. When owners are absent, snap traps are more humane, because a live trap would hold the occupant until it starved.

To create another type of simple, inexpensive, and usually effective live trap, bury a large tin can (half gallon or larger) in the ground, with the top exactly flush with the surface. Mice and shrews often stumble into such a pitfall trap; it does not harm them, and the smooth sides prevent escape. Escape is even more difficult if a lip is placed around the inside of the top, but a lip reduces the size of the hole through which the animal can fall. You

Figure 1.
Homemade trap.

can bait this trap with rolled oats, walnut meats, sunflower seeds, or some other attractant, but it may be effective even without bait. Cans sunk in the deep sand of a desert canyon may capture a great variety of small mammals. Be sure to remove the trap cans before leaving the area.

Any live trap should be placed to maximize the likelihood of capture and to minimize the danger to whatever might enter it. Choose a shady spot protected from rain and sun, and never leave a trap unattended for more than four or five hours. Live traps set in late afternoon should be examined several hours after dark and again at daybreak. Even diurnal species must not be exposed to direct sunlight.

A small animal captured in a live trap may be examined while restrained in a glass gallon jar and then released at the capture site. The release should be prompt, and handling of the animal should be minimal. Their bones are fragile and easily broken, and some animals can inflict painful bites. (See also the section "Mammals and California Society: Wild Mammals as Carriers of Pathogens").

Why capture live mammals? The study of small species begins with their identification, which usually depends on a critical examination of structural details. With a knowledge of some major external features and the locality of capture, you may not need details of dentition and cranial aspects. Individual mammals captured can also be marked for later recognition. A deer mouse, for example, can be restrained in a thick sock while a color (such as food coloring) is applied to the head or back. This mark remains until the next molt and enables you to recognize that individual if seen again. You can mark a local population in a variety of color combinations so that many individuals can be recognized. The survival of these unnaturally marked rodents indicates that the coloring does not make them especially vulnerable to predators.

Regulations may limit or prohibit the capture of some or all wild mammals. Prior to trapping, you should become familiar with local restrictions. Some mammals are protected because they are scarce (see table 2 in the section "Mammals and California Society: Conservation of Mammals"); others are potential reservoirs of disease. Trapping is usually prohibited in county, state, and national parks.

Although wild mice and other small mammals are naturally secretive, it is possible to create situations in which you can observe their behavior at close range without trapping. Meadow voles, which make discrete trails through grass, and deer mice often make their nests under boards lying on the ground, and a quick hand may capture a startled individual when the nest roof is suddenly removed. Be cautious when grabbing dazed creatures, however, for they can bite. You can place boards on the ground, and eventually mice will nest beneath them. A small "nest box" sunk in the ground may also eventually be occupied by wild mice. Provided with a loose lid, the box allows you to observe the presence of young in the spring and food stores in the autumn. Wild mice may tolerate minor disturbances, especially if your movements are slow and quiet.

At a campsite, where a resident population of scavenging rodents is likely, a spotlight covered with red paint or translucent red plastic will not disturb them. Mice may be already conditioned to search for scraps of leftover food, and a little encouragement will bring them to a dimly lit site. (Campsites should not be baited with food where bears range.) By marking individuals, you may learn about individual behavior traits.

Small mammal tracks can also be observed at such feeding stations. Sweep the ground clean of twigs and small stones to prepare a clean slate for the foot imprints of mice and other small mammals (fig. 2). A thin layer of wheat flour, lightly dusted over the smoothed dirt, will leave the tracks distinctly outlined. If they are sufficiently clear, you can make plaster casts of them by pouring fluid plaster of Paris on the imprints. With a collection of casts, you can learn both the consistent and the variable features of tracks.

You can learn to recognize some of the more distinctive mam-

Figure 2. Mammal tracks. (a) Raccoon *(Procyon lotor);* (b) Virginia Opossum *(Didelphis virginiana);* (c) Muskrat *(Ondatra zibethicus)* (note tail mark); (d) Black Bear *(Ursus americanus);* (e) Striped Skunk *(Mephitis mephitis);* (f) Badger *(Taxidea taxus);* (g) Elk *(Cervus elaphus);* (h) Mule Deer *(Odocoileus hemionus);* (i) Mountain Lion *(Panthera concolor);* (j) Bobcat *(Felis rufus);* (k) Beechey Ground Squirrel *(Spermophilus beecheyi);* (l) Long-tailed Weasel *(Mustela frenata);* (m) Gray Fox *(Urocyon cinereoargenteus);* (n) Coyote *(Canis latrans)* (note large outer toes of hind foot). R.F., right forefoot; R.H., right hind foot; F., forefoot; H., hind foot. Each scale bar equals 2.54 cm (1 in.).

mal tracks by observing animals as they move about. The origin of the tracks is then certain, and moreover, you can associate the spacing of the tracks with the gait of the individual that made them. Also, with some knowledge of tracks, you may frequently get a clue as to which species left feces, or scat. Scat is far more difficult to identify than tracks, for the scat varies in size and shape according to the diet and age of the mammal that made it (fig. 3).

A feeding station can also be used for larger species. Remember, however, that the same fare that draws a Raccoon or fox will also attract a skunk. Both the Spotted Skunk *(Spilogale putorius)* and the Striped Skunk *(Mephitis mephitis)* are fearless, and their extremely offensive scent, which can be squirted 3 to 4 m (10 to 13 ft), is very difficult to remove from fabric. Rats are also drawn to feeding stations, especially in suburban areas, and are generally undesirable for many reasons.

Bats cannot be enticed to a feeding station, but they occasionally do congregate and can then be easily observed. In arid regions, bats gather at water troughs, where they drop to the surface of the water to drink. Moderate light usually does not frighten them away and helps you identify them. Bats may also be found roosting in small or large groups in caves or mine shafts. Take pains to avoid disturbing resting bats, for movement of one may cause others to take flight. Sometimes you may find the nighttime roost of one or several bats, perhaps under the eaves of a porch roof; by examining the insect fragments under the roost, you can learn the diet of the bats. In relatively open habitats, you can follow bats at night by picking them off the roost and gluing Cyalume capsules to their fur.

The more sophisticated student of mammals may wish to follow individuals by radiotelemetry. Students have attached small radio transmitters to mammals ranging from rats and bats to deer and bears. These transmitters emit distinctive signals that can be picked up at least several hundred meters (a few thousand feet) away and sometimes more than several kilometers (a few miles). A small citizen band receiver with a directional antenna can identify signals from individual mammals. Using such a device, you can eventually determine a mammal's home range and home site, as well as where and when it forages.

Other observation techniques may apply to particular mammal species. Muskrats *(Ondatra zibethicus),* for example, can be watched from a car parked next to a slough or irrigation canal.

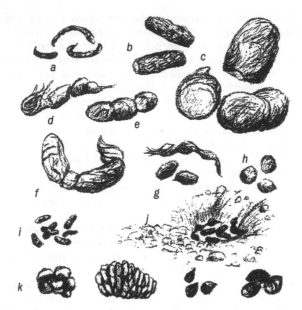

Figure 3. Mammal scats. Species of cats frequently deposit their rather uniformly segmented feces in so-called scrapes, often near rocks; the irregularly segmented scats of foxes and the Coyote are often found along paths and roads. Note the terminal curl in the scats of these canids. (a) Long-tailed Weasel *(Mustela frenata)*; (b) Raccoon *(Procyon lotor)*; (c) Mountain Lion *(Panthera concolor)*; (d) Gray Fox *(Urocyon cinereoargenteus)*; (e) Bobcat *(Felis rufus)*; (f) Coyote *(Canis latrans)*; (g) Yellow-bellied Marmot *(Marmota flaviventris)*; (h) Black-tailed Jackrabbit *(Lepus californicus)*; (i) Desert Wood Rat *(Neotoma lepida)*; (j) Beechey Ground Squirrel *(Spermophilus beecheyi)* scat in latrine; (k) four forms from the Mule Deer *(Odocoileus hemionus).* Not to scale.

They are active in early morning and late afternoon and can be attracted to a small raft about 1 m (3 ft) square, anchored in the open. Muskrats voluntarily take their food to such floating objects and sometimes cover small rafts with the shells of freshwater clams on which they have fed. The rafts can also be baited with carrots and other succulent vegetables.

Some of the shiest mammals can be observed while young, when they spend much time in play and before fear has developed. When young foxes first venture from the burrow, they readily play in the open. You can enjoy watching a litter of kits as they chase one another, catch grasshoppers, wait for their parents to bring food, or just sleep. You should watch from a respectful distance with the aid of binoculars or a spotting telescope to avoid alarming the parents.

Many mammals, especially carnivorans, can be brought into view with a device known as a predator call. These "whistles" produce the sound of a stricken rabbit and have a magnetic effect on foxes, the Coyote *(Canis latrans)*, the Bobcat *(Felis rufus)*, and raptorial birds. Even bears may respond. Predator calls are most effective when the caller is concealed in a blind.

The clever field naturalist will think of many more ways to learn about the habits and behavior of wild mammals. The basic ingredients are patience and the ability to remain quiet for long periods. Always avoid disturbing wild mammals. An alarmed animal may become careless and expose itself, and possibly the observer, to danger.

Various methods may be used to study dead animals. The species accounts in this book include comments on the structure and shape of the skull and the nature of the teeth. In many kinds of mammals, the teeth are quite distinctive. They are also extremely hard and durable and, therefore, endure as fossils when many other bones have been crushed or broken. Logically, then, dentition—the number and form of the teeth—deserves special attention. The species accounts give the number of teeth in each of four positions in both the upper and the lower jaw. For example, 2/1, 1/0, 1/0, 3/3 means that each side of the skull has two upper and one lower incisor, one upper and no lower canines, one upper and no lower premolars, and three upper and three lower molars. If you find an unfamiliar dead specimen, you can save the skull for later identification. When it has dried, small beetles of the dermestid family (Dermestidae) will do a fine job of removing the meat, and the critical features of the skull and teeth will become apparent (fig. 4). These beetles are naturally attracted to a skull that is hung outdoors from a tree or clothesline, out of reach of dogs, cats, and small children. The heads of larger mammals, after the removal of most of the flesh, can be boiled in a weak bleach solution for a short period for final removal of the flesh from the bone.

The species accounts also include measurements, in millime-

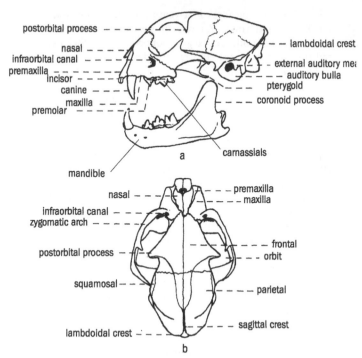

Figure 4. Skull of Bobcat *(Felis rufus)* showing important diagnostic features.

ters, of the main parts of the body, usually the total length (TL), the tail (T), and the hind foot (HF); sometimes the ear (E); and for bats, the forearm (F). The ear is commonly measured from the lower and outer notch to the tip; the hind foot is measured on the bottom surface; the tail is measured from its origin at the back to the tip of the last tail vertebra (fig. 5). The weight is given in grams or kilograms. If you find a freshly killed animal, these measurements will greatly facilitate identification. Be careful to avoid contact with fresh blood and external parasites.

You may wish to preserve freshly killed specimens that are in good condition. Many texts give instructions for preparing specimens for subsequent identification and study. One of the best is *Methods of Collecting and Preserving Vertebrate Animals* (Ander-

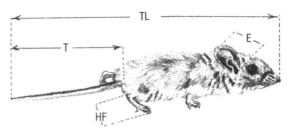

Figure 5. Measurement of mammals: TL, total length; T, tail (to tip of last vertebra); HF, hind foot (heel to end of longest claw); E, ear (notch to tip).

son 1948). Another is *A Manual of Mammalogy with Keys to Families of the World* (De Blasé and Martin 1981).

Museum collections of mammals provide data on past population distributions, current geographic ranges and their changes, and details of species morphology. They also provide material for chemical analysis. These collections are essential to the serious study of mammals and greatly aid our understanding of the evolutionary process. In addition, they form a basis for teaching students about mammalian forms, skeletal anatomy, and degrees of variation. Baseline data are critical in view of our inadequate knowledge of the majority of species. Although some readers may lament the presence of wild animals in a museum collection, these specimens provide a permanent database for future reference.

Equipment Suppliers

Advanced Telemetry Systems, 470 First Avenue N, Isanti, MN 55040 (monitoring equipment)

American Cyanamid Company, Bound Brook, NJ 08805 (Cyalume)

Bat Conservation International, P.O. Box 162603, Austin, TX 78716; www.batcon.org (bat detectors, bat houses, and bat house kits and building plans)

Communications Specialists, Inc., 426 West Taft Avenue, Orange, CA 92865 (monitoring equipment)

Custom Electronics, 2009 Silver Court West, Urbana, IL 61801 (small radios and receivers; there are numerous other suppliers)

H.B. Sherman Traps, 3731 Peddle Drive, Tallahassee, FL 32303

(good-quality traps for small mammals; larger live traps are often advertised in sporting magazines)
Trail Master, Goodson, and Associates, Inc., 10614 Widmer, Lenexa, KS 66215 (infrared equipment)

Features of Mammals

Everyone recognizes a mammal. The salient features are quite obvious. The nearly always live-born infants are nourished by milk from their mother's mammary glands, and all mammals possess hair, at least at some time in their life. However, many less obvious but not less important characteristics separate mammals from other vertebrates.

The following brief discussion of mammalian biology establishes a basis for interpreting the behavior of wild species. We apologize for the occasional but necessary use of unfamiliar or technical terms. Most are defined at their first occurrence in the text, and all are included in the glossary.

In mammals, the lower jaw, or mandible, consists of a pair of bones, the dentary bones, which are usually fused in front; the mandible articulates with the skull, mostly with the cranial squamosal bone. The skull articulates with the atlas, the first bone in the spinal column. Almost all mammals have seven cervical vertebrae; Neotropical anteaters and manatees have fewer, and sloths have from five to nine.

The mammalian heart is four chambered. A pulmonary circulation takes blood to and from the lungs, and a systemic circulation takes blood to and from the rest of the body. The separation of the blood into two circulatory systems is a very efficient arrangement that allows a high rate of metabolism. This, in turn, permits an excess of metabolic heat and accounts for mammals' characteristic warm-bloodedness and relatively constant body temperature, or homeothermy. The lungs lie within a thoracic cavity, protected by ribs. The bottom of this cavity is closed by a circular muscle, or diaphragm, the contraction of which enlarges the lung cavity, expanding the lungs and drawing in air through the trachea and bronchi.

The mammalian middle ear consists of three ossicles, or ear bones; two of these are modifications of bones from the lower jaw of reptiles. These bones are frequently enclosed by an auditory bulla, which is a rounded, bony prominence on the ventral part of the skull of most mammals. The roof of the mouth consists of a hard palate, which separates the air passing through the nostrils from the air and fluids in the mouth. This arrangement allows an infant mammal to breathe as it takes milk from its mother.

Mammalian teeth are usually heterodont; that is, they vary in shape, including incisors, canines, premolars, and molars. This feature is not unique to mammals; some synapsids (early reptile-like animals) were heterodont. Moreover, some mammals, such as dolphins and porpoises, have teeth that are homodont—all or mostly similar. Mammalian teeth are usually diphyodont, which means they occur in two generations. The front teeth (incisors, canines, and premolars) appear early in fetal development (except in marsupials) and are eventually replaced by permanent teeth.

In many mammals (including some bats, rodents, carnivorans, primates, and insectivores) a penis bone (the os penis, or baculum) lies within the penis, giving it some rigidity. The shape of the penis, or phallus, itself is characteristic for each species. In some species, females have an os clitoris. Presumably, these structures provide a limited physical isolating mechanism, but hybrids between different species are not unusual.

Although the abundance of new words may seem bewildering at first, the association of names and structures will soon become familiar and simplify the use of the keys. Many of these structures are typical of only certain groups of mammals and provide the basis for recognizing families and orders. You will gradually recognize the types of differences that characterize the many families and genera within our fauna, and approximate identification will become second nature.

ORIGINS OF MAMMALS

The transition from reptilelike animals to mammals, which began in the Permian (late Paleozoic), some 220 to 225 million years ago (table 1), was gradual and involved a number of physical and physiological modifications, including shifts in feeding, locomotion, metabolism, and sensory perception. For example, placement of the limbs more directly below the center of the body enabled more rapid movement.

The quadruped creatures from which mammals derived are called synapsids. Synapsids are sometimes referred to as mammal-like reptiles, which is misleading, for they scarcely resembled the modern reptiles we know today. When synapsids evolved, the line between them and reptiles had already become distinct. By the Middle Jurassic, perhaps 140 to 150 million years ago, some of these creatures had developed mammalian features. These primitive mammals lived in most areas of the world. Pinpointing the time at which early synapsids could be called mammals is difficult, but a beast called *Phascolotherium bucklandi,* of the Middle Jurassic in England, may be the earliest mammal. (The genus *Morganucodon,* sometimes said to be the first mammal, is now considered to be nonmammalian.) The earliest eutherian (placental) mammals diverged from the marsupial line in the Early Cretaceous. A fossil found in northeastern China and showing many structural details represents a small arboreal mammal of 125 million years ago.

Synapsids had a complex lower jaw composed of four loosely joined bones, one of which, the dentary bone, had teeth; the others likely functioned in hearing, probably of low-frequency airborne sounds, and also formed the connection between the lower jaw and the skull. With the gradual appearance of mammalian traits, major changes occurred in the lower jaw and its attachment to the skull. The hearing bones, behind the dentary, became part of the skull; two of them formed two of the ossicles in the middle ear, and one, the ectotympanic bone, became the outer part of the auditory bulla. The dentary bone formed a connection with the squamosal bone in the skull; that connection is a key mammalian

TABLE 1. Geologic Timetable, Indicating Sequence of Major Geologic and Evolutionary Events

Era	Period	Time of Beginning (mya*)	Epoch	Characteristics
AGE OF MAMMALS — Cenozoic	Quaternary	01	Holocene	Little Ice Age Medieval Optimum
		2	Pleistocene	Major Glaciation
	Tertiary	3	Pliocene	Formation of the Isthmus of Panama Large Carnivores
		5	Miocene	Continental cooling, increasing aridity, grassland formation, many grazing mammals
		23	Oligocene	Mountain building Advanced primates
		34	Eocene	Appearance of modern mammals
		55	Paleocene	Appearance of families of existing mammals
AGE OF REPTILES — Mesozoic	Cretaceous	65		Dinosaur demise; earliest monotremes, marsupials, and placental mammals; dinosaur diversity; flowering plants
	Jurassic	136		First Synapsida; early mammals; first birds
	Triassic	190		seed-bearing angiosperms
	Permian	225		Initial separation of continental plates First reptiles.
Paleozoic	Pennsylvanian	280		
	Mississippian	320		
	Devonian	345		First amphibians
	Silurian	395		First land plants; jawed fishes
	Ordovician	430		Early fishes
	Cambrian	500		Many invertebrates

*mya = million years ago

feature. The existence of a hard palate, which allows the simultaneous intake of air through the nostrils and food through the mouth, suggests that the young took milk from a nipple and also that there is sufficient O_2 delivery while eating to support homeothermy; the hard palate may also have served to buttress

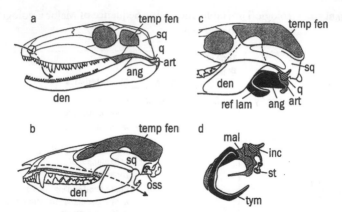

Figure 6. (a) Skull of an Early Permian synapsid (based on *Varanosaurus*). The dotted and solid lines entering the nostril show the passage of air into the mouth cavity. Indicated also are the temporal fenestra (temp fen), and, behind it, the position of the squamosal (sq) and quadrate (q) bones. The lower jaw consists of the large dentary bone (den), which bears teeth, and behind it the angular (ang) and the articular (art) bones, which are believed to have received low-frequency vibrations that were transferred to the quadrate in the skull. (b) Skull of a modern marsupial *(Didelphis marsupialis).* The dotted line indicates the passage of air through the nostril, over the hard palate, and into the tracheal passage; oss, middle ear ossicles. (c) Homologous elements in the jaw of an Early Triassic synapsid *(Thrinaxodon);* ref lam, relected lamina. (d) Middle ear of a modern mammal; inc, incus (derived from the quadrate); st, stapes; mal, malleus; tym, tympanic. From Hopson (1994), courtesy of the Paleontological Society.

the upper jaw. As synapsids became mammal-like, they lost their parietal foramen—a hole in the top of the skull—which in many reptiles today allows light to reach the parietal eye (fig. 6).

Synapsids developed various feeding habits; some were insectivores, some carnivores, and some herbivores. These feeding patterns were reflected in their teeth, some of which were clearly differentiated into incisors, canines, premolars, and molars. Some early forms were diphyodont. Synapsid teeth were covered by prismatic enamel, and some had branching roots; both features are characteristically mammalian. In some of these very early mammals, the lower jawbone had holes, indicating that nerves and blood vessels passed through them. This suggests moveable lips, which would have assisted in nursing and perhaps in keeping food in the mouth while chewing. But no clear, uniform series of physical changes occurred. In some early mammals, the postdentary bones still joined the skull at the quadrate bone. In *Sinocodon*, a Jurassic genus, the dentary bone joined the skull at the squamosal bone, as in modern mammals, but the loose bones behind the dentary still remained, as they did in early synapsids.

By the Late Cretaceous, some 65 million years ago, just prior to the decline and disappearance of the dinosaurs, mammalian differentiation was well under way. Marsupials, monotremes (egg-laying mammals), and eutherians such as ungulates and carnivores had evolved. Most Cretaceous mammals were mouse sized or rabbit sized; large forms did not evolve until after the passing of the dinosaurs. Cretaceous mammals had long since become warm-blooded, or endothermic, which allowed them to move about at night, without the need to obtain heat from solar radiation. Even today, most mammals are either nocturnal (active at night) or crepuscular (active at dawn or dusk, or both).

The Age of Mammals

Major physical changes in the continental outlines and the positions of the continental plates altered ocean currents and climates and allowed mammals to move into new areas. Movements of continental plates probably started with the beginning of the Earth, and they continue today. The Mesozoic changes were especially important, for they occurred when mammals first differentiated from synapsids.

The Cenozoic (approximately equal to the Tertiary), sometimes called the Age of Mammals, was also a time of major climatic changes, mountain building, and extensive differentiation of mammals. It is divided into five major epochs: the Paleocene, the Eocene, the Oligocene, the Miocene, and the Pliocene. Until the Eocene, Antarctica was joined to both South America and Australia, and northwestern Europe was joined to North America (Greenland). At this time the Earth was much warmer than it is today, and many kinds of mammals lived in the polar regions.

The Paleocene (65 to 55 million years ago) was a time of cool climates with little seasonality as we know it. Most of the modern orders of mammals appeared in the fossil record during this time but may well have differentiated in the Late Cretaceous. Seed-bearing plants underwent great development, continuing the differentiation that had begun in the Late Jurassic, and the abundance of seeds must have made possible the development of many seed-eating small mammals. Small rodents and rabbits appeared in North America, probably from Asia, and marsupials reached South America, Antarctica, Australia, and Madagascar. In the Eocene (55 to 34 million years ago), the climate became warm and humid, with a wet-dry seasonality. In the early Eocene, eastern North America and western Europe were still connected and had many mammals in common.

Marked global cooling and a wet-dry seasonality characterized the Oligocene (34 to 23 million years ago). At the beginning of this epoch, redwoods grew eastward to Colorado, and palms flourished in northern California. The cooling and a decline in rainfall accounted for an increase in grasslands, which led to a diversification of grazing mammals. Camels and the early relatives of horses in North America radiated during the Miocene (23 to 5 million years ago). At that time a further decline in precipitation and global temperatures apparently led to a further increase in grasses and, with them, grazing mammals, those species that feed on fresh grasses and forbs at their feet. A corresponding decline occurred in browsing mammals, those species that feed on leaves and tender twigs of shrubs and trees. (The distinction between these two types of herbivores is not ironclad; some species, such as the Mule Deer, both browse and graze.) The presence of herbivores, including antelope, made possible the development of various kinds of meat-eating mammals at about the time the Sierras and the Rocky Mountains were being built. Meanwhile, a Bering connection joining Alaska to Siberia served as a land bridge over which elephants and other animals entered North America.

In the Pliocene (5 to 2 million years ago), mountain building continued and may have accounted for increasing aridity and cooling in North America. Camels and horses continued to be abundant and diverse. The Pliocene ended with drastic cooling, culminating in the formation during the Pleistocene of glacial ice over much of North America and Europe, but not northeastern Asia.

The Pleistocene

The Pleistocene, or Ice Age, a period of nearly 2 million years, was a time of dramatically alternating climatic changes that greatly increased the diversity of our mammalian fauna. Some 20 glacial periods, 10 of which were severe, blanketed approximately 30 percent of the land surface with ice that was, in some regions, thousands of feet thick. Today, only 10 percent of the land mass is ice covered. During times of glaciation, the Great Basin was a humid region of many freshwater lakes, and many mountain slopes were deeply buried in ice. In the last major glaciation, snow lines in the Sierra Nevada and the Cascade Range were 1,000 m (3,300 ft) below current levels. In the Clear Lake region of California, pollen cores indicate the maximal lowering of temperature to have been 7 to 8 degrees C, and the immediate post-glacial temperatures to have averaged about 1.5 degrees C warmer than today's average. Such drastic changes in terrestrial temperatures have not occurred in the Holocene, the period following the Pleistocene and continuing through the present.

Ocean currents along our coast during glacial epochs were warmer than they are today. Extensive ice formation lowered sea levels by 100 m (330 ft) or more, so Alaska was joined to eastern Siberia, blocking the cold current from the Arctic Ocean that today chills the California coast.

In the cold, wet glacial periods, some northern mammals moved southward into our area; in the interglacial periods, however, when the climate ameliorated and ice melted, faunas moved northward and also into the mountains, which, in effect, became ecological islands where the isolated animal populations became distinct. During these glacial-interglacial fluctuations, the members of many species were alternately separated from one another and reassembled as they moved along north-south "pathways" parallel to the major mountain ranges. These environmental changes, with their successive isolation and mixture of populations, undoubtedly favored rapid genetic changes and account for the rich and varied mammalian fauna we have today. By comparing morphological and molecular features (DNA and other molecular markers) of different populations, it becomes possible to compare relationships among species. The phylogeography of western chipmunks (*Neotamias* spp.), for example, presents a complex puzzle for an enterprising student.

The Pleistocene mammals in California included many large forms, most of which became extinct at the end of the Ice Age. In western North America, Pleistocene camels were twice the size of modern camels. Large herbivores meant large predators: the Jaguar *(Panthera onca)* ranged not only throughout California but northward to Wyoming; the Dire Wolf *(Canis dirus)*, much larger than the modern wolf, ranged throughout North America; the Lion *(Panthera leo)*—now confined to the Old World—occurred northward to Alaska; and the sabertooth *(Smilodon)* was a common predator in California. With the disappearance of the extensive ice sheets, virtually all of this "megafauna" disappeared, as did some small mammals. The Woolly Mammoth *(Mammuthus primigenius)* disappeared in North America after the last glacial maximum, although it persisted in Siberia for some 4,000 years after the close of the last major glaciation.

Aborigines have been blamed for these extensive extinctions of mammals, because they occurred at nearly the same time that humans invaded North America. This widely accepted concept of "Pleistocene overkill" was developed by Professor Paul Martin of the University of Arizona. The Ground Sloth *(Nothrotheriops shastense)*, a slow-moving creature common in California in the Pleistocene, may well have been killed off by aborigines. Mass extinctions of both large and small mammals may also have been due to extensive climatic changes, however. Many large mammals of the Pleistocene (such as mammoths and mastodons) had long gestation periods and may have been unable to survive the harsh seasonality that we experience today. It has been postulated that the Pleistocene was relatively equable, with only moderate seasonal shifts.

Postglacial Climate

The climate of the last 10,000 years is known in more detail than are the climates of previous epochs, and climatologists have some knowledge of the causes of recent climate shifts. Recent climate deserves special consideration because it has affected the current distribution of many mammals.

The climate of the Holocene has been quite variable, partly because of variations in energy from our sun. Global temperatures, as measured from satellites, decline about 0.2 degrees C for each 0.1 percent decline in solar radiation. Maximal solar radiation occurs when sunspots, which affect the sun's magnetic field, are most numerous, and sunspot frequency usually follows an 11-year cycle. The relationship between the sunspot cycle and global temperatures is well established. The sun is also known to have longer cycles of intense radiation.

Changes in solar energy are also correlated with levels of carbon-14, once considered to be constant in the atmosphere: the creation of atmospheric carbon-14 is suppressed in times of high solar activity. This is important in the study of mammals because radiocarbon dating is used extensively in the dating of fossil midden heaps of wood rats (*Neotoma* spp.), Pleistocene fossils, and tree rings. Plant remains in wood rat midden heaps indicate that vegetational shifts have quickly followed Holocene climatic changes: fossil pollens from the early Holocene show the Sonoran Desert was moist, with abundant oaks and grasses.

A period of global warming, the Medieval Warm Period, occurred during a spell of intense solar radiation. The warmer temperatures are confirmed by greater annual growth of tree rings in the California Sierra Nevada. The Medieval Warm Period, also called the Holocene Climatic Optimum, extended from approximately the ninth century to the fourteenth century. This was about the time of the Anasazi settlements in the American southwest, as well as the era of farming in Greenland. The climate was warmer than that at the beginning of the twentieth century. The Mojave Desert had a persistent lake some 400 years ago, presumably because of warmer temperatures and increased rainfall.

The well-known Little Ice Age, from the mid-1400s to about 1850, was a time of global cooling. Residual small glaciers from this period persist in the Sierra Nevada today. From 1645 to 1715, a time of a total absence of sunspots, was a period of intense cold called the Maunder Minimum. This cold spell ended Viking farming in coastal Greenland, and droughts prevailed in the summers. A nearly 200-year drought in the Sierra Nevada lowered the level of Lake Tahoe sufficiently to allow trees to grow along the rim; their submerged stumps remain today. Reduced growth in the rings of bristlecone pines from the White Mountains of California reflect the strong effect of the Little Ice Age in western North America. Detailed accounts of the Little Ice Age have been provided by Ladurie (1971), Grove (1988), and Fagen (2000).

In Europe, the occurrence of mammals was more fully recorded than it was in North America; the Roe Deer *(Capreolus capreolus)* and the Polecat *(Mustela putorius)* disappeared from Finland during the Little Ice Age, returning after it, in the late 1800s. Almost certainly there were movements of California mammals in response to the Medieval Warm Period and the Little Ice Age, and some enterprising mammalogist may yet discover and date Holocene faunal movements in the New World. Currently, the FAUNMAP program, an electronic database maintained by the Illinois State Museum (www.museum.state.il.us/research/faunmap), documents the changing ranges of mammal species of the United States through the middle and late Pleistocene.

REPRODUCTION

For any species, maintaining its numbers is a struggle. Reproduction embraces a variety of events relating to the production and establishment of a new generation.

Seasonality

The breeding season of a species includes the period from mating to birth. In many small, short-lived mammals, such as mice, embryonic development (gestation) is brief; birth (parturition) may occur three weeks or less after mating. In most larger mammals, such as deer and bears, reproduction occupies months.

Breeding cycles usually have an annual rhythm (reproductive seasonality). Most biologists believe that the breeding season is adjusted so that the young are born at the season most favorable for their survival and rapid growth. Most mammals bear their young in spring, the time of melting snow, vegetative growth, and increasing warmth. Some small species, such as mice, also mate in spring. Other small mammals, notably bats, may mate in fall and produce young the following spring, and some carnivorans mate in summer or fall and give birth the following winter. Such diverse schedules of mating and parturition usually constitute adaptations to various other (usually nonreproductive) aspects of the species' life cycle.

Reproductive Factors or Cues

To reproduce, an animal must reach sexual maturity; accumulate an adequate amount of energy, usually in the form of body fat; and receive certain environmental cues. Body fat may be accumulated in two ways. Large mammals, such as deer, gradually add fat during the nonreproductive season, building a reserve of energy for the early stages of pregnancy and breast development. These species are called capital breeders because their reproduction draws on their stored capital (body fat). In contrast, many small species build reserves from their food only shortly before mating. These species are called income breeders. In both capital and income breeders, females also increase their food intake during pregnancy and after giving birth.

Until the body has a mature sexual system, it cannot respond to sexually stimulating signals from the environment. In addition, virtually all adult wild mammals in temperate areas remain sexually quiescent until they encounter certain environmental stimuli, which are mostly seasonal. The daily change in day length (or photoperiod) cues many annual events in most plants and animals. Each day from late December until late June, the length of daylight increases and the length of darkness decreases; this pattern is reversed between late June and late December. In many mammals, long periods of darkness suppress the gonads (testes and ovaries). When the animal is in darkness the pineal gland, a small gland in the upper median part of the brain, secretes a hormone called melatonin, which depresses gonadal activity, thus inhibiting reproduction. As nights become shorter in late winter and spring, secretion of melatonin declines, allowing testes and ovaries to increase in size and activity, and breeding commences. Because day length increases as darkness decreases, it was long supposed that light stimulates reproduction in mammals, as it does in birds. Careful studies on a number of species of mammals, however, have confirmed the inhibitory effects of darkness and melatonin. These effects are equally strong in diurnal and nocturnal mammals.

Other seasonal changes may also cue the sexual activity of some mammals. Laboratory observations on several kinds of voles (Arvicolinae) revealed that after exposure to short days, the gonads shrink or regress, but not to the degree seen in wild voles. This difference suggests that although the photocycle may affect reproductive seasonality in voles, other environmental cues may also do so. In fact, the reproductive season of some voles, at least, is not always closely associated with seasonal changes in day length. Instead, it is often related to plant growth. In the Great Basin, for example, the Montane Vole *(Microtus montanus)* breeds when fall rains produce a lush growth of annual herbs.

Some desert rodents, too, especially kangaroo rats (*Dipodomys* spp.) and pocket mice (*Perognathus* and *Chaetodipus* spp.), are apparently stimulated by the fresh growth of the annual plants on whose leaves they feed. In years when fall rains produce luxuriant plant growth, kangaroo rats and pocket mice are sexually active by late winter, but in dry years they may forgo reproduction entirely. The arrival of fresh plant material is suspected of stimulating their gonadal growth. A similar possible effect is seen in pocket gophers (*Thomomys* spp.), voles, and the Black-tailed Jackrabbit *(Lepus californicus)* in the Central Valley of California. These mammals tend to breed for longer periods in irrigated pastures than on adjacent nonirrigated land.

This stimulatory effect of plants on wild mammals has been duplicated in the laboratory. A number of plants, especially certain legumes and growing grasses, contain substances similar or even identical to the sexual hormones produced by vertebrate gonads. Other materials, found in sprouting grain, are stimulatory to the reproductive system but chemically different from vertebrate hormones. These phytoestrogens may well provide the additional reproductive stimulation that accounts for the surges of breeding in some desert mammals during years of lush plant growth, and their absence may result in the occasional failure of entire populations to breed in dry years.

Hormonal Control of Reproduction

An internal system of chemical controls regulates the reproductive process. The endocrine glands produce chemicals called hormones that circulate in the blood plasma throughout the body. These hormones have specific effects on certain target organs but not on most other tissues. In some cases, hormones stimulate other endocrine glands to produce different hormones; in other cases, they influence nonendocrine tissues.

One of the most important endocrine glands is the hypothalamus, which is situated in the midventral part of the brain. In response to environmental stimuli, the hypothalamus produces a hormone that is carried by a short blood vessel (portal system) to the anterior lobe of the pituitary gland, or hypophysis, which lies directly below it. This hormone, gonadotropin-release hormone

(Gn-RH), stimulates the anterior pituitary to release two gonadotropins—follicle-stimulating hormone (FSH) and luteinizing hormone (LH)—which are carried throughout the body.

The gonadotropins increase the circulation of blood to the gonads and also stimulate their growth and metabolic activity. The female gonads (ovaries) produce two hormones, estrogen and progesterone. Both are carried throughout the body in blood serum. They stimulate the growth and metabolic activity of the uterus and breasts and also make the female more receptive to male sexual advances. A male readily recognizes their presence in the urine of a female. These hormones also have important effects on the hypothalamus and the anterior pituitary. At low levels they tend to depress the release of Gn-RH from the hypothalamus and FSH and LH from the anterior pituitary; this is sometimes referred to as a negative feedback effect. At high levels, however, these same hormones stimulate the hypothalamus and the anterior pituitary so that more LH is released. High levels of LH cause the ovary to release one or several eggs (ova) in a process called ovulation.

The Estrous Cycle

In many female mammals, preparation of the uterus and spontaneous release of one or more ova are synchronized with sexual acceptance of a male. This period of readiness for mating is called estrus, or heat. If mating occurs, sperm meet the ova in the upper oviduct (between the ovary and the uterus); sperm enter one or more ova, an event known as fertilization; and the fertilized ova move down the oviduct to the uterus. One or more fertilized ova attach to the lining of the uterus, a process called implantation. Together each growing ovum and the uterine wall develop a placenta, tissue that promotes nutrition of the ovum. If mating does not occur, the unfertilized ova pass into the uterus and eventually die. Then the entire process, called the estrous cycle, may repeat itself.

In a minority of mammals, including voles, it is the act of mating that stimulates ovulation. Such induced ovulation has the advantage of timing the release of ova to the arrival of sperm, thus reducing the chance that ova will be lost. In species with induced ovulation, the period during which the female will accept a male is much longer than in other species.

Gestation

After successful mating are four possible sequences of events. In the first, fertilization and implantation take place quickly. The fertilized egg starts to divide immediately, and embryonic growth proceeds at a steady rate until birth. This direct development is the most common pattern in Pacific Coast mammals. It occurs also in the Opossum (*Didelphis virginiana*), although, like most marsupials, the Opossum does not have a well-developed placenta.

The second pattern—seen in the Marten (*Martes americana*), Fisher (*Martes pennanti*), Spotted Skunk (*Spilogale putorius*), and pinnipeds, among others—is delayed implantation (embryonic diapause). Fertilization and initial cell division occur shortly after mating, but cell division then ceases, and implantation may be postponed for weeks or even months. The adaptive significance of delayed implantation is not always clear, but in some cases it is. For example, the Black Bear (*Ursus americanus*) mates in summer—a period when the adults are not solitary—but gives birth in midwinter, thus allowing an extended period for growth before the young bear's first winter sleep.

A third pattern, delayed fertilization, is seen in vespertilionid and rhinolophid bats, in which the gonads of the two sexes do not mature in the same season. The testes become active in late summer, and sperm mature in fall. Mating occurs in fall and occasionally in winter, and the sperm are stored in the oviduct, remaining viable during winter while both sexes hibernate. Females ovulate when they emerge from hibernation in spring, and the stored sperm then fertilize the ova. Interestingly, ovulation and fertilization are synchronized among female bats that have overwintered together; this is probably because they are subjected to the same cues, although it is not known what these cues are. Females of some species also move to old barns, empty attics, or hollow trees to establish nursery colonies, where birth of the young is synchronized.

Many California bats also migrate, and in migratory species

the sexes may be separate for much of the year. Delayed fertilization might have arisen as an adaptation to the separation of males and females in spring. Delayed fertilization is a firmly fixed phenomenon in many bats of temperate regions and undoubtedly originated in ancient geological time. The adaptive significance of delayed fertilization may lie deep in the past, when seasonality, climate, and migratory patterns were probably very different.

A few bats follow a fourth pattern, delayed development. For example, in the California Leaf-nosed Bat *(Macrotus californicus)*, which neither migrates nor hibernates, mating, fertilization, and implantation occur in rapid sequence in fall. A period of delay follows implantation, then development proceeds at a normal rate; gestation takes some nine months.

Condition of Young at Birth

Many infant mammals are altricial—born nude and helpless, usually within a protective nest. Young mice (*Peromyscus* spp.), voles, shrews (*Sorex* spp.), and squirrels (Sciuridae) are blind and naked, with the earflap folded over the ear opening. The Opossum is born in a virtually embryonic state after a gestation of only 12.5 days. In contrast, other species have a longer gestation after which the precocial young are born well formed, with eyes open and ears erect. Young Porcupines *(Erethizon dorsatum)*, for example, have a long gestation and are born well developed, even with quills. Jackrabbits, like all hares (*Lepus* spp.), are fully furred and alert at birth and can feed on tender grass even before their fur is dry. Likewise, deer and antelope are precocial. There are intermediates between the altricial and precocial conditions; for example, cats are born with some fur but with their eyes and ears closed.

The speed of postnatal development varies among species and is probably an important adaptation. Among ground squirrels (*Spermophilus* spp.), for example, hibernating species, which have a brief activity period, grow more rapidly from birth than does the nonhibernating Antelope Ground Squirrel *(Ammospermophilus leucurus)*.

Litter Size

Each species of mammal has its characteristic litter size, but always with some variation. Humans usually produce one infant from a single ovum, but there are occasional twins or triplets. Similarly, most bats produce one young annually, but there are some twins and even quadruplets among the tree bats (*Lasiurus* spp.). Among mammals that regularly have several young in a single litter, there is usually substantial variation. This variation has received much attention from ecologists, for presumably there is an underlying pattern of adaptive significance. Litter size in herbivorous mammals varies with plant growth, increasing in years with richer growth.

There are perhaps three basic patterns of variation in litter size: variation among species (interspecific), local variation within a single species (intraspecific), and geographic variation within a single species. Litter size tends to compensate for death rate (mortality) in the long term, but birth rate (natality) and death rate are not always in balance. On the one hand, a large litter size and overall high birth rate (two or more litters) may compensate for a high mortality rate; on the other hand, a low mortality rate (great longevity) may be an adaptation to low productivity. These concepts, however, are more easily formulated than confirmed; theories on natality and mortality are based more on assumptions than on facts.

The tendency of species to have large or small litters has been described by the concept of K- and r-selection: K-selected species, such as the Black-tailed Deer (*Odocoileus hemionus*), are usually long-lived, occupy stable environments, and can survive by producing small litters, whereas r-selected species, such as voles, are short-lived, suffer a high mortality, have an abundant but fluctuating food supply, tend to breed at an early age, and produce large litters when their resources are abundant. This concept was originally proposed by the geneticist Dobzhansky in 1950 and was expanded by MacArthur and Wilson in 1967. The notion of K- and r-selection is actually rather simplistic, for most species lie somewhere in between. Moreover, a species may be K-selected under certain environmental conditions, such as poor plant production in a given year, but r-selected when resources are abundant. The concept is useful, but only if we realize its limitations.

Milk and Lactation

Mammals nourish their young with a secretion from the mammary glands of the female. Nursing (or lactation) is a uniquely mammalian feature and should be carefully studied by anyone wishing to compare the life patterns of different species.

Mammary tissue begins developing in the growing embryo, apparently from modified sweat glands, and sexual differentiation is apparent very early. Although present in both sexes, mammary glands normally function only in females. With the approach of sexual maturity and ovarian cyclicity, hormones from the ovaries and the anterior pituitary stimulate further mammary growth, or mammogenesis. In contrast to the breast tissue of the human female, however, which develops most significantly during puberty, mammary tissue in most other mammals undergoes most of its development during pregnancy. The placenta, through its release of a hormone called placental lactogen (PL), plays a major role in this development. The amount of PL secreted is directly related to the number of placentas, which normally equals the number of embryos. This results in a degree of mammogenesis appropriate to provide milk for the litter; mammogenesis is greater in females producing larger litters. After the young are born, the mother increases her food intake so that she can produce an adequate amount of milk.

The mammary gland is composed of secretory hollow spheres called alveoli , which drain into a branching system of ducts (fig. 7), which in turn lead into the nipple. In ungulates, the ducts lead into a collecting chamber, the cistern, housed in the udder; the milk leaves the cistern through a false nipple, or teat, at the time of nursing. About each alveolus are myoepithelial strands, fine muscles that contract under hormonal stimuli. This complex structure is supplied with nerves, especially from the nipple, leading to the hypothalamus in the brain.

Because the role of hormones in mammogenesis in wild mammals is poorly known, we must study the process in laboratory and domestic species. In the laboratory rat, the ducts develop under the stimulation of estrogen, and alveolar growth is induced by progesterone and estrogen from the ovaries, prolactin from the anterior pituitary gland, and PL. Moreover, insulin (from the pancreas) promotes the stimulatory action of the ovarian hormones.

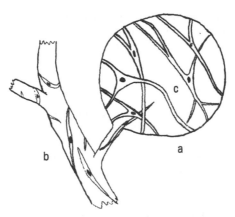

Figure 7. (a) Mammary alveolus; (b) adjacent duct of the mammary gland, showing (c) the covering of contractile cells, which force milk from the alveolus to the ducts and eventually to the nipple.

The roles of these hormones vary from one kind of mammal to another and also through the duration of both pregnancy and lactation. As the embryo or embryos develop, progesterone and PL levels rise and become increasingly important in mammogenesis. At the same time, the level of estrogen, and perhaps its role in mammary development, gradually declines.

The hormones that induce the growth of mammary glands also prevent the synthesis and flow of milk until the infant is born. At the time of birth there are major changes in circulating hormones; these include a drastic decline in progesterone, elimination of PL with the loss of the placenta, and a rise in estrogen from the ovaries. In addition, stimulation of the nipple by the infant is essential to the full function of mammary tissue. Thus, milk production and flow are responses to the altered hormonal balance following birth and also the stimulation of suckling.

Suckling is instinctive for all mammals, and the infant usually finds a vacant nipple shortly after birth. In every California species the number of nipples is greater than the mean litter size, so finding a nipple is seldom a problem. Moreover, mammary development is greater in females with larger litters because a larger number of embryos results in a greater amount of PL. Thus, a female with a large litter is prepared to nurse it.

Because the nipple has a rich supply of sensory nerve endings, it is very responsive to touch, and suckling initiates nervous signals to the hypothalamus. These signals cause the hypothalamus to produce and release oxytocin, which flows into the posterior pituitary, where it enters the circulatory system. Oxytocin influences the myoepithelial strands about the alveoli to contract, forcing accumulated milk into the ducts and the nipple.

This nervous stimulation of the hypothalamus, the resultant release of oxytocin, and its stimulation of the myoepithelial strands constitute a well-known neuroendocrine loop. The message from the nipple to the hypothalamus is the neural part, and the release and effects of oxytocin are the endocrine part. The contraction of the myoepithelial strands ejecting milk into the ducts is a pleasurable experience for a woman nursing her baby. Suckling of her brood most likely provides physical pleasure to the lactating female mouse as well.

In domestic cattle, milk flow, or letdown, is known to be stimulated by the sight and sound of a milking machine, and the high-pitched squeals of infant mice produce the same effect in lactating mice. Thus, it appears that several kinds of stimuli may cause the release of oxytocin. Fear, however, may inhibit milk flow. Fear-induced failure in milk flow is known in both domestic mammals and nursing women and may partly explain the difficulty of rearing wild mammals born to females captured when pregnant.

Small amounts of milk may drip from the nipple even in the absence of an infant. The nursing infant, however, accounts for most of the removal of the milk. The infant withdraws milk by two actions. First, by gripping the nipple between its tongue and palate, the infant removes the milk by a stripping action, much as a person strips milk from the teat of a cow by hand. Second, sucking reduces pressure within the mouth, so that the infant literally sucks from the nipple as if it were a straw. These two very different actions together constitute suckling.

The first material made by the mammary gland is a thin, almost colorless fluid called colostrum. Although colostrum is low in lactose (milk sugar), it is rich in sodium, calcium, and fat-soluble vitamins. Perhaps the most important of its components are immunoglobulins, which provide immediate protection from disease microorganisms. In some mammals, such as rabbits and humans, these antibodies are also transmitted across the placenta, but in most species colostrum provides the initial protection from disease organisms. Colostrum flows for only about one day.

The synthesis of milk (lactogenesis) follows that of colostrum, and in most species of mammals the infant subsists on milk until weaning—the time at which it is capable of obtaining adequate nourishment from solid food alone. Normally the young begin to take solid food before weaning, but some mammals (seals, for example) stop nursing long before they take their first solid food, living on their body fat in the meantime.

The composition of milk varies widely among different species and also with the stage of lactation. Variations occur in the four major constituents of milk: fat, protein, sugar, and water. In some domestic species, lactose levels may be high early in lactation, then decline gradually while fat and protein increase. There seems to be no general rule to explain these variations. Presumably the nature of the milk is adapted to the species as well as to the stage of dietary independence.

Specific patterns of milk intake exist. The infant Opossum virtually fuses to a nipple but still has an irregular pattern of suckling, as indicated by the sporadic visibility of milk through its translucent skin. Rabbits nurse but once a day. Most mammals probably have rather regular frequencies of nursing, but there is little real information on nursing rates, growth rates, or caloric intake of the young or the lactating mother.

There are very likely adaptive features to lactation, and an enterprising student could make major contributions to behavioral ecology by exploring early food and growth patterns of native mice, bats, and squirrels. How does a lactating bat, for example, adjust its need for increased food intake to the evolving nutritional demands of its young and still follow its usual foraging pattern? It is indeed ironic that milk, lactation, and seasonal changes in mammary tissue—the central features that unite the class Mammalia—have been well studied in only a small number of wild species.

SOCIAL GROUPS

The nature and permanence of the pair bond or group assemblage vary widely among mammals and significantly affect how they care for their young.

In polygynous species, one male mates with more than one female. In Sea Lions (Otariidae) or Wapitis *(Cervus elaphus),* a bull may have control of a large number of females, whom he jealously defends; this is called defense polygyny. Such control is limited, for the females usually have no hesitation in mating with other males. Breeding success, the spread of a male's genetic attributes or fitness, is quite variable in polygynous species, for some males may mate with many females, whereas other (bachelor) males may not mate at all. In polygynous mammals, parental care is solely the responsibility of the female.

In monogamous species, such as the Parasitic Mouse *(Peromyscus californicus),* the parents form a permanent pair bond. In these species biparental care persists until the young are independent; the male is likely to help feed and protect them. Paternal care in pair-bonded species may follow directly by way of territorial behavior of the male, but this behavior is more likely to result in exclusion of potential competitors rather than in care as such.

Species in which only the leading, or alpha, pair breed may engage in cooperative breeding. In this pattern, known in some species of the dog family and also in voles, care is provided by alloparents: siblings, an older generation of young, but this is clearly not parental care, though something nearly identical. Sometimes female alloparents, such as female foxes born the previous year, even become pseudopregnant and develop hormonally to the point of providing milk for their younger siblings. In another form of cooperative breeding, two female foxes will den together and mate with a single male and may share the care of their young.

Some mammals are promiscuous, with mating a matter of proximity. A female mouse, for example, may defend a specific territory, and several males may mate with her successively. In this case the female is polyandrous. In other species of mammals, territoriality may not develop until the approach of birth, and before this a group of males may breed randomly with a group of females. In this situation, called polygynandry, the males are polygynous and the females polyandrous.

Commonly males compete to mate with one or more females. Such male-male aggression is frequently symbolic, designed to establish the superiority of one male. In some species, however, such as Wapiti, combat may result in the death of one of the males. Females may also suffer from exceptionally aggressive males: the male Sea Otter *(Enhydra lutris)* holds the female's head and nose when mating, and resulting injuries sometimes kill the female.

Parental care is parental behavior that enhances the growth and survival of offspring. Although in a broader sense, maternal care includes prenatal nourishment, this discussion focuses on postnatal care.

In mammals, parental care virtually always includes lactation, although the young of a few species, such as hares *(Lepus* spp.) and Porcupines *(Erethizon dorsatum),* might survive solely on solid food from birth. A diverse pattern of associations of suckling and parent-offspring contact is seen. A foal feeds frequently, for example, and is seldom far from the mare, whereas young rabbits take milk but once a day and may be left unattended for hours.

As the infant takes milk, it also receives heat from the mother—not only through radiation and contact but also from the milk itself, which is at very nearly the inner or core temperature of the mother and warmer than the body of the infant. Moreover, in addition to providing materials for growth, milk contains calories that are converted to heat. Thus, in some mammals at least, milk and suckling behavior raise the body temperature of the newborn until it can thermoregulate, or control its own body temperature.

The significance of heat from both milk and body contact with the mother has been shown in captive rats and mice. The body temperature of newborn young declines between suckling periods, with a substantial saving of energy. As the growth rate slows, the infant develops a better thermoregulatory ability, concurrent with growth of fur and increase in body size. The thermoregulatory aspect of maternal behavior is undoubtedly very important for such altricial mammals as mice, shrews, and bats, but perhaps less so for such precocial species as Porcupines, hares, and ungulates.

Because the nipples lie on the ventral part of the body in almost all mammals, the mother's body constitutes a cover for her nursing offspring. In this way, the mother protects her brood not only from hostile aspects of the physical environment but also from potential predators.

Clearly, lactation is the basis of parental care in mammals. The early mother-offspring contacts have important effects on the behavioral development of the young—and, to some extent, the young elicit maternal behavior in the mother. The hormonal basis for lactation is naturally bound to other reproductive events, especially birth. Thus, the breasts develop during gestation, and the activities of the newborn young stimulate the release of milk.

Maternal Aggressive Behavior

The mother of newborn young is usually aggressive toward other adults. This well-known behavior should dispel the myth that aggression is a trait solely of males. The hostility of females with suckling young occurs in most, if not all, orders of mammals. Maternal behavior is produced by the hormones oxytocin and estrogen. Indeed, custodial and aggressive maternal behavior may be a constellation of responses to the same hormonal stimuli. Oxytocin and estrogen administered to virgin laboratory rats result in prolonged maternal behavior, such as nest building and picking up of newborn young. The rise of these two hormones when the young are born presumably accounts for the appearance of maternal instincts. Inasmuch as suckling induces a prompt secretion of oxytocin in the mother, the infant is responsible for preserving the maternal behavior of its mother. Maternal aggressiveness is strongest just after young are born and tends to decline as they grow. It disappears with the experimentally early removal of suckling young.

Although in most mammals a female with young is indifferent or even hostile toward the young of other females, exceptions seem to occur among certain colonial bats. In colonies of the Pallid Bat *(Antrozous pallidus),* a cluster of 10 very small young (naked and with eyes still unopened) may be guarded by a single female, which seems to have temporary custody of the nursery.

Robert L. Rudd and his associate Albert J. Beck noticed another cluster of somewhat older young (perhaps two weeks of age) that were guarded by two adult females; a still older group (about four weeks of age) was found to be segregated, each individual young with its mother. These observations suggest that in colonial bats, female aggression and intraspecific hostility may somehow be suppressed when the offspring are very small. This pattern is facilitated by simultaneous ovulation and birth within a given colony.

The duration of parental care in mammals is loosely tied to the duration of lactation. It is also linked to the age of dispersal, which normally occurs prior to sexual maturity. Thus, parental care is brief in a vole, which may be sexually mature at four weeks of age. Young Beavers (*Castor canadensis*), however, remain in their parents' lodge for nearly two years, though overt parental care probably ceases long before dispersal. These events—lactation, dispersal, and sexual maturity—although sequential, differ from one species to another and probably constitute adaptations to many sorts of environments.

POPULATION FLUCTUATIONS

Several hundred years ago chroniclers documented dramatic changes in the numbers of small mammals, such as lemmings, in Scandinavia: they suddenly appeared in high densities and just as suddenly disappeared. The total time from one population peak to the next was approximately three to four years. Because of the apparent regularity of these changes, they have been called "cycles." The early accounts included the mythical "swimming out to sea" of lemmings and even the beaching of certain small whales. Population cycles have been intensively studied in the past 50 years, but our understanding of them is still imperfect.

Densities of wild animals are determined, in part, by carrying capacity (CC), the environmental features (food, cover, and other characteristics) needed to support a population of a given size. Carrying capacity varies widely, however, with both environmental fluctuations and the nature of the population. Almost continuous changes in rainfall and temperature alter vegetation in an irregular fashion. These changes in climate are sometimes drastic, as in the warm Medieval Optimum and the cold Little Ice Age. Habitat productivity also varies seasonally and over shorter periods. Moreover, seasonal CC is different for different animals. For example, it is greatest in spring for voles (Arvicolinae), which require growing vegetation, but increases in fall for seed-eating mammals such as tree squirrels *(Sciurus* and *Tamiasciurus)*. In winter the nutritional requirements of a mammal such as a deer, especially a pregnant female, may increase as the environmental productivity declines; cold weather places greater demands on the body. Wild mammals alter their food selection with the seasons, further complicating the concept. Moreover, CC varies with the age structure of a population, for young animals make much greater nutritional demands than do adults.

To understand population fluctuations it is necessary to ask several questions. How do these fluctuations vary among species? What promotes demographic stability in those species that do not fluctuate greatly in numbers? Do predators limit population density, and, if so, at what levels? Are populations self-limiting, or density dependent? The answers to these and other basic questions are not always readily apparent.

Population Changes in Prey Species

Populations are more unstable in typical *r*-selected species such as voles than in K-selected species such as deer. Because most mammals lie somewhere between K-selection and *r*-selection, their numbers do not usually fluctuate violently. Instability is very conspicuous in voles, and they have received a great deal of attention. Under favorable conditions of food, voles frequently exceed their CC. They often occur in habitats dominated by one or only a few types of food and are especially vulnerable because they depend on grasses and small forbs for both food and cover. A rise in abundance of voles and failure of their food and cover frequently occur simultaneously, apparently due to heavy pressure exerted by the voles on the plants.

In contrast, the Deer Mouse *(Peromyscus maniculatus)* occupies a habitat with a great variety of foods and has the capacity to quickly change its diet with changes in food supplies. Thus the Deer Mouse can eat pine seeds, Douglas-fir *(Pseudotsuga menziesii)* seeds, fungi, insects, and other delicacies. To be sure, both kinds of rodents experience changes in numbers, but the densities of Deer Mice are usually much more stable.

Increases in prey species, especially *r*-selected species, are to be expected over time, because these species usually have a strong reproductive capacity. The potential suppression of their populations by predators is less well understood. Normally, large predators require large prey and may appear to suppress high densities of that prey. For example, large predators seem to help keep populations of deer (a K-adapted species) within their carrying capacity. Predators, however, do not definitely limit the growth of prey species.

Some mammals greatly increase in numbers following the removal or reduction of their predators. For instance, the California population of Black-tailed Deer *(Odocoileus hemionus)*, reduced dramatically during the gold rush and following years, rebounded after the Mountain Lion *(Panthera concolor)* was eliminated in many areas. Today, the lion population has recovered very well, whereas that of the deer is at carrying capacity and may now in fact exceed the carrying capacity of its environment in some areas, in the sense that its habitat can support its current numbers only at the expense of the deer's health and stamina. This species is a browser, eating, among other foods, the leaves of trees as high as it can reach; such a browse line indicates that deer populations have exceeded their carrying capacity.

Despite the apparent correlation between removal of predators and an increase in the deer population, it is not known if predators, such as the Mountain Lion and the Coyote (*Canis latrans*), take enough deer to suppress or greatly affect healthy populations. Although predators can certainly take substantial numbers of deer, there is little evidence that they can reduce deer to low population levels. Some researchers believe that Coyotes, major predators of fawns in spring and summer, contribute to stability in deer populations. Many factors, however, such as other prey species and predator density, habitat and forage availability, and disease, to name a few, may play a role.

Ironically, predators are usually most effective in lowering population densities that have already begun to decline. Great numbers of voles, for example, may so destroy their habitat that decline becomes inevitable. Although a high vole population attracts both mammalian and avian predators, its decline does not seem to be due to predation. The late Professor Oliver Pearson of the University of California at Berkeley found that predators (feral cats, Gray Foxes [*Urocyon cinereoargenteus*], Raccoons [*Procyon lotor*], and others) take approximately 5 percent of voles when populations are increasing but a much higher percentage when they are declining. Vole populations are density dependent when they are increasing but inversely density dependent when they are declining. That is, as vole densities increase, food becomes a factor in limiting their numbers, but as they decline, predators accelerate their decline. Voles have several traits that collectively reduce stability. They reach sexual maturity and reproduce at an early age; they ovulate only at the time of mating; they can mate shortly after parturition, so that a female may be nursing her first litter while carrying her second; and their mean life span is relatively short (less than a year), so that any environmental factor that retards reproduction quickly reduces the entire population.

Population Changes in Predator Species

Both avian and mammalian predators are attracted by high densities of their prey. Also, when prey are abundant, predators find food more easily and may produce more young. The Red Fox *(Vulpes vulpes)* is known to produce litters of six or more kits when prey are abundant but only three when prey numbers have declined. Also, more fox kits survive when their foraging parents have abundant food. The same pattern is seen in Coyotes when populations of the Black-tailed Jackrabbit *(Lepus californicus)* fluctuate.

Some predators may be density dependent as well, although the regulation of their populations may also be influenced by territoriality and other factors. A sharp reduction in numbers of a Coyote population, for example, results in a dramatic increase in the size of litters. For the most part, population changes in predators are poorly understood, but there is a general tendency for predator levels to decline following a decrease in densities of their prey and to rise when more food becomes available.

THERMOREGULATION

As pointed out earlier, the production of a substantial amount of heat (endothermy) enabled early mammals to become nocturnal and highly mobile. A major result of endothermy was the preservation of a relatively constant body temperature (T_b). Body temperature does fluctuate, usually within narrow limits: that of nocturnal squirrels, such as Flying Squirrels (*Glaucomys sabrinus*), is relatively low in the daytime, whereas that of diurnal squirrels, such as chipmunks (*Neotamias* spp.), declines at night.

Most mammals maintain a rather high T_b of 34 to 41 degrees C through a high metabolic rate and the low conductivity of their outer covering—fur in the case of most terrestrial mammals and fat in whales, dolphins, and porpoises (Mysticeti and Odontoceti). Maintaining a balance between creation and loss of body heat is important, and the problem is aggravated by great changes in ambient temperature (T_a). Control of T_b, or thermal stability, requires both physiological and behavioral responses.

Maintaining a relatively constant T_b is more difficult for a small mammal because its surface area is greater relative to its body mass. As the mass-to-surface ratio increases, the rate of loss of body heat declines. This difference is frequently expressed as an exponent, which is a physiological adjustment for metabolic body mass. The exponent is a measure of the increase in metabolism relative to increase in body mass. Commonly the exponent .75 is used, but .67 is more accurate. These exponents are derived from measurements taken on resting and fasting animals at thermoneutral temperatures; they do not apply to field conditions.

The zone of thermoneutrality is a range of temperatures within which an animal does not need to make behavioral or physiological adjustments to keep metabolic heat production in balance with heat loss. There is an upper limit to T_a above which an animal cannot adjust its own temperature, and death is the usual result. This is called the critical thermal maximum. This concept is studied in the laboratory but does not relate to most situations in nature.

Mammals can accelerate heat production when exposed to excessive cold. A common physiological mechanism is nonshivering thermogenesis (NST). This is the production of heat from brown adipose tissue (BAT), a specialized fat tissue richly provided with capillaries and branches of the autonomic nervous system. BAT occurs in both hibernators and nonhibernators. It is frequently found in the interscapular region of rodents. NST is a common mechanism of warming in such animals as voles *(Microtus* and *Clethrionomys)* and some mice *(Peromyscus)*. In laboratory studies, NST increases as day length declines. Many mammals (e.g., the Deer Mouse *[Peromyscus maniculatus])* also increase heat production by increasing their basal metabolism when T_a declines. Finally, heat is produced by shivering, an involuntary contraction of muscles induced by a drop in skin temperature.

In cold environments, mammals also need mechanisms to retain the heat they generate. The skin and furry covering, or pelage, of most mammals not only function as camouflage but also reduce heat loss. Pelage insulates a mammal from excessive cold by trapping air among its hairs. Changes in pelage frequently result from changes in the annual photocycle: hairs may become longer and denser when a mammal is exposed to short days. In some voles *(Microtus)* the fine density of the underfur increases without a change in the guard hairs, and in deer (Cervidae) the empty spaces within the hair increase, providing greater insulation.

Countercurrent exchange is another common physiological device for modifying heat loss. The large ears of elephants and jackrabbits contain countercurrent arterioles and venules. These arterioles can be constricted, reducing heat loss by reducing blood flow to the surface of the body. Countercurrent exchange is also done through the feet of such animals as the Arctic Fox *(Alopex lagopus)* and Caribou *(Rangifer tarandus),* and through the fins and mouths of some whales.

Huddling together tends to reduce body surface relative to body mass, thus reducing heat loss through radiation. This behavior is frequently seen in bats, and its benefits have been demonstrated many times in the laboratory. Cave bats frequently roost in clusters, and this habit tends to reduce heat loss. In the Pallid Bat *(Antrozous pallidus),* it also reduces both oxygen consumption and water loss. Adult mice and voles frequently share nests in winter, reducing heat loss and energy use. The Antelope

Ground Squirrel (*Ammospermophilus leucurus*) also reduces energy expenditure in winter by huddling.

In warmer environments, skin may protect a mammal from excessive exposure to sunlight. Melanin in the skin serves this function for many desert mammals, such as the Antelope Ground Squirrel. Pelage also protects a mammal from excessive heat, up to a point; after the pelage warms the animal usually moves into the shade, where the pelage may lose heat by radiation.

Breathing, urination, radiation of body heat, evaporative cooling, and drinking cool or cold water are common mechanisms of ridding the body of excess heat. Exhaled air and urine leave the body at T_b or close to it and account for some loss of body heat. Most mammals have turbinate bones that are richly provided with capillaries. This built-in air-conditioning system warms and humidifies entering air and cools and tends to dry exhaled air.

Body heat is lost, more or less continuously, by radiation. Dilation of arterioles in the skin, such as in the ears of jackrabbits, accelerates this heat loss.

Evaporative cooling lowers the temperature of the skin. Bats, for example, have a high evaporative water loss. The effectiveness of evaporative cooling depends on relative humidity: less humid air can absorb more water. Evaporative cooling is thus most effective in desert mammals. However, such cooling can be expensive in desert species, for only a limited loss of body fluids can be tolerated. Dilation of arterioles in the skin promotes evaporative cooling. Sweat glands also assist in this cooling, but not all mammals have sweat glands; they occur in horses and sheep and are abundant in humans. A form of evaporative cooling is panting, in which water is lost from the oral tissues and heat and water are lost from the lungs. Many mammals immerse themselves in water, which then evaporates when they emerge. Moreover, while the animal is in the (presumably cool) water, it loses heat to the water through conductance.

A mammal can also modify heat loss by a variety of other behavioral mechanisms. The simplest and most obvious is to modify exposure. Many species, for example, may seek shade. Many species of mice and shrews escape hot dry air by retreating into underground burrows. The Kit Fox (*Vulpes macrotis*) is not really physiologically suited to its dry, hot desert environment. It uses radiation and panting but also remains underground in periods of great heat stress.

The behavioral thermoregulation of infants, especially those that are altricial, may differ from that of adults. Infant mice and voles, for example, generally tolerate a certain degree of hypothermia. They receive warmth by brooding by the mother, and also the father in some species; by ingestion of the mother's warm milk; and by huddling with their siblings. They develop a capacity for thermoregulation at one or two weeks of age. Precocial young, such as jackrabbits and snowshoe hares (*Lepus* spp.), are furred and possess rather efficient thermoregulation.

SEASONAL DORMANCY

A wild animal is under continuous pressure to maintain an energy balance. During hostile seasons, when maintaining this balance is particularly difficult, a decline in bodily activity or metabolism may occur. If such a period of inactivity occurs during winter, it is called hibernation; if it occurs during summer, it is called estivation.

Hibernation and estivation involve a complex of physiological changes from the usual normothermic state. Body temperature drops, and breathing, heart rate, and metabolism all slow. Oxygen consumption, cell respiration, and kidney activity profoundly drop. Eventually, however, the individual returns to a normothermic state. In contrast, the individual usually does not recover from hypothermia, an unnatural decline in T_b that is usually caused by a decline in food or T_a.

Several physiological changes prepare a mammal for hibernation. These changes are correlated with fluctuations in temperature and rainfall, but they also follow an endogenous rhythm in the animal that is not precisely annual. For example, the Golden-mantled Ground Squirrel (*Spermophilus lateralis*), when kept under constant conditions of light and temperature, has a natural or free-running rhythm of 11.5 months. Exogenous factors fine-tune the endogenous rhythm.

As the season for winter sleep approaches, the hibernator becomes rather sensitive to declining T_a. When the body's temperature reaches a certain setpoint, it may become dormant for a few hours—sometimes for several nights successively—or even for a day or more. At this time the individual appears unable to proceed into a deep dormant state. During the dormant periods the heart rate slows to a range of two to eight beats per minute, and breathing may be reduced to twice per minute. The body temperature drops to a point just slightly above that of the surrounding air; a ground squirrel (*Ammospermiphilus* and *Spermophilus*) in its underground nest, for example, may have a body temperature of 6 degrees C when the air is at 4 degrees C. As the season progresses, the setpoint appears to be successively lowered until a metabolic decline leads to a profound drop in heart rate and general activity. The dormant periods lengthen, so that by midwinter a ground squirrel may remain dormant or torpid for 10 days to two weeks.

Thus, in most mammals hibernation is not a long spell of continuous dormancy but a series of rather brief dormant periods interrupted by arousals. These periodic arousals begin with an increase in body temperature, which is shortly followed by increases in heart and breathing rates; these changes continue until the normal (active) rate of metabolism has resumed. Periodic arousals appear to be caused by endogenous stimuli, possibly water loss or accumulation of metabolic wastes in the blood, and may last only a day or a few hours before the animal returns to dormancy. Invariably the individual urinates at this time, and a little food may or may not be eaten. Some mammals, such as the Black Bear *(Ursus americanus)* and the Golden-mantled Ground Squirrel, do not eat between the onset of hibernation in late summer or early fall and emergence the following spring. Others—certain chipmunks *(Neotamias* spp.), for example—store large amounts of food and do eat during periodic arousals. As spring approaches, the frequency of periodic arousals increases and the periods of dormancy become briefer.

Prior to hibernation or estivation, the animal accumulates sufficient body fat, or white adipose tissue (WAT), to sustain metabolism during dormancy. Although fat itself is not an endogenous cue, adequate fat is closely correlated with hibernation in ground squirrels, pocket mice *(Perognathus* spp.), jumping mice *(Zapus* spp.), and bats (Chiroptera). The Golden-mantled Ground Squirrel increases its total fat by 300 percent prior to hibernation. Similarly, the Western Jumping Mouse *(Zapus princeps)* increases its dry body weight by about half. Stored body fat in a female bear nourishes not only her but also her embryo, which is born in midwinter, and supports the production of milk, which feeds the cub until the mother emerges some months later.

In addition to WAT, there are special pads of BAT that are richly provided with nerve endings and a blood supply. During periodic arousals BAT quickly generates heat, which is rapidly carried to such critical regions as the brain, heart, and lungs. As a result, the anterior part of the body warms first. BAT is stimulated by the hormone noradrenaline, which comes from the sympathetic nervous system. During periodic arousals, when the heart and lungs work at an increased speed, WAT may provide additional warmth. When the hibernating mammal emerges in springtime, stores of BAT and WAT are depleted but not entirely exhausted.

Many small mammals lose body mass during winter, thus lowering metabolic needs. As part of this process, in some species of shrews *(Sorex)* and voles *(Microtus)* the brain shrinks in winter, also reducing metabolic requirements, for the brain needs a relatively large amount of energy. This is known as the Dehnel effect, named for the investigator who first noticed winter shrinkage of brains in shrews. This shrinkage is also controlled by the annual photoperiod.

The environmental (or exogenous) cues for hibernation are unknown but probably include the annual photocycle. In the Golden-mantled Ground Squirrel, adult males enter hibernation in August or early September, while the weather is warm and food is plentiful. Females are slower to enter dormancy, perhaps because they must accumulate some extra fat for the manufacture of milk in springtime, and lactation probably delays this accumulation in late summer. The young are last to enter seasonal dormancy. Townsend's Ground Squirrel *(Spermophilus townsendii)* and Belding's Ground Squirrel *(Spermophilus beldingi)* enter dormancy in July. They are, however, extremely fat. Temperature and snow cover appear to affect emergence from hibernation in spring. Ground squirrels emerge earlier on south-facing slopes, where the snow melts early and the sun warms the soil, than they do on north-facing slopes. Hibernators generally emerge earlier in years with an early spring.

Several groups of California mammals are well known to hibernate, and this physiological pattern clearly developed at different times in their evolution. We know this because, among closely related groups, there are both hibernating and nonhibernating species. Why some species hibernate and sometimes close relatives do not is not known. Among the Heteromyidae, for example, kangaroo rats *(Dipodomys* spp.) are not known to hibernate, but several species of pocket mice *(Perognathus* and *Chaetodipus)* are. The kangaroo mouse *(Microdipodops* sp.) accumulates WAT in the tail and becomes dormant for at least part of winter. Most species of ground squirrels and chipmunks hibernate, but tree squirrels *(Sciurus* and *Tamiasciurus* spp.) do not. Some carnivorans, such as the Black Bear, have a deep winter sleep, whereas others, such as the various kinds of weasels (Mustelidae), do not. This adaptation is obviously not related to mammalian phylogeny.

Almost all California bats hibernate, usually moving into caves in fall. They are fat at this time, and this seems to make them very sensitive to the lowered temperatures they encounter in caves. Studies have shown that dormant bats in aggregation maintain higher body temperatures than do bats of the same species roosting individually. Whereas cave bats arouse from dormancy spontaneously within the chilled air of a cave or mine shaft, tree bats remain dormant until air temperatures reach nearly 20 degrees C; this behavior probably precludes their hibernating in caves. Bats of many species seem to lose body water during hibernation, though metabolism of stored fat releases some water into the body. Upon their periodic arousals, bats frequently ingest water droplets from their fur.

The hibernation of the Black Bear is rather unusual. Bears may remain in their winter dens for six or seven months, during which time they not only do not urinate or defecate but also apparently do not arouse. A dormant bear maintains its body temperature at about 31 to 32 degrees C. Its heart rate may drop to 12 beats a minute, with a consequent 50 percent drop in metabolic rate. Although one may call this state "hibernation," it is physiologically very different from hibernation in ground squirrels and chipmunks. Not surprisingly, the bear preserves a core temperature not greatly lower than its active temperature, for parturition occurs during the coldest part of winter, and from that time the offspring take milk.

A hibernating mammal clearly saves energy. Although periodic arousals may consume large amounts of energy— nearly 90 percent of the energy used during hibernation—there are great savings during the dormant periods. The savings are relatively greater in small species. Professor Kenagy at the University of Washington calculated that the Golden-mantled Ground Squirrel in hibernation saves some 80 percent of the energy that would be required for normal activity.

Hibernation reduces reproduction; most hibernators have only a single litter annually. In compensation for the shorter time available for growing, the young grow extremely rapidly.

SENSES

Imagining the world in which wild mammals find themselves is difficult, for their sensory specializations are quite different from ours. Unlike birds, mammals almost always have an excellent olfactory sense (although whales [Cetacea] do not). Mammals also usually have poorer vision than birds. Beyond this, mammals' senses are specialized in ways suited to their activity patterns and habitats; abilities that are not frequently employed tend to deteriorate. Nocturnal mammals usually have well-developed senses of smell and hearing but may have much poorer vision than we do. Moles and pocket gophers (*Thomomys* spp.), which spend much of their time in the darkness of their burrows, also have poor vision. On the other hand, tree squirrels (*Sciurus* and *Tamiasciurus* spp.), which are diurnal, possess excellent vision suited for arboreal activity and the visual detection of predators.

The sensory stimuli a mammal can perceive are generally correlated with those it can generate. Insectivorous bats (Chiroptera), for example, which can detect sound waves well in excess of 20,000 cycles per second (cps), or 20 kHz, can also produce these high-frequency sounds. Similarly, the olfactory ability of most mammals is paralleled by a variety of scent-producing structures; together these two systems form the basis for much mammalian communication and behavior. For the most part, mammals live in a world of odors—a world of many sorts of self-produced and alien smells.

Olfaction

Most species of mammals have a large olfactory nerve tract. The nasal passage is expanded and folded in a way that increases its sensory surface. In addition, in some species a branch of this passage leads to a pair of small openings in the roof of the mouth. These openings are lined with odor-sensitive tissue that can detect smells entering the mouth. This structure is called the vomeronasal organ. The oral detection of odors can be observed when an animal such as a deer (Cervidae) or horse (Equidae) raises its head and lifts its upper lip to expose the vomeronasal organ to airborne odors. This position is known as *Flehmen*.

Olfaction plays numerous important roles. Locating water in the desert is, to some degree, a matter of recognizing water vapor in the air. Probably food is located primarily by its odor. Even to our relatively insensitive olfactory endings, most plants have distinctive odors, and undoubtedly most herbivores recognize a broad spectrum of plants, some of which may be only mildly aromatic. Mammalian predators use odors to detect their prey, and the prey may be warned by the scent of a predator. Even young ground squirrels (*Ammospermophilus* and *Spermophilus* spp.) recognize and avoid the scent left on the trail of a rattlesnake.

Scent Marking

A variety of dermal glands produce odors with which a mammal can establish signposts within its territory. Urine also contains distinctive scents, and many mammals customarily discharge small amounts of urine to renew their olfactory signposts. Odors in urine reveal not only the species and sex of the mammal but also its sexual condition and individual identity.

Most research on scent marking concerns dogs (Canidae) and laboratory animals. Among dogs, olfaction is a major means of social organization. Foxes (*Urocyon* and *Vulpes* spp.), wolves (*Canis lupus*), Coyotes (*C. latrans*), and domestic dogs all secrete substances into urine and vaginal discharge that communicate aspects of territoriality, reproductive state, and social dominance. Adult males of these and other species place urine throughout their territories and constantly explore scents left on rocks, pathways, or tree trunks; thus, they continuously search for intruders while renewing their own boundary markers. A male can also recognize the scent of a female and detect whether she is in estrus. This odor can be carried many miles in the air, a phenomenon well known to anyone who has kept a female dog in heat.

Other types of scent marking are also common. The importance of scents is well known to trappers of furbearing mammals, who attract mammals to traps with the scent of previously trapped animals. Wood rats (*Neotoma* spp.) have ventral glands that produce a secretion that rubs off on rocks and branches when they run. The licking of newborn young is an essential activity of many mothers and provides a means of later recognition in at least some species. In dirt bathing, which has been observed in the Beechey Ground Squirrel *(Spermophilus beecheyi),* some kinds of kangaroo rats *(Dipodomys* spp.), and pocket mice *(Perognathus* spp.), the dorsal gland seems to be applied to the ground; this may constitute an effort to mark an area. In some species, such as the Porcupine *(Erethizon dorsatum),* for reasons not well understood, the male may urinate on a female, a phenomenon known as a urinal shower.

Deer and elk (Cervidae) and antelope possess tarsal glands and sometimes facial glands that produce scents with which they mark their territories. Deer and elk may apply secretions from these glands directly onto the trunks or branches of trees. The secretions of facial glands are seasonal and may increase with gonadal activity. Tarsal glands in the Mule Deer *(Odocoileus hemionus)* secrete odors that may serve for individual identification.

The Mule Deer, or Black-tailed Deer, and the Pronghorn *(Antilocapra americana)* appear to make extensive use of olfactory signaling. The male Pronghorn marks by scraping a small area with the forefeet, urinating on the pawed ground, and then defecating on the urine. The reason for this behavior is unknown; because this species moves about from one area to another in small groups, odors probably do not have a territorial significance. Deer and Wapiti *(Cervus elaphus)* may mark themselves with urine, either on the belly or in the area of the tarsal glands. In particular, bull Wapiti do this when in rut.

The olfactory signals of mammals usually are detected by both sexes and all ages, in contrast to insect pheromones. The word pheromone was first applied to odors produced by one insect to elicit a specific response, usually sexual attraction, in another of the same species but of the opposite sex. A pheromone is usually a single compound that causes a very precise behavior. In contrast, the odors synthesized by mammals usually consist of several compounds, frequently modified by bacteria, and cause various responses depending upon the age, sex, and experience of the recipient. For these reasons some students of animal behavior prefer not to call the olfactory signals of mammals pheromones.

The Role of Olfaction in Reproduction

Odors can affect mammalian reproduction in ways that go beyond announcing the presence and reproductive state of the individuals occupying the area. Odors affecting reproduction can be divided into releasers, which cause prompt behavioral responses, and primers, which alter the hormonal state of the individual receiving the odor, thus modifying its sexual development and performance. The role of releasers in most species is to attract males to a female in estrus, but in some species (such as sheep), females in estrus are attracted to a male. Primer odors of conspecific individuals may stimulate sexual activity in members of the opposite sex and suppress it in members of the same sex. Although much of the current information on these effects is based on the behavior of laboratory populations of the House Mouse *(Mus musculus)*, there are similar responses in other species of mice *(Peromyscus* spp.) and voles *(Microtus* spp.).

Generally there are five well-studied primer effects in laboratory mice. The Bruce effect is the failure of a blastocyst, or fertilized egg, to implant when a mouse in early pregnancy is exposed to the odor of urine of a male other than the one with which she has mated. Apparently odors from an alien male modify hormonal levels so that implantation fails.

Odors in the urine of sexually active adult males can have two stimulatory effects on the reproductive performance of nonpregnant females. When female mice are subjected to male urine, their estrous cycles become shorter and more frequent. This effect has been shown not only in laboratory mice but also in a vole (Prairie Vole *[Microtus ochrogaster]*) of the central United States and may also exist in California voles. Female mice, moreover, reach sexual maturity earlier when exposed to the odor of male urine. These effects do not occur if the males have been castrated. Clearly the odor and not male behavior causes the effects because they are absent in females whose olfactory nerve tract has been destroyed.

In contrast, the urine of females has two suppressant effects. When two or more female mice are kept together without a male, their odors tend to retard the frequency of estrous cycling and also to retard sexual maturity in young females. One may speculate on the significance of these phenomena in wild populations.

Most of these primer effects are, to date, known only in mice and, except for the Bruce effect, would seem to require a preponderance of one sex or the other. To suggest that they lack an important role in wild populations, however, would be premature.

Most mammals have accessory reproductive glands, called preputial glands, that release materials into urine. Preputial glands are under the control of gonadal hormones and are usually larger in males than in females. Among laboratory mice of known social structure, the preputial gland is largest in the dominant males, which also do more of the urine marking. Sexually experienced females respond most strongly to the odor from the preputial products; response is weak in virgin females and absent in pregnant females. Thus, it appears that preputial secretions have an important role in social organization and sexual activity in mice.

In some kinds of wild mice, the male nuzzles or licks the genital region of a female in estrus and seems to receive vaginal scents through the vomeronasal tract. In some species, these olfactory signals may be essential to stimulate the male to mate. In hamsters, for example, which are related to deer mice (*Peromyscus* spp.), voles, and wood rats, if the male's vomeronasal tract is anesthetized, he is rendered sexually incapable.

The Pig *(Sus scrofa)* is renowned for its ability to detect truffles in the ground, and sows possess a special skill for recognizing the unusual fragrance of truffles growing as much as 1 m (3 ft) beneath the surface. This talent stems from the production of a peculiar substance in truffles that is also produced in the testes of the boar and secreted in his saliva. The musklike odor of this material attracts the sow. Human testes produce the same material, and it is secreted in the underarm perspiration of adult males.

Vision

Visual signals require light. Because most mammalian groups are largely nocturnal, they communicate primarily by sound and odor; visual communication is less common than it is in birds. In some cases, however, the significance of visual signals among mammals is apparent. Visual signals have been intensively studied in elephants and primates, and such communication may have further, as yet undetected roles in California mammals.

Except for flying squirrels (*Glaucomys* spp.), all squirrels are diurnal, and it is logical to assume that they communicate by visual as well as vocal means. Visual intraspecific signals occur in chipmunks (*Neotamias* spp.) and ground squirrels, but the significance of their various tail flicks is poorly understood.

Many visual signals in mammals are passive. Color patterns, for example, are visual transmissions, although their meaning is not always apparent to us. The stark black-and-white pattern of skunks, obvious even in dim light, announces their potential danger. The color patterns of many tree squirrels may not carry an obvious meaning, but the fur of ground squirrels and chipmunks is usually colored so that these animals are not clearly visible when at rest in their normal habitat. In some species, these color patterns may aid in species recognition. The erection of dorsal manes in the Coyote and the Gray Fox *(Urocyon cinereoargenteus)* and the flaring of the rump patch in the Pronghorn *(Antilocapra americana)* are powerful intraspecific signals.

An animal's posture, too, may be a visual signal. A submissive individual, such as a puppy, can indicate its social ranking by lowering its head or rolling on its back, whereas an aggressive animal might raise its head or bare its teeth. Although vocal signals may also be given, the visual aspects are very important. One of the most startling posturing signals is the handstand of the Spotted Skunk *(Spilogale putorius):* when alarmed, this skunk sometimes stands on its forefeet so that its black-and-white dorsum is entirely directed toward its potential enemy. Distinctive posturing also accompanies courtship in some mammals, such as deer and elk.

Most California mammals retreat from bright light. Nevertheless, light serves important functions for mammals. For example, the light/dark cycle is essential to the timing or entrainment of daily rhythms. Also, some nocturnal species, such as bats, may employ environmental light in navigation. For instance, the Big Brown Bat *(Eptesicus fuscus)* uses the glow of the evening sky in determining the direction in which it begins its evening foraging; the directional movement of captives can be altered by changing the source of light.

Vocalization

Sound production is very common among most vertebrates, and virtually all mammals use sound under certain conditions. Although sound in mammals is more varied and extensive than it was thought to be a few years ago, it is still poorly understood in comparison to sound in birds, on which researchers have assembled detailed information. Sound is important partly because it does not take much energy to produce. The metabolic expense of a bark is minimal compared to that of producing scents.

Many mammals produce ultrasonic sounds (frequencies above 20 kHz). A broad range of ultrasonic frequencies is available, but sounds at these frequencies dissipate quickly. For this reason, ultrasonic frequencies are used for communication only over very short distances. Because humans can only detect ultrasonic vocalization using extremely sensitive instruments, the full extent to which mammals use such frequencies may not be known for some time.

Small young of some rodents emit ultrasonic squeals of alarm that quickly draw the mother to the nest. This maternal response is strong in nursing females, which are attracted even to young not their own.

Laboratory rats and mice also produce characteristic sounds during courtship and mating. These copulating calls reach 22 kHz in rats and approximately 70 kHz in mice. The pitches vary by gender and have the effect of stimulating mating. They are, moreover, under the control of the testicular hormone testosterone and disappear after removal of the testes. In the Little Brown Bat *(Myotis lucifugus)*, there seem to be none of the premating tactile or visual overtures characteristic of most mammals. As the male mounts the female, however, he utters a highly distinctive copulation call that induces acceptance by the female.

Various species of insectivorous bats emit lower-pitched calls of distinctive frequency and duration. Carrying farther than the ultrasonic calls, lower frequencies are probably more important for intraspecific recognition. Vocal communication among Pallid Bats *(Antrozous pallidus)* is believed to help flying individuals locate their roosts and also to maintain the association of females and fledged young.

Simple calls, such as the clear two- or three-note whistle of the Pika *(Ochotona princeps)*, may be given by both sexes and may be distress or alarm calls as well as territorial announcements. Many kinds of mammals emit alarm calls. These are always short and usually consist of a single note. Similarities exist among the alarm calls of different species; the alarm call given by a Beechey Ground Squirrel *(Spermophilus beecheyi)* is recognized as a danger signal by the California Quail *(Callipepla californica)*.

Some mammals also utter prolonged songs. Male Pikas, for example, do so during the breeding season. These songs are individually distinctive and may provide the basis for sexual as well as individual identification. The ability to produce songs has also been reported for species of whales, some voles *(Microtus),* the House Mouse *(Mus musculus),* and the Grasshopper Mouse *(Onychomys leucogaster* and *O. torridus).* The latter emits a complex of sonic patterns at both moderate and very high frequencies.

Apart from precise vocal signals with specific meanings, some mammalian sounds appear to express rather general emotions. These sounds are commonly accompanied by a characteristic body movement, and the vocal and visual signals clearly reinforce each other. Low-pitched sounds usually indicate hostility or aggression. Typical examples are the growls and other guttural sounds that are accompanied by grimaces or a show of teeth—characteristic mammalian expressions of hostility. In contrast, the high-pitched squeals of a submissive puppy often go along with a crouching or supine position. Such vocalizations can be heard in many orders of mammals.

Echolocation

Some mammals locate food and detect obstacles through the echoes from ultrasonic or other highly pitched sounds (as in bats), or from lower sounds within the normal range of human hearing (as in whales and porpoises [Cetacea]). Such echoes reveal the distance and size of the object from which the sound waves bounce; if the object is moving, they may also indicate the speed and direction of movement. Echolocation has been most intensively studied in bats, which can maneuver and capture flying insects in the dark, but it is also known to occur in shrews, whales, porpoises, and several species of birds.

Echolocation depends on the relationship between wavelength and frequency. Sound in air travels at a constant 344 m/s (or 34.4 cm/ms), except for minor variations introduced by air temperature. The constant speed of sound means that wavelength decreases as frequency increases. Bats are capable of emitting very high frequency sounds, up to 200 kHz, and their specialized ears can hear echoes from these frequencies. (The inner ear of a bat is not in direct contact with the bones of the skull; this may be a mechanism to prevent the bat from hearing sounds of its own body, such as those from jaw movements.) The very short ultrasonic sound waves can bounce back from a very small object, such as a mosquito. In contrast, a sound wave at a lower frequency is longer than, for example, a moth, and thus would not bounce back from the moth. In addition, because high-frequency sounds dissipate rather quickly, sound emissions from one foraging bat are not likely to jam those from another. Thus, ultrasonic emissions are well suited for the echolocation of small objects. The production of sound by bats is complicated by harmonics, sound waves the frequency of which can be divided by the frequency of the fundamental sound.

Typically the high-frequency emissions of North American bats vary over a brief period of time: the emission frequently ends at a lower pitch than it begins, a phenomenon known as frequency modulation (FM). In bats of the family Vespertilionidae, the FM pulse may begin near 100 kHz and drop to near 20 kHz. FM enables echoes to indicate the movement of their source. For example, if a moth is moving away from a bat, the lower-frequency echoes emitted nearer the end of the pulse take longer to reach the ears of the bat. The bat's complex brain can determine its distance from the moth; if the sound is heard by both ears, the bat can also determine its direction. FM may provide some clue to the size of prey as well as its position.

Ultrasonic emissions are pulsed so that echoes return from objects during the quiet periods between emissions. The bat's ear muscles contract during signals, so that it does not hear its own sound, but relax in time to receive the echoes. As a bat approaches its prey, echoes travel over a shorter distance; the bat then reduces the duration of the impulse, so that the echo continues not to interfere with the emission. Thus the emissions must be very brief, because some objects from which echoes might return might be very close. Echolocation pulses may be as brief as 5 ms. To avoid confusion of signals, a hunting bat might emit pulses lasting only from .2 to .5 ms.

The volume of the emission varies with the size of the prey, greater volumes being used for small flying insects and low volumes to detect insects that are larger or stationary. Species called gleaners use low volumes to detect the echoes and sounds coming from stationary insects. Gleaners usually have greatly enlarged, moveable ear pinnae that enhance their capacity to determine the direction of the prey's sounds.

Sound travels much faster under water, and whales and porpoises emit impulses that can be heard by human ears. They also detect sounds produced by prey.

MIGRATION AND MOVEMENTS

Three distinct types of movements are characteristic of wild mammals. Movements that involve an eventual return to the place of origin are called migrations, regardless of their length. We commonly think of migrations as being annual, but some are seasonal. Some are even daily: because daily foraging activities frequently end in a return to a nest or resting site, they qualify as migrations. Nevertheless, we have come to consider such short-term excursions as movements within home ranges. Even very brief movements, however, may constitute migrations, and at times it is hard to distinguish between daily foraging, dispersal, and clear-cut annual migrations. Finally, all animals appear to have a life stage in which they depart from the home site and assume residence in another area; this sort of movement is called dispersal. Thus, the origin of a migration is seldom the exact place of birth, although it is sometimes very close.

Migrations

Probably in the earliest human societies, aborigines followed migrating ungulates, and a knowledge of the factors affecting the times of arrival and departure of these species was essential to the well-being of the hunter groups. Today, the migration routes and schedules of herds of wild American ungulates, such as the Black-tailed Deer *(Odocoileus hemionus hemionus)* and Mule Deer *(O. h. columbianus),* provide part of the basis for determining the times and locations of legal hunting of these animals. The short-range and long-range movements of bears are still poorly known, but radiotelemetry has begun to show the movements of individual male and female Black Bears *(Ursus americanus)* as well as where bears are prone to den for winter. This information helps game managers minimize damage from these potentially dangerous carnivorans and plan the times and places for public hunting of bears.

Throughout the Sierra Nevada, deer (Cervidae) migrate every fall from the higher elevations. In Tulare County they spend summer between 2,000 and 3,000 m and move downhill to between 500 and 1,200 m in winter. In a study by the Department of Fish and Game, one doe was followed by radiotelemetry for a year. She had a summer home range with a radius of approximately 1 km (.5 mi). In early October she moved to a lower elevation, where she again stayed within an area of 1 km in radius. In late May she returned to the summer range, a distance of about10 km (6 mi) in a direct line, in less than three days. In this study some individuals, marked by bells, moved annually to summer and winter ranges roughly 30 km (18 mi) apart.

The North Kings deer herd in eastern Fresno County has discrete migratory paths. Except during migration, movements are slight and adjusted to weather and forage conditions. Some individuals make exploratory excursions to higher elevations in late winter, when the winter range and forage are deteriorating. Movements are irregular in spring, for deer sometimes pause for several days or even weeks along their narrowly prescribed migration paths. The uphill migration takes the deer to several elevations, and individuals tend to return to the home range they occupied the previous year. A differential ascent results in bucks reaching higher elevations, above most of the does. They return to lower elevations with early storms in October, using the same routes as in spring, but the distance traveled and the time taken depend on the severity of fall weather.

Bats are migratory, but there are conspicuous differences among species, and within a species, migratory patterns may differ between the sexes. These variations are characteristic of species, but the patterns seem unnecessarily diverse under modern climatic conditions.

Tree bats *(Lasiurus* and *Lasionycteris)* make the most extensive movements of any North American bats. Tree bats are found far into Canada in summer, and the Hoary Bat *(Lasiurus cinereus)* is found from the north end of Hudson Bay southward to central Mexico; its migratory southern destination is unknown. Tree bats are powerful fliers and move far to the south every fall. Not uncommonly, they are found on ships 160 km (100 mi) or more from the continent, and nonmigratory insular populations of *Lasiurus* occur in the Hawaiian Islands, the Galápagos, the West Indies, and Bermuda.

In contrast, the various species sometimes referred to as cave bats (*Myotis, Eptesicus, Pipistrellus,* and other genera) migrate from a cave or mine shaft (the hibernaculum) in spring to a summer range and return to the hibernaculum in fall. These migrations are not necessarily north-south in direction, and some individuals may not cover much ground. These migratory patterns have been extensively studied in the United States and Europe by banding bats when they are hibernating and recapturing them in their summer quarters. By hibernating, these species leave a hostile winter climate without leaving a geographic region.

Migratory patterns of the Guano Bat *(Tadarida brasiliensis),* one of the cave bats, have been studied by E. Lendell Cockrum of the University of Arizona. Some populations of this bat summer from Arizona eastward to Texas and migrate south into Mexico for winter, flying more than 1,600 km (1,000 mi) each way. California populations, however, seem to make only local movements in spring and fall, and some individuals in southern Nevada and western Arizona may winter in southern California. Females move north in spring and assemble in large nursery colonies, where the young are born. Most males remain in the south, for example, one summer colony in the Mexican state of Chiapas consisted of 40,000 males and no females.

Migratory patterns probably evolved so that species could escape adverse environmental features and pursue beneficial ones. There seems to be a powerful genetic component to these patterns. Nevertheless, they are clearly sensitive to minor environmental changes today, as the preceding examples indicate. One can speculate on the environmental pressures behind the evolution of these rhythmic movements. In the case of the Mule Deer, the logic seems fairly clear. Fresh food occurs throughout summer only at higher elevations, but deep snow in winter precludes grazing there. Also, during the downhill migration gonads are growing, and the sexes share in the winter range, where mating occurs. The annual migration of this mammal is clearly adjusted to both the occurrence of fresh forage and the need for joining of the sexes.

Migration can take many different forms. The annual movements of cave bats such as *Myotis* to and from their hibernacula involve radiation from the cave in many different directions with no single geographic trend. This movement is profoundly unlike the general north-south movements of a *Lasiurus* or a *Tadarida;* the latter two movements themselves appear to be rather different, with *Tadarida* entering a cave for winter and *Lasiurus* remaining at least somewhat active in the southern part of its range. Migrations in these three groups quite possibly evolved independently.

Remember, however, that modern migratory patterns may have evolved under markedly different climatic conditions. We may still be in the Pleistocene, with drastically fluctuating climates. In the rather recent past, from about 1450 to 1850, the Little Ice Age saw lowered temperatures and reduced agricultural productivity in the midlatitudes. In contrast, prior to about 1000, there was a period of mild temperatures, and successful agricultural establishments existed in Greenland. Patterns of migration likely evolved to allow survival under a rather broad climatic spectrum. The Pleistocene saw the disappearance of many mammals, small as well as large, and those species surviving today are those sufficiently flexible to accommodate to the sudden and drastic environmental changes of that period. This may be one reason that some wild mammals may not seem perfectly adapted to conditions that exist at this point in geological time. For example, the diversity in migrations is not easily explained as a set of finely tuned adaptations to current conditions.

Home Ranges

Most small mammals, such as mice, squirrels, and rabbits, forage daily over more or less the same area, each individual (or sometimes family) remaining essentially on the same ground every day. This area provides adequate food and cover for the individual and is called its home range. If its area of residence is invaded by another of its own kind, the resident mouse or squirrel may be hostile and drive out the alien, or the alien may take over. The part of the home range that is defended against intruders is called a territory. Territories may be quite apparent to observers of such mammals as squirrels and sea lions, which are obvious in displaying their social relationships, but bats appear to forage harmoniously over the same ground. The distinctions between territories and home ranges are not always clear, and there are important differences in these concepts among various species.

Although a theoretical function of male territoriality is the exclusion of other males from his mate or mates, this effort is frequently only partly effective. A bull sea lion or fur seal may defend his surroundings against intrusion by other bulls, but the cows may move from one territory to another during their fertile periods. A dominant male Douglas's Squirrel *(Tamiasciurus douglasii)* is often less than 100 percent successful in preventing the encroachment of subordinate males into his territory, and females may mate with the intruders.

Although most studies of home ranges of small mammals are based on live trapping and releasing animals, radiotelemetry has also been useful. Because radiotelemetry reveals the actual location of an animal during its natural meanderings, details of its home range as well as seasonal changes can be carefully plotted by anyone willing to spend the necessary time in the field.

Some mammals usually stay very close to home. The Beechey Ground Squirrel, *(Spermophilus beecheyi),* for example, forages close to its burrow entrance; it rarely moves far, nor does it remain away for long. This squirrel was studied in great detail at the Hastings Natural History Reservation in the Carmel Valley by the late Jean Linsdale and his associates. Like many rodents, the Beechey Ground Squirrel makes most of its brief trips along well-worn paths. Young individuals may stray no more than 5 m or so from the burrow entrance, but adults frequently move out more than 30 m. An abundance of fresh grass and small forbs allows the squirrels to feed close to their burrow entrance; as they deplete this food supply, they move farther away. Professor Linsdale observed that almost all daily movements were less than 100 m one way and that the few long journeys of 320 m or more involved a change in residence.

Similarly, wood rats have rather small home ranges. The Desert Wood Rat *(Neotoma lepida)* seldom goes more than 7 to 10 m from its home. Because home ranges overlap only slightly, there seems to be a clear element of territoriality in their boundaries. Home ranges of the Dusky-footed Wood Rat *(Neotoma fuscipes)* have been meticulously plotted by radiotelemetry. As with small mammals generally, adult males have larger home ranges (averaging more than 2,200 m^2) than adult females (1,900 to 2,000 m^2) or juveniles (about 1,700 m^2). The home range is not fixed, for when vegetation dries up after the end of the spring rains, some wood rats move widely, presumably in search of food. Females nursing young remain close to their nests, however, and have home ranges of less than 1,500 m^2. Other seasonal changes in home range may also relate to reproduction: male and female home ranges tend to overlap during the breeding season and be mutually exclusive at other times.

Radiotelemetry suggests that Kit Foxes (*Vulpes macrotis*) in the San Joaquin Valley hunt in family groups. They may forage over the same ground as other family groups, but not at the same time. Thus, they consistently use a home range but seem not to defend territories against other Kit Foxes. They do not move extensively; a Kit Fox may spend its entire life within a 5 km^2 (2 mi^2) area.

Radiotelemetry and tagging of Black Bears (*Ursus americanus*) by researchers in the Department of Fish and Game reveal that in summer, an adult bear ranges over an area of from 12 to 25 km^2 (5 to 10 mi^2). Movements increase in fall, but there is no well-marked migration. Autumnal movement may be undertaken in search of food or a place to hibernate.

Among many larger herbivores, social organization modifies the nature of home ranges. An ungulate cannot disappear in a hole in the ground but rather depends on running to escape predators. A herd using common ground can employ multiple eyes, ears, and noses to detect enemies.

In the Roosevelt Elk (*Cervus elaphus roosevelti*), a home range is an area occupied by a herd of cows and their offspring. At all seasons it is located near a good supply of food. The bulls remain separate except during the breeding season. In the Pronghorn (*Antilocapra americana*), the herd instinct also prevails, and the herd moves to the most favorable food supplies. On less productive land, it changes its home frequently.

Individual home ranges of Black-tailed Deer are the same from year to year. At higher elevations they are larger: the forage is less abundant, and the seasonal plant growth later, and the higher elevations may make greater metabolic demands than do habitats below 1,800 m. At the highest elevations a deer might range over a 10 km^2 (4 mi^2) area, whereas below 1,800 m, the home range is usually less than half that size.

Homing

An important factor in both daily movements and migrations over long distances is an animal's homing ability. Although little is known about the sensory input that indicates to a mammal its geographic location, homing or displacement experiments have shown that many kinds of bats can find their way home from up to several hundred miles away in a few days. Students have transported banded bats various distances and directions from their home caves and then waited for their return. The Guano Bat (*Tadarida brasiliensis*) can return from up to nearly 640 km (400 mi) away, traveling about 32 km (20 mi) a night. The Pallid Bat (*Antrozous pallidus*) has been known to cover up to 51 km (32 mi) in one night, and the Little Brown Bat (*Myotis lucifugus*) can return 96 km (60 mi) the same night it is released, covering about 6 km/hr (4 mi/hr).

As interesting as these experiments are, they do not reveal the means by which a bat finds its way home. The homing ability has an important role in the annual cycle of many kinds of bats. The annual movement away from a cave in spring may be hundreds of kilometers, and individual bats return to the same caves the following fall. Some of these distances are so great as to preclude the likelihood of familiarity with landmarks en route.

To find its way home from an unfamiliar region, an animal first must be able to determine its new position. This process is called orientation. Anyone who has been lost in a forest is aware that most humans have a poorly developed sense of orientation. Second, the animal must have the ability to determine direction. This process is called navigation. Different kinds of animals use various sorts of navigational cues, which vary with their sensory ability. Orientation and navigation together enable the animal to return home from an unfamiliar region.

Although many bats clearly have the ability to navigate, our understanding of this ability is far less advanced than it is regarding birds. Some students have released bats with covered eyes and bats with uncovered eyes at varying distances from their diurnal resting sites and observed the speed with which they return. Generally, bats with their eyes covered seem to return less frequently and less quickly than those that can see. Similar experiments have found that at distances of 16 km (10 mi) or more, the lack of vision impairs the homing of some species of bats. This pattern suggests that some bats have a familiar home range with a radius of approximately 16 km (10 mi) and can find their way home by echolocation, but that at greater distances vision is required for effective navigation.

Homing ability may be rather limited in some mammals that do not naturally make extensive movements. Among Deer Mice *(Peromyscus maniculatus)*, for example, individuals displaced only 100 m from their home range may not return; there are always some individuals that assume a new residence near the point of release or in another area. On the other hand, a Great Basin Pocket Mouse *(Perognathus parvus)* managed to return to its burrow overnight from a distance of roughly 2 km (about 1 mi). Generally, the differences between the homing abilities of mice and bats undoubtedly reflect not only a difference in motivation but also a fundamental lack of ability of many if not most mouse species to orient and navigate over long distances.

Dispersal

Dispersal, the permanent departure of an individual from its birthplace, occurs at some point in the life cycle of all plants and animals and seems to be an innate drive in all species. Dispersal almost always occurs prior to sexual maturity, although in some species it takes place after mating but prior to birth of the young (or oviposition, in the case of egg-laying animals). This innate drive of the young may be assisted by the increasing hostility of the parents as their offspring approach adulthood. Beaver *(Castor canadensis)* colonies, for example, typically consist of an adult male and adult female, the young of the year, and young of the previous year. As the older offspring reach sexual maturity, at about two years of age, the adult male drives them from his territory. Dispersal, however, is seen early in some mammals and is not always associated with parental behavior.

Most individuals disperse from the natal area to another well within the geographic range of the species. Dispersal tends to reduce inbreeding and enhance the genetic heterogeneity and vigor of the population. Dispersal may take a mouse only a few meters from its birth site, but a young Coyote *(Canis latrans)* may travel many kilometers before settling in a new area.

The tendency of dispersing mammals to remain in the environment to which they are adapted is well illustrated by the spread of the Muskrat *(Ondatra zibethicus)* in California. This aquatic rodent was brought into the state for its value as a furbearer and was raised captive in the 1920s. Previously it occurred only along the margins of the Colorado River and along the eastern slope of the Sierra Nevada in California, although it is widespread along slow streams and marshes in much of North America. Because the value of Muskrat skins does not justify the cost of rearing them to maturity, disappointed managers of Muskrat farms released them in the Central Valley and a wild population soon became established. Subsequently the Muskrat dispersed along irrigation canals and other slow-moving watercourses throughout the valley. Its distribution within central California, however, remains almost totally restricted to the immediate margins of irrigation ditches and sluggish streams. Dispersal along any other route is exceptional.

At the edge of a species' geographic range, dispersing individuals may move into areas not already occupied by that species. In this situation, dispersal may promote the expansion of the geographic range of the species. Most commonly, such dispersal is not permanent. If peripheral regions were favorable for the existence of the species, they would already have been populated. Nevertheless, around the margin of a species' geographic range, dispersal constitutes a continuous or pulsating pressure for expansion. A fluctuation in climatic conditions (or human-induced changes in the environment) may alter a marginal habitat to make it suitable. Because climatic changes occur on a small scale every year, small shifts in the geographic ranges of mammals do occur. When climatic changes are major and extended, such as those that occurred during the Pleistocene, there may be vast shifts in the geographic ranges of mammals. Entire faunas may move. Moreover, different species may move in separate directions, for their environmental demands are seldom the same.

Shifts in geographic ranges of mammals due to natural changes in the environment occur slowly and are not frequently witnessed during the human life span. Human-induced changes causing such shifts may be rapid and profound, however. In California, the Cotton Rat *(Sigmodon hispidus)* has spread from its original home along the Colorado River into the Imperial Valley with the advent of irrigation. This region was previously desert, a hostile and deadly environment for the Cotton Rat. But when irrigation created a suitable habitat, the species spread into the Imperial Valley through dispersal.

Clearly, animal movements are not random and disorganized. Throughout an individual's life, there are phases of dispersal, occupancy of a home range, defense of a territory, and sometimes migration. The life cycle can be viewed as a series of successive types of movements, each with its own special function, and comparisons among species aid in appreciating the roles of movements in nature.

MAMMALS AND
CALIFORNIA SOCIETY

Since the arrival of humans in the New World, populations of wild mammals have been critical to our welfare. Prey species, such as deer (Cervidae), rabbits (Leporidae), and ground squirrels (*Ammospermophilus* and *Spermophilus*), provided food and clothing for the first Indians and today continue to draw many people to the outdoors. Furbearers were the stimulus for much early exploration and continue to occupy the efforts of a small number of trappers. Sightings of wild mammals provide countless hikers, campers, and picnickers with a thrilling and highly valued experience, and the economic importance of these outdoor enthusiasts to makers of outdoor gear, optics, cameras, and books should not be underestimated. Likewise, there is a thriving (and growing) industry of whale watching and other nature-oriented touring.

On the negative side, some species are injurious to crops or domestic stock, and others carry diseases transmissible to humans. Ground squirrels, for example, not only take some forage and seeds but also are reservoirs of plague and other diseases and usually unpopular with landowners (although they are favorites of park visitors). The majority of California mammals, however, are innocuous or provide humans with economic, scientific, and aesthetic benefits. Some species are more charismatic than others, but there is a growing attitude that every species is important, as is diversity.

Although extinction is a normal process, human activities (e.g., habitat destruction, water pollution, and the introduction of nonnative species such as feral cats) have accelerated the rate at which species are disappearing. The expansion and increased intensity of human activities bring greater pressure on populations of wild vertebrates. Agriculture and environmental pollution can alter floras and arthropod populations, which are essential to the well-being of wild mammals. The previous widespread use of the insecticide DDT, now outlawed in the United States, probably accounted for a decline in numbers of bats (Chiroptera) by reducing their food supply. Some bats are highly beneficial because they consume enormous quantities of mosquitoes and other insect pests; their populations are now gradually recovering.

Even when wild animals are not directly destroyed, habitat alterations may reduce the carrying capacity of the land—its ability to provide adequate food and cover. When the carrying capacity for a given species is reduced, its numbers tend to decline, or individual body size may be reduced. In some situations, habitat alteration may increase the carrying capacity for other species. For example, the removal of a large stand of firs and pines, allowing the development of a rich cover of grasses, forbs, and shrubs, increases the carrying capacity for deer, ground squirrels, bears (Ursidae), and chipmunks (*Neotamias* spp.) while decreasing that for the Northern Flying Squirrel *(Glaucomys sabrinus)* and Douglas's Squirrel *(Tamiasciurus douglasii).* Adjustments to habitat changes are sometimes slow and may result in changes in geographic ranges. Highway traffic is always destructive to wildlife. Perhaps the two species of skunks are most commonly seen dead on the road, but automobiles are not selective and destroy numbers of rare mammals as well.

No native species of California mammal has become extinct in historic times, but certain well-marked subspecies (races) appear to have passed into history. Species protection and habitat protection and restoration in California have been reasonably successful; however, California has more endangered species than any other state. (This is partly because it has a greater variety of species to begin with.)

Conservation of Mammals

The vast majority of mammals can adjust to small environmental changes: their abundance and local distributions may fluctuate, but extinction is not usually an immediate danger. Some species, however, because of specific environmental requirements or sometimes because of slow reproductive rates, are especially vulnerable to environmental disturbances. Habitat changes can either reduce the numbers of such species or greatly restrict their geographic range. To preserve populations of wild species at levels compatible with concepts of multiple use of land, regulatory agencies have given scarce species particular recognition and protection.

The California legislature in 1970 passed the Endangered Species Act, which provides for recognition of rare and endangered species by the Fish and Game Commission. The Department of Fish and Game (DFG) has reviewed these species in a series of documents entitled *At the Crossroads*, published biennially beginning in 1974. Annually, the Department also publishes a review of these species, *The Status of Rare, Threatened, and Endangered Animals and Plants of California*. Congress enacted legislation to protect endangered animal species in 1966; in 1973 it enacted the federal Endangered Species Act, which covers plants as well as invertebrate and vertebrate animals.

Federal law currently defines an endangered species as a form faced with an immediate danger of extinction in at least a significant part of its geographic range. A threatened species is one that is rare and approaching endangered status. The state also identifies some species as endangered and designates others as rare—that is, threatened. Some mammals identified by the state as rare or endangered are not given any special status by federal regulatory agencies. Table 2 lists the California mammals designated by state and federal authorities as rare, threatened, or endangered. The DFG has also designated a large number of mammals as species of special concern: species that, because of declines in habitat, food sources, and other vital requirements, may become endangered. Finally, it has designated some species as fully protected, meaning that they may not be taken or possessed at any time, except (with a state permit) for necessary scientific research. A complete list of these species and subspecies can be found on the internet at www.dfg.ca.gov/hcpb/species/species.shtml.

Mammals listed in Table 2 are totally protected. In many instances, suitable habitats on public land are set aside for one or more of them. Some of them have very small populations, however, and their survival is uncertain. As numbers dwindle, vulnerability increases. Some populations of Bighorn Sheep *(Ovis canadensis)*, for example, are dangerously small and may be unable to survive increasing predation by Mountain Lions *(Panthera concolor)*, poaching, and competition from Burros *(Equus asi-*

TABLE 2. Rare or Endangered California Mammals: 2003

Mammal	State	Federal
Buena Vista Lake Shrew *(Sorex ornatus relictus)*		Endangered
Island Fox *(Urocyon littoralis)*	Threatened	
San Joaquin Kit Fox *(Vulpes macrotis mutica)*	Threatened	Endangered
Sierra Nevada Red Fox *(Vulpes vulpes necator)*	Threatened	
Southern Sea Otter *(Enhydra lutris nereis)*		Threatened
Wolverine *(Gulo gulo)*	Threatened	
Guadalupe Fur Seal *(Arctocephalus townsendi)*	Threatened	Threatened
Steller Sea Lion *(Eumetopias jubatus)*		Threatened
Sperm Whale *(Physeter catodon)*		Endangered
Sei Whale *(Balaenoptera borealis)*		Endangered
Blue Whale *(Balaenoptera musculus)*		Endangered
Fin Whale *(Balaenoptera physalus)*		Endangered
Humpback Whale *(Megaptera novaeangliae)*		Endangered
Northern Right Whale *(Eubalaena glacialis)*		Endangered
California Bighorn Sheep *(Ovis canadensis californiana)*	Endangered	Endangered
Peninsular Bighorn Sheep *(Ovis canadensis cremnobates)*	Threatened	Endangered
Point Arena Mountain Beaver *(Aplodontia rufa nigra)*		Endangered
Nelson's Antelope Ground Squirrel *(Ammospermophilus nelsoni)*	Threatened	
Mojave Ground Squirrel *(Spermophilus mohavensis)*	Threatened	
Morro Bay Kangaroo Rat *(Dipodomys heermanni morroensis)*	Endangered	Endangered

Giant Kangaroo Rat *(Dipodomys ingens)*	Endangered	Endangered
San Bernardino Kangaroo Rat *(Dipodomys merriami parvus)*		Endangered
Fresno Kangaroo Rat *(Dipodomys nitratoides exilis)*	Endangered	Endangered
Tipton Kangaroo Rat *(Dipodomys nitratoides nitratoides)*	Endangered	Endangered
Stephens' Kangaroo Rat *(Dipodomys stephensi)*	Threatened	Endangered
Pacific Pocket Mouse *(Perognathus longimembris pacificus)*		Endangered
Riparian Wood Rat *(Neotoma fuscipes riparia)*		Endangered
Salt Marsh Harvest Mouse *(Reithrodontomys raviventris)*	Endangered	Endangered
Amargosa Vole *(Microtus californicus scirpensis)*	Endangered	Endangered
Riparian Brush Rabbit *(Sylvilagus bachmani riparius)*	Endangered	Endangered

nus). Once a population gets too small, it may face problems even if it recovers. For example, the Northern Elephant Seal *(Mirounga angustirostris),* whose numbers were drastically reduced by over-hunting in the 1800s, has made a spectacular comeback (from 100 individuals 100 years ago to more than 100,000 today), but because these animals are all derived from the same 100 ancestors, their limited gene pool may create problems if the species faces changing environmental conditions in the future.

The DFG not only maintains preserves for wild species but also provides for conservation programs, including research on rare species of mammals. This work is supported in part by fees from hunters, trappers, and fisherpeople, many of whom are active conservationists aware of the need to maintain viable populations and habitat. Other financial contributions come from the sale of native species stamps, personalized license plates, and passes to wildlife preserves, and from the Endangered Species Tax Check-off Program.

Many nongovernmental organizations dedicate themselves to preserving charismatic species such as sea mammals. Less glamorous animals, such as bats, have benefited from the popularizing efforts of a small group of people, and today, children returning from summer camp are likely to bring back bat roosting boxes instead of bird houses. The Nature Conservancy, supported by private contributions, has been very successful in protecting wildlife by purchasing and preserving critical lands on which endangered plants and animals exist and also in working with state agencies and private landowners to restore ecologically unique ecosystems. Other prominent conservation groups, such as the National Audubon Society and the Sierra Club, may also contribute financially to species survival and to habitat preservation.

Legal protection of wild mammals and their environment is usually based on scientific logic, but sometimes attitudes and decisions— including designations of protected species—reflect ideological, political, economic, and emotional factors. For example, in 2001, a representative to the International Whaling Commission from Japan declared that Minke Whales *(Balaenoptera acutorostrata)*, a popular food species in that nation, and one seen off California's coast, were "cockroaches" and "voracious gluttons" that were stripping the oceans of fish. One can only wonder how fish populations survived the onslaught of Minke Whales for so many millions of years. At the other extreme, when it was proposed that a herd of 29 common Black-tailed Deer *(Odocoileus hemionus)* in a suburban park in central California be euthanized, activists insisted that it instead be translocated to a wilderness park, an unknown area with food plants foreign to them. After four months in the wilderness, 23 of the successfully transplanted deer were dead. Some fell victim to Mountain Lions and Coyotes *(Canis latrans)* (of which they had no experience), and to car, train, and fence collisions. Others died of unknown causes, possibly also predation, and were subsequently scavenged. The only confirmed survivor at the end of one year walked 40 km (25 mi) in a straight line to an urban area near its place of birth. The estimated cost of the translocation was $85,000, not including volunteer, staff, and administrative costs to the DFG and the park district.

Some rare or endangered mammals occur on private lands, and their presence and protection may involve varying degrees of expense to the landowners. Large mammals, such as the Tule Elk *(Cervus elaphus nannodes)*, move freely from public to private land, at times breaking fencing and eating forage intended for cattle. Small rodents may damage crops or compete with grazing stock. This problem may not be costly, and it can be argued that rodents actually benefit grasslands by aerating the soil and dispersing seeds. But where economic interests are concerned, conservation is often hotly contested.

It is widely believed that land can never be cultivated or developed if any endangered species is found on it, but that is not always so. Ideally a rare species should be left unmolested, but when economic interests take precedence, the DFG has a long-standing policy of accepting a certain amount of habitat destruction. Sometimes it does this in exchange for mitigation, which is, in effect, a compromise. A landowner may agree, for example, to donate part of the property to open space, donate or restore property elsewhere and relocate the animals, or donate a small percentage of profits to other conservation efforts. In actual practice, the discovery of listed species on property results more often in delays and extra cost than in stoppage.

Landowners are legally bound to honor the designations of rare and endangered forms, but they are not compensated for the trouble or expense of complying with the law. Conservationists and the general public usually agree on the need for special protection for scarce animals, but they do not always pay the cost of this protection. Although these socioeconomic problems are usually minor, they are real and commonly neglected.

Occasionally, a landowner disregards the law and purposefully plows a field to rid his land of a listed species. One case in southern California received much press, because the landowner, who purchased and plowed the land knowing of the presence of an endangered kangaroo rat (*Dipodomys* sp.), received much support from activists opposed to the Endangered Species Act. The prosecution of such cases does not help the species that are displaced or killed.

Common Conservation Methods

Protection of either a species or its habitat is intended to prevent its further decline or even extinction. No one state or nation can protect the habitat of a whale, for the sea is a multiple-use environment. It is relatively simple, however, to outlaw the capture of certain marine species. In contrast, preservation or restoration of a habitat such as the North Bay salt marshes ensures a home for the Salt Marsh Harvest Mouse (*Reithrodontomys raviventris*) and a host of other relict species.

Habitat protection is generally the most simple and successful means of protecting rare species and has the advantage of preserving whole biological communities; land to which animals are relocated, whether natural or restored, may lack food species or other features critical to the survival of the transported species. Success in habitat preservation depends partly on preserving parcels of adequate size. The maintenance of wildlife corridors (links between shrinking areas of prime habitat) is increasingly seen to be important. For example, clear-cutting may create islands that the Wolverine *(Gulo gulo)* cannot leave for lack of a continuous forest canopy.

Restoring habitats is a more complex challenge than preserving them. It often requires the elimination of nonnative plants and animals—a time-consuming and expensive task. Released exotic animals, such as rabbits and cats *(Felis sylvestris),* can cause extensive habitat destruction and loss of native species. For example, the Island Fox *(Urocyon littoralis),* which prefers chaparral and woodland, is threatened by exotic herbivores that eat and uproot native vegetation; this fox is also out-competed by feral cats and plagued by the cats' parasites. A native species (the Golden Eagle *[Aquila chrsyaetos],* a DFG species of special concern) is being removed from the fox habitat as well, to ensure the survival of the more threatened animal, and is being replaced by the formerly native Bald Eagle *(Haliaeetus leucocephalus),* which does not prey on the fox. The introduced nonnative race of the Red Fox *(Vulpes vulpes),* now common in the state, also has detrimental effects on the native fauna: it is a predator of the endangered Clapper Rail *(Rallus longirostris)* and a competitor of the native Kit Fox *(V. macrotis).*

Protection of a species is most feasible for those that occupy a small geographic area, for depredations against them are more easily controlled. Some species of kangaroo rats, for example, occur on limited ranges, and their presence or absence is not difficult to determine. Furbearers, such as the Marten *(Martes americana)* and the Fisher *(M. pennanti),* are also relatively easy to protect because the sale of furs is recorded and traffic in pelts becomes public knowledge.

The success of programs supporting the growth of rare species populations depends on the reasons for the species' rarity and the expertise of the people involved in the programs. Relocation is the most effective conservation strategy for such large herbivores as the Wapiti *(Cervus elaphus),* but it risks mixing genetically diverse stocks. Moreover, it is not guaranteed to last. One herd of Tule Elk *(C. e. nannodes)* released on dry Mount Hamilton traveled miles to settle near a reservoir and nearby ponds with at least some tules. Captive breeding and restocking are most expensive and require a population viability analysis (PVA) estimating the likelihood that a reintroduced population of a given size will survive. Usually data are inadequate for a PVA. Release of captive bred herbivores is likely to be successful if the PVA is promising, but release of captive bred carnivorans requires that the prey population be adequate and that the released animals be capable of hunting, which they may not be. PVAs are clearly useful for endangered species, such as the Stephens' Kangaroo Rat *(Dipodymus stephensi)* in Riverside County.

Mammals Recently Extinct in California

Within historic times, four species of mammals—the Gray Wolf (sometimes just called Wolf) *(Canis lupus)*, Grizzly Bear *(Ursus arctos)*, Jaguar *(Panthera onca)*, and Bison *(Bison bison)*—have disappeared from our state. Each survives elsewhere, however, so their elimination from California is properly called extirpation, not extinction. One can express regret that these magnificent beasts no longer exist in our state, but were they to roam freely over our lands today, they would cause some serious problems. They all demand large pieces of land, and all have habits that conflict with human activity.

The Gray Wolf occurred along the eastern edge of the state and in the Central Valley. Early travelers mentioned Wolves in many parts of California, but some of these animals may have been Coyotes. In his diary of his gold rush experiences in 1850, J. Goldsborough Bruff frequently wrote about wolves, which he distinguished from Coyotes on the basis of both size and color. Bruff noted that Wolves occurred both near Sacramento and in Calaveras County. In the early part of the twentieth century the Bureau of Biological Survey employed trappers and hunters to eliminate the Wolf from cattle ranges. The Wolf disappeared from Nevada in 1923. In California, the last Wolf was captured southwest of Tule Lake in 1924. The last Wolf in Oregon was killed in 1974. More recent reports of Wolves in the Sierra Nevada are based on an animal of Asiatic origin, presumably an escaped captive or pet. As the descendants of wolves released in Yellowstone National Park in Wyoming spread to other states to the west, however, it is reasonable to expect their appearance in northwestern California in the near future.

The Grizzly Bear is the New World representative of the Brown Bear *(Ursus arctos)*. The Grizzly once occurred widely throughout the western United States and was a constant threat to humans and domestic stock. Its total lack of fear combined with its destructive habits rendered it a dangerous member of our fauna. Up to the 1850s or 1870s, the Grizzly roamed the Central Valley, Coast Ranges, Cascade Range, and Sierra Nevada. Generally, it reached its greatest concentrations in the Central Valley and in chaparral at low elevations. Vivid accounts of this huge bear are to be found in *California Grizzly* by Tracy Storer and Lloyd P. Tevis Jr. This bear was genuinely abundant, and there were numerous places where travelers risked meeting groups of 10 to 20 or more Grizzlies. Because its centers of density coincided with ranching activities, it was persecuted whenever encountered. The last Grizzly in California was killed in the early 1920s, and the last one in Oregon was killed in 1931.

The Jaguar is the largest cat in the New World and, like the Grizzly, an awesome creature to have as a neighbor. A large Jaguar may exceed 110 to 120 kg (242 to 264 lb.) and can bring down prey up to the size of a grown horse or a large bull. Unlike the Grizzly, however, the Jaguar is rarely, if ever, a threat to human life. It is, in fact, a rather shy creature and is seldom seen without the aid of hunting dogs. The Jaguar roamed the South Coast Ranges of California between San Francisco and Monterey up to at least 1826. The last known individual was killed in Palm Springs about 1860. Outside of California, the Jaguar once ranged northward to Colorado, but it must now be regarded as extremely rare in the United States and is federally listed as endangered. Today Jaguars are sometimes found in the states bordering Mexico, but it is not clear if these are occasional strays or residents. They have been conspicuously destructive to domestic stock.

The Bison once lived in the northeastern corner of California, eastern Washington, and southeastern Oregon, and the adjacent regions to the north and northeast. It was most common about Malheur Lake in Oregon and in the area between Eagle and Honey Lakes in California. The Bison disappeared early in the postglacial period, prior to the arrival of Europeans. It died out in

California about the time the Native Americans acquired horses but before they had firearms. Europeans, the source of the horses, thus apparently had only a secondary role in the demise of the Bison in California.

Predation on Domestic Livestock

Most attacks by predatory mammals on domestic stock are on sheep. Cattle suffer rather small losses, and poultry are usually well protected. Most domestic sheep are more or less unprotected from predators; in winter they are concentrated in fenced pastures at lower elevations, and in summer many are moved to open ranges on public lands in the mountains. Sheep suffer an annual mortality of some 5 to 10 percent from all causes, including predation. The Coyote is the major predator of domestic sheep in the western United States. In California Coyotes take some 10 percent of lambs annually, fewer on ranches with domestic guard dogs.

The Coyote is the most versatile carnivoran in North America, and its adaptability renders it capable of great harm as well as genuine benefit. Although Coyotes may, at times, feed largely on mice, ground squirrels, and rabbits, they also kill substantial numbers of deer and sheep. It is realistic to view their predation on deer as ecologically desirable, but the killing of domestic sheep can be very costly to the rancher. In recent years, Coyotes have increasingly preyed upon pet dogs and cats in suburban and even urban environments. Even humans have been attacked; in Los Angeles, Coyotes killed a small child.

Detailed studies of the diets of Coyotes indicate some interesting differences between individuals. Coyotes that managed to escape from old-style, unpadded leghold traps frequently did so at the cost of a foot. Such animals could be recognized by their distinctive tracks and were referred to as "peg-leg" Coyotes. These individuals were apparently handicapped in their pursuit of natural quarry and were more prone than others to attack domestic sheep. A recent study indicated that sheep killers are chiefly alpha males. Other Coyotes, possibly some that are aged and losing vigor, seem to develop the habit of preying on easily captured domestic animals, including pet dogs and cats. When the troublesome individuals are captured, predation usually ceases. Specialists in predator control therefore believe that the most efficient protection from damage by Coyotes is directed at the specific individual inflicting the damage. The ban on the use of leghold traps (except under special circumstances involving endangered species) has led to the use of Havahart-type traps for the control of noxious individual Coyotes.

Feral Dogs

Dogs are unlike other predators in California. Although they may seem feral at times, often they belong to someone who feeds and cares for them. Unlike domestic cats, dogs usually do not survive and reproduce in the wild, and in this sense they are not truly feral. Depredation by dogs is largely the fault of irresponsible dog owners.

Though the responsibility for damage done by dogs lies with their owners, this responsibility can be difficult to establish, because many uncontrolled dogs are also unlicensed. The appraisal of damage inflicted by dogs is further complicated by the reluctance of ranchers and conservation, animal control, and law enforcement officers to release information on dogs they have dispatched in the act of molesting domestic stock or wildlife. Thus, depredation by feral dogs is very difficult to measure with any real precision. Nevertheless, knowledgeable officials agree that serious problems exist in many parts of the state.

Depredations most frequently occur where wildlife or domestic stock are concentrated near human habitations, and where groups of dogs roam uncontrolled at night. Sheep ranching areas are an example. Dogs account for an estimated 4 to 45 percent of the depredations on domestic sheep; the degree varies with the concentration of sheep near human habitation. In some regions, damage attributed to dogs may equal that assigned to Coyotes.

Uncontrolled dogs also kill an estimated 1,300 deer annually in California; this is a very small part of the total deer population, but it could be locally important. A population of deer that migrates to a community of foothill homes in winter, for example, is vulnerable. As human settlements encroach into foothill areas, this damage will increase. Also, instances of roaming dogs killing antelope when fences and deep snow handicapped the antelopes' movements and prevented their escape have been noted.

The greatest danger posed by loose dogs, however, is to human health. The risk of rabies is ever present in a society with a high population of free-ranging dogs, and the risk is great where the disease is endemic in skunks and Raccoons *(Procyon lotor)*. Although most of the 1 to 2 million dog-bite victims annually have trivial wounds, a dog bite is unpleasant, and the frequent follow-up of the rabies treatment is both expensive and extremely painful. Moreover, the wound itself can be serious. The death of a young woman in San Francisco in 2001 is a tragic reminder of the potential danger of uncontrolled dogs.

Although feral dogs are not truly part of the mammal fauna of California, they do constitute a population of free-ranging mammals and are frequently far more dangerous than our native mammals.

Feral Cats

Cats are the animals most commonly abandoned in the countryside and city parks; they readily breed in the wild and establish truly feral populations. Such populations can be found in most habitats, except the very hot deserts and higher mountains. When left to look after themselves, cats, being strictly carnivorous in nature, feed on a variety of lizards, mammals up to the size of a young rabbit, and birds up to the size of a two or three week old pheasant. They are skillful and effective hunters. Capturing and neutering them accomplishes little if they are then released to continue destroying wildlife. Pet cats allowed to roam outdoors at night also do an incredible amount of damage to wildlife; studies in Great Britain have shown that they kill tens of millions of birds and small mammals each year.

Furbearing Mammals

From the earliest periods of the European invasion into North America, furbearers drew humans to the western states. The

Pacific Coast States are blessed with a variety of mammals with beautiful fur, mostly carnivorans and rodents. Some species are probably less common than in former years, and others are probably at their original population levels, if not even more common. Several species are introduced mammals in our fauna—for example, the Muskrat *(Ondatra zibethicus)*, the Red Fox, and the Opossum *(Didelphis virginiana)*.

Early History

The principal furbearers in the early days were the Sea Otter *(En-hydra lutris)* and the Guadalupe Fur Seal *(Arctocephalus townsendii)*. Trappers fanned out from eastern North America in search of Beaver *(Castor canadensis)* and began operating in California early in the nineteenth century, also taking River Otters *(Lutra canadensis)* and Mink *(Mustela vison)*. Hunters contributed skins of deer and elk, which are not included among furbearers today. The gold rush in the middle of the nineteenth century diverted the energies of potential trappers, but professional trapping probably never ceased.

For the latter half of the nineteenth and the first half of the twentieth centuries, prices and demand for pelts were low and trappers pursued chiefly high-quality furbearers. Eventually, early regulations limited the capture of these animals to levels designed to prevent their disappearance from our fauna.

Current Fur Trapping

In the early 1970s an increase in fur prices stimulated commercial trapping, and the sale of wild furs sporadically became an important source of income for a relatively small number of people. The total value of furs taken commercially reached all-time highs in the 1970s. In the 1977–1978 season, the sale of those furs in California brought in $1,159,126. Passing the $1 million mark for the first time, this sum reflected an increase in licensed trappers and animals taken, as well as the high price of furs. The value of Raccoon, Coyote, and Gray Fox *(Urocyon cinereoargenteus)* skins came to more than $100,000 for each species; Muskrat pelts brought more than $200,000; and the Bobcat *(Felis rufus)*, with fur of mediocre quality, brought more than $400,000. In the 1978–1979 trapping season, the total value of furs of wild mammals sold in California was $2,399,565; the value of Bobcat pelts alone was more than $1,130,000. The average price of a Bobcat skin had risen from $106 in 1977–1978 to $190 in 1978–1979. (In the early 1940s, a good Bobcat pelt brought the trapper $6.)

By 1998–1999, the price per Bobcat pelt had dropped to $30.55, and the total revenue to trappers for all species had declined to $34,212, with the Muskrat representing 70 percent of that total. Although the wearing of furs had become politically incorrect, prices (except for Bobcat pelts) had not changed much since the 1970s, perhaps because of demand in foreign markets. The number of animals taken had dropped sharply, however, largely as a result of the ban in California on body-gripping traps imposed by Proposition 4, passed in November 1998. For example, in the 1997–1998 season, 1,127 Coyote skins sold for a mean price of $10.49, but in the following season only 301 Coyotes were taken, at an average price of $10.80. Similarly, in 1997–1998 a total of 1,059 Bobcats were taken; in 1998–1999 the take dropped to 190 with virtually no change in price.

The Bobcat achieved prominence because today it is the only spotted cat whose fur can legally be sold. Despite the poor quality of the fur and the very thin skin, the demand for Bobcat fur remains high. Fortunately, the Bobcat is a prolific animal and its populations hold up well under trapping pressure. During most of the 1990s, the annual take of Bobcats was more than 1,000. Most of the commercial capture of Bobcats was done with the aid of dogs; only 28 percent was done by trapping. The total take of the Bobcat in the 1990s was less than 20 percent of the 14,400 take allowed annually by the Office of Scientific Authority (of the U.S. Fish and Wildlife Service).

Aesthetics and the Significance of Commercial Trapping

The majority of Californians today who are interested in furbearing mammals care more about the thrill of observing one than shooting or trapping one for its pelt. Even casual outdoor enthusiasts speak with excitement of having a glimpse of a Marten or Coyote. In former times, trapping provided extra money for low-income people, but today trappers are more likely to get pleasure from capturing wary animals than from augmenting their incomes. The low prices of furs today make trapping a very poor source of supplemental income, but large numbers of Muskrats continue to be taken.

Several furbearing mammals, particularly the Mountain Lion, the Coyote, the Black Bear *(Ursus americanus)*, and the Muskrat, can be extremely destructive to agriculture. The major Coyote depredations are against domestic sheep, especially in the north coast counties and in the winter lambing region of the Sacramento–San Joaquin Delta. Muskrat and Beaver burrows through small levees and irrigation ditches cause levee deterioration and occasional water losses. Beavers sometimes cause major levee breaks. Other furbearers can be local or sporadic pests. The Badger, for example, is infrequently encountered, but one Badger digging in pursuit of a ground squirrel can ruin a levee and thus counter the good it does as a destroyer of ground squirrels. And skunks, who have no real friends at all, are sometimes major reservoirs of rabies.

Furbearers are important predators, however. The same Coyote that kills lambs in winter preys on ground squirrels, mice, and possibly a few ducks and Muskrats from spring until fall. The skunk subsists mainly on soil insects and mice, with an occasional bird's egg in spring. The Mink may destroy many young Muskrats. Mountain Lions, Bobcats, and Coyotes kill some fawns. This outrages deer hunters, but predation is perhaps preferable to starvation, the fate of deer populations totally protected from predators.

Game managers, who are generally conscientious professionals, must take all these factors into consideration. At the same time, the public exerts various and sometimes conflicting pressures on the Director of the DFG; the Fish and Game Commission members, who are appointed and subject to political pressure; and the wardens, who most commonly face the public in the field.

Some conservationists would prefer total protection for all furbearers, including the Bobcat. Population figures over the decades indicate that trapping has little effect on Bobcat populations. Although total protection would very likely increase Bobcat densities, it would also increase attacks on domestic stock. The Bobcat, fortunately, remains a common carnivoran, which one may encounter on mountain trails, cutover forest lands, coastal headlands, and mountain pastures. Throughout North America the Bobcat, like the Coyote, has been persistently persecuted, yet both have been able to thrive under this pressure.

Many carnivorans have been protected for years and are today as numerous as they probably were when California was first settled. Some are common or even locally abundant. The Marten and the Fisher, two of our most prized furbearers, are frequently seen in favored habitats, with the Fisher being by far the more secretive species. They are most common in remote areas, and their numbers could perhaps tolerate regulated trapping. The River Otter, another valuable but protected furbearer, is common along many California watercourses. Although it is not commonly observed, its tracks and its slides indicate local abundance of this graceful predator. It is difficult to believe that some would begrudge this fine animal the few trout it takes. The Wolverine, Kit Fox, montane Red Fox, and Ringtail (*Bassariscus astutus*) also have some protection.

The Mountain Lion has increased and spread and now is found in suburban areas around San Francisco Bay and in some areas on the Sacramento Valley floor. Road kills in the mountains have become more common in recent years. The Mountain Lion is a valuable predator of the Black-tailed Deer but sometimes enters suburban areas to feed on house cats, and attacks on hikers, although very rare, do occur, at times with fatal results. Passage of the Wildlife Protection Act of 1990, a ballot initiative that not only banned the sport hunting of Mountain Lions in California but requires the state to spend no less than $30 million per year for 30 years on habitat protection, demonstrates the ability of private citizens to effect change in environmental law.

Wild Mammals as Carriers of Pathogens

We tend to think of an ecosystem as comprising the particular taxa in which we happen to be interested. Any given ecosystem, however, contains more species of integrated parasites than of free-living forms. Very often the parasites have evolved with their hosts and do not greatly damage them. The same parasites, however, can make humans very ill.

Some wild mammals host important pathogenic organisms that are transmissible to humans. These pathogens include viruses, bacteria, protozoa, and parasitic worms. Several of them are common enough to justify special precautions in handling mammals, especially rodents, or sometimes in simply being near populations of wild mammals. Among the more important are those that cause rabies, plague, tularemia, hantavirus, Lyme disease, and relapsing fever, all of which but the last can be fatal to humans.

In handling a wild mammal, take the utmost care to avoid contact with its blood. We recommend the use of disposable rubber gloves to avoid contact of blood with your skin. Commonly we have minor scratches in our skin, and these may allow entrance of a disease-causing microbe. After handling a wild mammal, always wash your hands with soap and warm water. When removing a tick from yourself or from a pet, never use your bare fingers. With a pair of fine forceps or fingers protected with tissue paper, grasp the tick where the mouthparts enter the skin, and pull with a steady force straight back; do not twist the forceps from side to side.

Rabies

Rabies encompasses several viral strains that occur in a broad variety of wild mammals, especially carnivorans, rodents, and some bats. The rabies virus is generally transmitted in saliva and for this reason is most likely to be acquired through a bite from a carnivoran, such as a skunk or a fox, but the virus may be in the air of bat caves. Rabies can be transferred from a pregnant female Guano Bat *(Tadarida brasiliensis)* to her embryo. It may occur in deer, cows, horses, and domestic dogs and cats. In most areas almost all dogs are vaccinated against rabies, but cats are generally unprotected and are occasionally infected by the disease. The typical symptoms in a wild mammal include loss of fear; the appearance of a usually shy species in a strange place, showing no fear, suggests rabies. An apparently tame fox walking around in circles should be captured by an animal control officer. A bat abroad in daylight should not be handled. Although most bats are small, innocuous beasts whose bite can scarcely break the skin, a strangely acting bat could be rabid.

Plague

Plague is a bacterial disease found throughout the world. It is common in many wild rodents, especially mice, rats, and some ground squirrels (*Spermophilus* spp.); in rodents, it is referred to as sylvatic plague. The plague bacterium is passed from one mammal to another by fleas. Commonly it creates a blockage in the gut of a flea, and an infected flea may regurgitate when feeding, thereby injecting the agent into another animal. If the animal dies, the fleas may leave its body to attack another mammal or may remain in the host's nest to infest the next individual that enters. Thus, the disease is difficult to eradicate. A person can become infected through flea bites or by eating the inadequately cooked meat of an animal ill with plague. Dogs appear to be relatively resistant, but plague is not unusual in domestic cats.

Plague has been found in many species of small mammals in California and is especially common in the Golden-mantled Ground Squirrel *(Spermophilus lateralis)*, as well as in other ground squirrels and chipmunks in the Sierra Nevada. Some mice *(Peromyscus* spp.) and voles *(Microtus* spp.) are sometimes reservoirs of the plague bacterium. Public health representatives poison rodents about resort areas to lessen the likelihood of vacationers encountering plague. Plague-infected rodents have been found about Lake Tahoe, for example, and plague has occurred in the Bushy-tailed Wood Rat *(Neotoma cinerea)* and the Deer Mouse *(Peromyscus maniculatus)* in Lava Beds National Monument. Plague has also been found in California carnivorans, including Black Bears, Coyotes, Bobcats, and wild pigs.

In humans, plague is called bubonic plague because it produces swollen lymph glands called bubos. The swellings are commonly under the arms or in the groin but may occur elsewhere. When plague enters the respiratory system, it is called pneumonic plague, the "black death." Pneumonic plague, which can be transmitted by coughing and is highly contagious, killed one-quarter of the population of Europe in the fourteenth century.

Between 1970 and 1980 13 cases of human plague in California were reported, with five fatalities, and it is very likely that some cases go unrecognized and unreported. Seventeen human cases were reported in California from 1980 to 1990; between 1990 and 2000 there were eight, one of them fatal. Exposure to plague is certain to increase with greater use of outdoor recreational areas.

Tularemia

Tularemia is a bacterial infection common in rabbits, hares, and many kinds of mice and voles. Humans can contract the virulent

Type A form of tularemia via the bites of the Pacific Coast Tick (*Dermacentor occidentalis*) and of the Deerfly (*Chrysops discalis*); by skin contact with the blood of a rabbit, squirrel, or other small mammal; or by eating incompletely cooked infected meat. We can acquire the less virulent Type B form of the disease by drinking water that has been contaminated, usually by rodent (e.g., Muskrat) urine. The disease is rarely, if ever, passed from one human to another. If untreated, tularemia is sometimes fatal in humans.

Tularemia is probably capable of infecting virtually all wild mammals, and it has been found throughout most of temperate North America and Eurasia. The disease is widespread in California but seems to be declining in importance. It has been most frequently contracted in Kern and Los Angeles Counties.

Relapsing Fever

Relapsing fever, or borreliosis, is a tick-borne bacterial disease associated with chipmunks, ground squirrels, and tree squirrels (*Sciurus* and *Tamiasciurus*). Ticks (*Ornithodoros hermsi*) can transmit the disease to humans through feeding or a coxal secretion. The symptoms (high fever, nausea, and severe headaches) disappear and then recur in about a week, giving this disease its name. Several such relapses may occur if the patient is not treated with an appropriate antibiotic. The ticks live in the nests of mammals. If a host, such as a chipmunk, has a nest beneath the floor, in a wall space, or elsewhere in a mountain cabin, the ticks may move to the bedding of the human occupants to feed. They feed fully in less than an hour, usually at night. Therefore, the victim typically is unaware of them. In areas where the Beechey Ground Squirrel (*Spermophilus beecheyi*) is common, Pacific Coast Ticks are likely vectors of the disease.

Infected ticks are known to occur around Lake Tahoe, Packer Lake, and Big Bear Lake in California, as well as in many other mountain areas above 1500 m (5,000 ft). Ticks infected with what appears to be a strain of the relapsing fever spirochete have been found at Eagle Lake by Professor Jerold Theis of the School of Medicine at Davis, but the survey did not detect the spirochete in any of 180 rodents (mostly squirrels and chipmunks) examined. Relapsing fever is not usually a serious problem, but in 1973 it was acquired by 62 visitors staying in vacation cabins on the north rim of the Grand Canyon.

Spotted Fever

Rocky Mountain spotted fever, an uncommon disease in California, is caused by a rickettsia (a kind of bacterium) and transmitted in California by the Pacific Coast Tick. This tick is commonly found on clothing after outdoor activity, but it rarely bites people, though it often attaches to dogs. Its natural hosts are Cottontails *(Sylvilagus)* and other small mammals. Ticks are both the vector and the reservoir of the rickettsia. The disease may cause nausea, abdominal pain, headaches, and a rash beginning on the palms or soles. It is fatal in about 5 percent of cases.

Tick Paralysis

This condition is not caused by a microorganism but is induced by a neurotransmitter blockage, which results from the injection of saliva following the attachment of a female tick. When the tick attaches at the rear of the scalp it can be unnoticed, and the symptoms, which mimic those of polio, may not be properly diagnosed. If the tick is removed early, the symptoms may disappear within 24 hours, but if the tick remains, tick paralysis may be fatal. Although human cases have never been found in California, the disease has been identified in domestic animals.

Hantavirus

A number of related viruses of the genus *Hantavirus* cause various kinds of hemorrhagic fever in humans. Hantavirus, in one form or another, occurs across the United States and as far south as Argentina. In the western United States one or more hantaviruses occur in native rodents, especially Deer Mice *(Peromyscus maniculatus)*. Hantavirus pulmonary syndrome (HPS) is caused by the Sin Nombre virus (a type of hantavirus), which is known to occur in Deer Mice as far north as Oroville but is particularly common in the eastern part of the state (e.g., Inyo County). HPS has been known only since 1993. It was identified in California in six people in 1999: two of them died two to five days after admission to a hospital. Twenty-nine cases of HPS, including 14 deaths, were reported (or later identified) between 1990 and 2000 in California. Death may result in 75 percent of human cases if the disease is not recognized and promptly treated.

HPS can result from inhaling the feces of an infected rodent. Sweeping floors and removing rodent feces from pantry drawers may be sufficient to create virus-infected dust. Possibly farm workers inhale the infected dust from hay.

Arenavirus

Another hemorrhagic fever is caused by an arenavirus called Whitewater Arroyo virus, recently found in the western United States. The reservoirs are several species of wood rats (*Neotoma* spp.), and the virus is believed to be acquired through inhaling dust and droplets of wood rat wastes. Previously undetected, the disease probably caused the deaths of several people who had symptoms of hantavirus but not the virus itself. In 2000, Whitewater Arroyo virus was identified as the cause of death of a 14-year-old girl in Alameda County and perhaps the cause of death of a 30-year-old woman in Orange County.

Lyme Disease

Lyme disease has come into prominence in recent years. It is caused by a spirochete *(Borrelia)* and transmitted by the bite of the Western Black-legged Tick *(Ixodes pacificus)*. Larval ticks, which hatch in late summer, attach to mice (especially *Peromyscus*) and, in greater numbers, to fence lizards *(Sceloporus occidentalis)* and alligator lizards *(Elgaria multicarinata)*, dropping off in several days. They molt into nymphs, which may feed on humans, other mammals, and lizards. Professor Robert S. Lane at the University of California at Berkeley found that the ticks lose the spirochete when they feed on lizards but retain it when they feed on rodents. Hosts of the adult tick may include humans, wood rats *(Neotoma fuscipes)*, cottontails *(Sylvilagus)*, Bobcats, feral Pigs *(Sus scrofa)*, kangaroo rats *(Dipodomys californicus)*, and particularly deer *(Odocoileus)*. The life cycle of the tick requires about three years.

Wildlife biologists and others working in leaf litter of hardwood forests are at greatest risk from April to July. Nymphal ticks are more common than adult ticks in such areas and are more likely to carry the spirochetes. Lyme disease is readily cured with antibiotics when recognized at the onset of symptoms; the classic sign is a bull's-eye rash that spreads gradually. The rash may not take this form, however, or it may be absent. Other symptoms are often flulike, including headache, fever, and malaise.

Giardia Disease

Giardia is a protozoan parasite in the gut of many wild and domestic animals and may survive in water for a long time as a cyst. Humans acquire the disease primarily by drinking contaminated water but can also acquire it through anal-oral contact with an infected person.

Trichinosis

Trichinosis is caused by the minute Trichina Worm *(Trichinella spiralis)*. In the wild, trichina worms occur in rats at garbage dumps and in their predators, usually bears, foxes, and Coyotes. Humans sometimes acquire the worms by eating poorly cooked bear meat, but the disease is not very common. Normally it may be asymptomatic, but if large numbers of encysted larvae are ingested at once, there may be severe inflammation of the striated muscles.

Dog Heart Worm

The Coyote is a natural reservoir of the nematode *(Dirofilaria immitis)* that causes heartworm in domestic dogs. Larval worms are passed from one host to another by mosquitoes. On the rare occasions that they are transmitted to a human (through a mosquito bite), they encyst in the lungs, where they mimic lung cancer.

Raccoon Worm

Raccoon nematodes *(Baylisascaris procyonis)* live in the guts of Raccoons. Their eggs are excreted and can survive for long periods in the soil. If a person comes in contact with Raccoon feces or with garden soil in which the eggs have settled, the now embryonated eggs may enter the human gut. They continue to develop in the small intestine, and the larvae migrate through the body. If one migrates to the brain, it is likely to be unrecognized, and death follows. The best prevention is to exclude Raccoons from your garden. Never feed household pets by leaving a pan of food outside; the food attracts skunks, Opossums, Raccoons, and sometimes foxes.

Histoplasmosis

This disease is caused by *Histoplasma capsulatum,* an airborne fungus, which is sometimes very common in bat caves. Fungal spores, when inhaled, may develop in the lungs and induce coughing, thoracic pain, and other symptoms that mimic tuberculosis. It can be fatal.

FOOD OF MAMMALS

California mammals exhibit many feeding patterns. Some species are extremely versatile dietary generalists. Others confine their feeding to a single category, such as plants, insects, or other vertebrates. Still others are extremely restricted specialists and may feed on only a single kind of food. Diversity in feeding is one way in which different species live in the same area compatibly, without interfering with one another in their search for food or, in some cases, without competing for food.

Few specialists are so narrow in their selection of food as to compete with no other species. Nevertheless, the specialist possesses some adaptation that enables it to find nourishment in food that others might disdain, thus achieving a high degree of independence in the constant scramble for food. A dietary generalist, on the other hand, potentially competes with other generalists. Dietary overlap in itself, however, does not constitute competition. Competition does not arise until a desired object, such as food, becomes limited in supply or availability. When there is not enough food for all the generalized feeders, competition may become a reality. Because a generalist is versatile, however, it may well be able to subsist on alternate food items when one becomes scarce. Competition is easy to imagine but very difficult to establish in nature.

The ability to shift from one sort of food to another enables the generalist to thrive in spite of variations in food supplies. This versatility is obtained at the cost of occasional extensive foraging, an expense not usually shared by the specialist. Thus, there is a trade-off. The specialist has its food supply continuously at hand and never needs to search for it, but when its food supply fails (which it seldom does), the specialist is doomed. The generalist must forage for its food, and foraging takes energy, so the energy in the food must exceed the cost of obtaining it. The generalist, however, always has something to eat. The generalist may seek the most nutritious food that can be obtained at the least expense of energy. We have, for example, known Deer Mice *(Peromyscus manicula-*

tus) to feed entirely on maggots in a rotting deer carcass. Such food contains little waste and requires no expense in foraging.

Intermediate conditions exist between the restricted diet of the specialist and the broad diet of the generalist. For example, two common California rodents eat woody plant fibers. The Porcupine *(Erethizon dorsatum)* spends most of its time sitting against the trunks of small conifers, especially pines, from which it strips the woody tissue beneath the bark. In springtime, however, the Porcupine regularly repairs to mountain meadows and feeds on grass. On the other hand, some voles *(Microtus* spp.) normally and by choice feed on the tender growth of forbs and grasses but have been known to eat the bark and roots of trees, even to the extent of killing orchard trees.

Specialized Diets

Before it can find an advantage in restricting its energy intake to a single kind of food, an animal must develop the capacity to obtain and digest a material that is not available to generalists. By selecting foods not sought after by others, a species takes a big step toward preserving its food for itself alone. Moreover, the specialist must concentrate on a kind of food that is not temporarily eliminated by variations in weather, for even a temporary food shortage might wipe out a population depending on it.

One of the best-known specialized feeders among California mammals is the Red Tree Vole *(Arborimus pomo)*, a small species that lives only along the coasts of Oregon and northern California. This little vole feeds on the soft tissue in the needles of Douglas-fir *(Pseudotsuga menziesii)*, an abundant substance not normally eaten by other mammals. Thus, it does not compete with other mammals for food. This advantage involves some risk, however: if the Douglas-fir should disappear, the Red Tree Vole would perish with it. Thus, it is an obligate specialist. Caterpillars of the Pine White *(Neophasia menapia)*, a butterfly in the family Pieridae, have been known to defoliate pine trees and sometimes also to attack the Douglas-fir. The distinctiveness of the Red Tree Vole indicates that it has survived as a specialist for a long time, despite occasional inroads of the caterpillars on its food supply.

Anatomical adaptations can give a species an advantage in

seeking certain foods without making it an obligate specialist. The Marten *(Martes americana)* and the Badger *(Taxidea taxus)* are two moderately specialized carnivorans. The Marten is light and agile and travels easily through trees, where it captures tree squirrels and birds. The Badger is heavy and stout, with powerful forelimbs for digging, and pursues a variety of small rodents, especially such soil-dwelling forms as pocket gophers (*Thomomys* spp.) and ground squirrels (*Spermophilus* spp.). The Marten and the Badger are not obligated to subsist on their normal prey; each would find adequate nutrition in the food of the other. Nevertheless, each is moderately specialized to pursue its normal food in its normal habitat: coniferous forests for the Marten and dry, arid areas of loose soil for the Badger. Clearly, these two carnivorans must compete only rarely.

Ruminants

Mammals lack the enzymes necessary to digest cellulose, a major part of the cell walls of grasses, forbs, and other plant materials. Instead, this complex sugar is broken down by microbiota of cellulose-fermenting bacteria and protozoa, which survive on the sugar products resulting from its digestion. The use of such microbiota is most elaborately developed in ruminants. These are grazing mammals such as deer (Cervidae) and sheep that have a four-chambered or ruminant stomach, a specialization for the digestion of these plant tissues. In the rumen (one chamber of the stomach), cellulose is digested to simple sugars that can be utilized by the bacteria and protozoa of the gut. Ruminants regurgitate partly digested plant foods from the rumen and, by "chewing their cud," further masticate food particles that have been chemically changed by the microbiota; this facilitates further digestion. When this material subsequently enters the small intestine, the mammal can absorb some of the products of microbial digestion, such as proteins. It also digests the microbes themselves.

Other herbivores, such as voles or the Porcupine, have cellulose-digesting bacteria and protozoa in the caecum or even in the large intestine. These mammals are called hindgut fermentors. Presumably, they derive less benefit from the digestive activities of their microbiota than do the more specialized ruminants; but in some hindgut fermentors (horses [Equus]), materials in the colon may be moved forward by antiperistaltic action so that nutrients may be absorbed in the small intestine.

The palatability of plants to herbivores is partly a function of the toxicity of certain plant compounds to the microbiota. Some plants contain oils and alkaloids that may immobilize or destroy the cellulose-digesting bacteria and protozoa.

Coprophagy

Several kinds of mammals, notably rabbits and hares (Leporidae), alternately produce soft feces and hard feces. Typically, soft feces are passed at night and promptly eaten as they leave the anus. This reingestion, called coprophagy, enables the mammal to assimilate some materials digested by the intestinal microbiota, including those in the large intestine. The process thus minimizes nutrient loss. It also conserves water. In arid environments, loss of body water is always a threat, and desert mammals develop several means of reducing dissipation of body fluids. Coprophagy is reported in pocket gophers (Thomomys spp.) and shrews (Soricidae); it may also occur in other groups.

Generalized Diets

Some carnivorans are actually bona fide generalists. For example, although the Gray Fox (Urocyon cinereoargenteus) seems to relish mice and small birds, it readily feeds on berries and other plant parts. It is at home in trees and takes birds' eggs in season. The Gray Fox is structurally and phylogenetically a carnivoran, but it is, in fact, a generalized feeder—an omnivore. Two other carnivorans, the Raccoon (Procyon lotor) and the Coyote (Canis latrans), are notorious for their versatility in feeding, although the Coyote eats much less plant material than does the Raccoon. The Black Bear (Ursus americanus), another carnivoran, actually feeds largely on plant materials.

Rodents are commonly considered to be plant feeders, and many species are said to subsist on seeds, but in fact most are true generalists and highly versatile in their pursuit of food. Deer Mice and their allies tend to eat whatever is abundant and nutritious. This adaptability can give the appearance of specialization, for if one type of food is abundant, Deer Mice may feed on it exclusively. In years when pines produce a very heavy seed crop, Deer Mice may subsist on pine seeds to the exclusion of other palatable foods. Some years, however, many trees and shrubs fail to produce seeds. Widespread shortages of pine seeds, acorns, manzanita berries, and other energy-rich foods from plants are not unusual, and seed scarcities occur among some plants in most years, perhaps caused by late frosts, insect attacks, or similar variables. At such times the generalist turns to other food items, such as insects and underground fungi. Many small mice feed heavily on insects, in most environments probably consuming a far greater volume of insects than birds do. (In addition, insectivorous birds are largely migratory and only present in midlatitudes for part of the year.)

A true generalist is a facultative specialist—an opportunist. Temporary specialization by generalists has frequently been observed at times of abundance of particular food items. In the summer of 1951 a sudden abundance of the California Tortoiseshell Butterfly *(Nymphalis californica)* provided an abundance of caterpillars and pupae. For a short time the caterpillars formed the main food of the Golden-mantled Ground Squirrel *(Spermophilus lateralis)*, which pursued these insects on the ground. The caterpillars climbed into bushes to pupate, and chipmunks *(Neotamias* spp.) fed on the hanging chrysalids; the Yellow-pine Chipmunk *(Neotamias amoenus)* ate little else while the chrysalids lasted.

Insectivores, as indicated by their name, feed largely on insects, but they, too, are somewhat flexible. Shrews *(Sorex* spp.) take quantities of conifer seeds in years of heavy seed production, and the Water Shrew *(Sorex palustris)*, which preys mostly on nymphs and larvae of aquatic insects, also captures small minnows, trout fry, and frogs. Although moles (Talpidae) feed on insects to a large extent, they also take large numbers of annelid worms and have been known to eat underground parts of plants.

Seasonal changes in food are apparent to anyone who examines the feeding patterns of a generalized feeder in detail. Available food supplies change not only from year to year but also from season to season. A Deer Mouse that eats pine seeds as they fall on bare ground in September and October may have to search for different foods when snow falls. Unlike chipmunks, which are notorious for laying up large amounts of food for later use, Deer Mice appear not to have such a well-developed habit. The winter diet of the Deer Mouse consists mostly of food such as insects and fungi that occur about the snowless margins of fallen logs or large boulders. With the approach of spring in the Sierra, seeds of the big leaf maple *(Acer macrophyllum)* may fall and germinate on the surface of the snow. These seeds are eagerly eaten by Deer Mice and are probably extremely nutritious. Later, as the snow disappears and the sun warms the ground, Deer Mice eat sprouting seeds and tender shoots of grasses and forbs, and green plant material forms much of their fare in spring.

Normally, one studies the food of mammals either by direct observation of feeding in the field or by examining the contents of the stomach or the cheek pouches. Animals provided with fur-lined cheek pouches frequently fill them with seeds as they forage, thus reducing the number of trips needed. Rodents such as pocket mice (*Perognathus* and *Chaetodipus* spp.) and kangaroo mice (*Microdipodops* spp.), however, may at times prey heavily on insects, and they rarely place live insects in their pouches. Thus, examination of cheek pouches presents a biased picture of their total diet.

The diets of wild mammals are in fact poorly known; for most species our knowledge comes from fragmentary anecdotal accounts. Many opportunities exist to discover patterns of feeding and dietary similarities and differences among related and sympatric species, as well as seasonal variations in food availability and use. Much is to be learned, for example, regarding dietary overlap among sympatric bats and the effect of foraging mice and rabbits on the plant cover.

IDENTIFICATION
OF MAMMALS

Classification and Names

In this book the taxonomic arrangement of orders, families, and genera is taken from *Classification of Mammals above the Species Level* (McKenna and Bell 1997). Species names are from *Mammal Species of the World* (Wilson and Reeder 1993). Most of these names are familiar to mammalogists, and the new combinations are based mostly on recent evidence of mammalian relationships. We have tried to indicate where such changes are controversial.

In contrast to a species, which is presumed to exist as a discrete entity in nature, a genus is a concept that exists only in our own minds. As knowledge of mammalian relationships accumulates, generic concepts are sometimes modified, and the allocation of species to genera may also be changed. Such changes account for unfamiliar combinations of generic and specific names. For example, a very old and familiar name, *Citellus*, the generic name for most of the ground squirrels, has been changed to *Spermophilus* owing to nomenclatorial technicalities.

These changes can be frustrating. We urge readers not to be hasty in adopting proposed taxonomic changes. Many of them may be suggested for invalid reasons and are quickly forgotten. Changes above the species level frequently result from shifts in attitude, and attitudes have a history of changing.

Over the past 30 years there have been great efforts to standardize common names of animals, especially vertebrates. We praise these efforts but cannot always condone the results. No rules exist for the use of common names. Generally we have used the common names employed by Jones et al. (1982) and Laudenslayer and Grenfell (1983). We do, however, believe that it is desirable to be flexible with common names and use those that are appropriate. For example, "Wild Boar" is a well-established name for wild populations of *Sus scrofa*, whether descendants of

European Wild Boar, feral domestic strains, or mixtures of the two. The name "Wild Boar" appears in the titles of many articles on this species, and it would be confusing to drop this name, as has been proposed, and adopt "Wild Pig."

The name change of our most common free-tailed bat, *Tadarida brasiliensis*, illustrates another aspect of the problem. Previously the scientific name was *Tadarida mexicana*, and the animal was sometimes called the Mexican Free-tailed Bat. Because it is acknowledged that *mexicana* is really a subspecies of *Tadarida brasiliensis*, the full scientific name is *Tadarida brasiliensis mexicana*. Some workers have therefore adopted the inappropriate "Brazilian Free-tailed Bat," creating confusion, not clarity. We have employed "Guano Bat," a much older name with historical significance. In another thoughtless adaptation of a common name from a scientific one, *Peromyscus californicus* has recently been called the California Mouse. Many mouselike rodents have a specific name derived from "California." We use this species' previous name, the "Parasitic Mouse," which is appropriate and has biological significance.

Many common names are admittedly quite misleading. The Ornate Shrew *(Sorex ornatus)* is a dull-colored creature, and the Broad-footed Mole *(Scapanus latimanus)* has forefeet no broader than those of other members of the genus. There are many other examples.

We have followed conventional practice regarding common names and hope readers will have no difficulty in recognizing species referred to by different, usually older, names. Confusion can be kept to a minimum by learning generic and specific names, which change less frequently. These are the names by which species are indexed in scientific literature, which it is our sincere wish that this volume will stimulate readers to investigate.

In this book we have capitalized common names that refer to individual species (e.g., Wolverine, Red Fox) and lowercased names that refer to groups rather than species (e.g., fox, wood rat).

Scientific Naming of Mammals

Any kind of organism is designated by a genus, or generic name, and a species, or specific name. These names are derived from either Latin or Old Greek, or they are Latinized proper names. This system of naming is called the binomial system, and it has been in universal use since Carl Linnaeus introduced it in 1758. The binomial, or scientific name, applies universally, throughout the world, regardless of how common names may vary geographically. Related species are placed in a single genus (whose name is always capitalized). For example, the various kinds of meadow voles are placed in the genus *Microtus.* The several California species are recognized by their specific names, such as the Long-tailed Vole *(Microtus longicaudus).* The Beaver *(Castor canadensis)* is the only member of the *Castor* genus in North America, but its name shows it to be a relative of *C. fiber* of Eurasia. The species name is written in lower case. Various genera of small mice are placed within the families Muridae, Zapodidae, Heteromyidae, or others; and the families of rodents are placed within the order Rodentia. Such an arrangement is intended to reflect the relationship of different groups, or taxa, of animals to one another; all higher taxa are capitalized but are not italicized.

Actually, the hierarchy of classification is far from perfect above the level of the order. Modern orders of mammals apparently became distinct in the Cretaceous period, and many mammal relatives are in orders long extinct. For that reason, a phylogenetic tree contains modern orders that are related through other taxa (part of the branches of the tree) that no longer exist, and the current taxonomic arrangement is more of a phylogenetic "bush" than a tree. Obviously, within this hierarchy are numerous taxonomic levels, and in a book on California mammals, a simplified hierarchy is quite adequate. The following scheme is useful to place our mammalian fauna in perspective.

Phylum	Chordata	
Subphylum	Vertebrata	
Class	Mammalia	
Subclass	Prototheria	Egg-laying mammals
Subclass	Theria	Marsupials and placental mammals
Superorder	Marsupialia (Metatheria)	Marsupials
Superorder	Eutheria	Placental mammals

The categories used in this book, including for the single marsupial in California and all the remaining placental mammals, are order, family, genus, and species.

Keys

The identification of a mammal depends on the condition of the specimen, the parts available, and the uniqueness of those parts in particular species. A skull from an owl pellet, for example, may be adequate to identify a shrew or vole but insufficient to identify a deer mouse. To assist in specific recognition, the keys usually include details of dentition and skull features as well as aspects of color and body proportions.

The keys are arranged so that major groups are progressively divided into lesser groups—families into genera, for example, and genera into species. Each key begins with a couplet of general contrasting characters that separate the major category into two subgroups. The specimen at hand should fit into only one half of the couplet, and the number at the right-hand margin leads to the next appropriate couplet. By going from one couplet to another, the user should be able to identify the specimen.

Keys are not flawless, however, and the terse contrasting phrases of a couplet cannot reflect individual variation. The keys, therefore, should be used together with the descriptions, plates, and range maps. After a specimen has been identified, it should be saved. This facilitates the identification of subsequently found individuals. Labeled with date and locality, moreover, it may expand the information contained in this volume.

Key to Orders of Mammals

Scope of This Volume

The following pages contain information on the wild mammals that are known to be found in California today or that approach our borders and might reasonably be expected to occur here. Not included are several introduced forms that are either very limited in their occurrence or exist in a state of semicaptivity, such as on ranches.

The amount of knowledge on some kinds of mammals, especially bats, shrews, wild mice, and squirrels, is unfortunately scarce. Most of these forms are not rare, however, and our relative ignorance of their habits should challenge enterprising mammalogists to discover more about them.

Range Maps

The range maps are guides to the geographic occurrence of California mammals. These maps are based on collection records and, sometimes, also on observations. They are, at best, only approximations of regions in which species may be expected. A species does not occur everywhere throughout its indicated range; usually it is restricted to a certain habitat. The River Otter

(Lutra canadensis) and the Water Shrew *(Sorex palustris)*, for example, remain close to the margins of rivers and creeks within their respective regions of occurrence. Moreover, human activities such as lumbering, mining, and agriculture alter the abundance and extent of distribution of many species. For this and other reasons, the geographic distributions of wild animals are seldom static. Thus, range maps, based partly on past records, are never precise. Despite their shortcomings, however, we believe that range maps are extremely useful when used in conjunction with keys, descriptions, and illustrations.

For species whose abundance and distribution have been drastically reduced, such as the Pronghorn *(Antilocapra americana)* and Wapiti *(Cervus elaphus)*, the current, not the historic, range is shown. We have omitted maps for species that occur throughout the state. We have also left out maps for introduced mammals; not only are the time and place of their introduction sometimes not known accurately, but their occurrence is frequently uncertain and often in a state of flux.

SPECIES ACCOUNTS

MARSUPIALIA (METATHERIA)

This order contains the opossums (Didelphidae) of the New World as well as the kangaroos and wallabies (Macropodidae) and phalangers (Phalangeridae) of Australia and New Guinea. These mammals produce young after a very brief gestation and little maternal nourishment during pregnancy. The newborn young are in a near embryonic condition; only the forearms, sense of smell, and mouth are well formed. The minute infant crawls to its mother's pouch, or marsupium, which contains the nipples. (Many South American and Australian marsupials lack a marsupium, though they have the other physiological and reproductive characteristics of the marsupial order.) There it attaches to a nipple, where it feeds for several weeks. The anterior set of milk teeth (incisors, canines, and all but the last premolar) do not develop in infant marsupials. This failure is an adaptation to the prolonged period of lactation, when the delicate nipple occupies the anterior region of the mouth.

The history of marsupials began in North or South America in the Early Cretaceous, when the two continents were loosely connected. When mammals began to emerge as a distinct group, some 150 million years ago, North America was the center of marsupial evolution. The oldest fossils are from North America. Because these fossils consist only of teeth or jaw fragments, it is difficult to imagine the animals' ecology. In some cases of solitary teeth, it is not apparent whether the original owners were marsupials or placental mammals. Early marsupials were small and probably either insectivorous or frugivorous (fruit-eating).

A Late Cretaceous fossil of a marsupial is known from Mongolia, and this species survived in Thailand until the middle Miocene. Another Late Cretaceous fossil is known from Madagascar. The first marsupials also entered Australia in the Cretaceous, via Antarctica. No placental mammals existed in Australia at that time, and marsupials radiated into many morphological and ecological types. Marsupials also reached to western Europe in the early Eocene, when Europe was joined to North America, and survived there until the Miocene. The pathway between North and South America—which may have been direct, through the Greater and Lesser Antilles, or both—was broken some 59 million years ago, in the early late Paleocene. Marsupials disappeared in North America in the Miocene and returned when the Isthmus of Panama was established in the late Pliocene.

Opossums (Didelphidae)

Opossums, a rather generalized form of marsupial, are the only type to exist in the United States today. Superficially ratlike, they are structurally unique in many ways. They have five upper and four lower incisors on each side, a very small braincase, a prehensile tail, and an opposable thumb on each hind foot. At birth the neonate has well-developed forearms, with minute claws, and presumably also has a good sense of smell. It crawls to its mother's pouch and attaches to an elongate nipple.

The generic name (*Didelphis* meaning "two-womb") refers to the complete separation of the uterus into two horns. Two lateral vaginas lead from the genital opening and join, making a loop, just below the two uteri. The penis is divided at its apex, as well. When mating, each half of the bifid penis enters a lateral vagina. A median vagina, however, functions as the birth canal.

VIRGINIA OPOSSUM
PI. 5, Figs. 8, 9

Didelphis virginiana

DESCRIPTION: A large, ratlike animal with many teeth, mostly small. Unlike rodents, it has large canine teeth, lacks a pair of chisel-like incisors, and has a thickened, prehensile tail; the female has a marsupium. Its fur is light gray and rather long and loose. Each hind foot has an opposable thumb. The female has a pouch in which the young are carried. The skull is distinctive, with a high dorsal ridge called the sagittal crest (fig. 9b). TL 700–900 mm, T 290–400 mm, HF 62–75 mm, E 47–57 mm.

Figure 8. Female Virginia Opossum *(Didelphis virginiana)* carrying her young with her; this obviates the need to return to a nest.

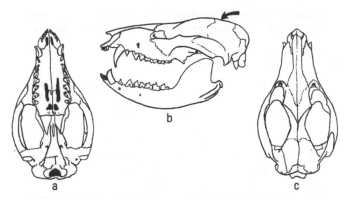

Figure 9. Skull of the Virginia Opossum *(Didelphia virginiana):* (a) ventral view; (b) lateral view (arrow to sagittal crest); (c) dorsal view.

Weight: 1.5–3.1 kg; females are about two-thirds the size of males. Dentition: 5/4, 1/1, 3/3, 4/4.

DISTRIBUTION: Occurs widely in cultivated areas at lower elevations, seeming to avoid mountains above 1,000 m. Its range extends from the eastern edges of the Sacramento Valley and San Joaquin Valley west to the coast and from San Diego north to Washington. It is also found in most of the eastern half of the United States, south through most of Mexico.

FOOD: The Opossum is a very generalized feeder. It forages for almost anything edible, plant or animal. Most of its food consists of soil-dwelling insects, but it also eats small mice, birds' eggs, nuts, berries and other fruit, and corn.

REPRODUCTION: This species breeds at any time, but most mating occurs from January to July. This is the only mammal in our area to reproduce without a well-formed placental attachment to the mother. A yolk-sac placenta produced by both the embryo and the mother's uterine lining contains sufficient yolk to provide energy for the infant's very brief embryonic life. Gestation lasts only some twelve and a half days. The minute young, about the size of a bumblebee, crawl to the mother's pouch and virtually fuse to a nipple. They remain in the pouch for about two months; for approximately one month more, they continue to nurse but also make excursions from the pouch and take some solid food. They are sexually mature in the first year, and a female may produce two litters of four to 10 young annually. Opossums are prolific but short-lived.

COMMENTS: The Opossum was first introduced at San Jose, California, in 1910, and into Oregon and Washington at approximately the same time. It readily adjusts to agriculture, from which it seems to obtain much of its food and shelter.

INSECTIVORA

The insectivores, order Insectivora, are part of the superorder Lipotyphla, which includes shrews (Soricidae), moles (Talpidae), hedgehogs (Erinaceidae), and their relatives. (Some Lipotyphla forms may not be closely related to one another.) The insectivores are perhaps the most primitive and oldest assemblage of placental mammals. The earliest known insectivore lived in the Middle Cretaceous in North America, and both shrews and moles are known to have lived during the Eocene in North America and Europe, which were connected at that time. Today the shrews are represented by a broad spectrum of families, mostly in North America, Eurasia, and Africa.

Insectivores are mostly mouse-sized mammals with small eyes and fine, sharply pointed teeth that serve well in their owners' probing search for food. The molars have sharp cutting edges that crush the exoskeleton of prey. Insectivores have simple intestines with no cecum. The pubic symphysis is reduced, with the margins of the pubes connected by cartilage. California has two families of insectivores, the shrews and the moles, which can be separated by the following key.

Key to Families of Insectivora in California

1a Eye orifice small but distinct, and eye clearly visible; pinna clearly present; forefoot somewhat narrower than hind foot; teeth with some reddish pigment; zygomatic arch absent . . . . Soricidae
1b Eye orifice minute or grown over with skin; pinna absent; forefoot broader than hind foot; teeth white; zygomatic arch weak and delicate but present . . . . Talpidae

Shrews (Soricidae)

Shrews are rather delicate, mouselike insectivores of a dull brown, reddish, black, or gray color. They can be immediately distinguished from mice by their long, moveable snouts; their upper incisors, which are sickle shaped and somewhat resemble the canines of carnivores; and, in Pacific Coast species, a wine red pigment on most of the surface of their teeth. The skull lacks the zygomatic arch present in moles. Some shrews utter high-pitched sounds and locate prey by echolocation. A Vagrant Shrew (*Sorex vagrans*) that entered one of our homes sat up on its haunches and uttered a shrill scream, perhaps a territorial signal.

Shrews have a distinctive life cycle. The young are born in spring and rarely become sexually mature before late the following winter. Sexual maturity is indicated by a rapid increase in body weight and the development of reproductive structures, and mating occurs in late winter or early spring. In many species, the female has only one litter in her lifetime. Shortly after the young are weaned, the adults die, so that by midsummer, virtually the entire population consists of young born during the previous few weeks.

Shrews have a high metabolic rate, perhaps reflecting their high surface-to-mass ratio, their high-protein diet, or both. They sleep frequently but are active both day and night.

Key to Genera of Shrews (Soricidae) in California

1a   Upper jaw with row of three unicuspid teeth (fig. 10)......
....*Notiosorex*
1b   Upper jaw with row of four or five unicuspid teeth (fig. 17a)
....*Sorex*

DESERT SHREW   *Notiosorex crawfordi*
Pl. 1, Fig. 10

DESCRIPTION: A small, gray or brown shrew with a short tail, relatively large ears, and faintly reddish teeth. It has only three unicuspids in the upper jaw (fig. 10); long-tailed shrews (*Sorex* spp.) have four or five. TL 81–90 mm, T 24–26 mm, HF 9–11 mm, E 8–9 mm. Weight: 3–5 g.

DISTRIBUTION: Resides in the southernmost part of California. This rarely encountered shrew is most likely to be found in areas of scattered scrub oak, juniper, sycamore, and cottonwood, and in more arid habitats of sagebrush and associated shrubs. It occurs east to western Oklahoma and south to central Mexico.

FOOD: In captivity, this shrew subsists on a broad variety of in-

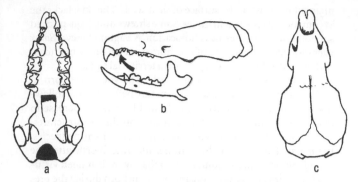

Figure 10. Skull of the Desert Shrew *(Notiosorex crawfordi):* (a) ventral view; (b) lateral view (arrow to unicuspid teeth); (c) dorsal view.

sects. In the wild, it presumably has a generalized diet of large terrestrial insects.

REPRODUCTION: Pregnancies occur from April to November. A litter has three to five young, and females may possibly have two or more litters a year.

Long-tailed Shrews *(Sorex)*

Long-tailed shrews belong to a genus *(Sorex)* of closely related and frequently very similar insectivores. Their identification rests largely on details of dentition; body color and tail color are useful but not always reliable. The details of the life history of some common species are rather well known. Some species, however, are known from very few specimens, and virtually nothing is known of their biology. The comments on these species are, therefore, regrettably skimpy.

Key to Species of Long-tailed Shrews *(Sorex* spp.)

1a Inner side of lower jaw with postmandibular canal (fig. 11a); inner (median) surface of unicuspid teeth without pigmented ridge (fig. 12a) . . . . subgenus *Sorex* (2)

1b Inner side of lower jaw without postmandibular canal (fig. 11b); inner (median) surface of unicuspid teeth with pigmented ridge (fig. 12b) . . . . subgenus *Otisorex* (3)

2a Third upper unicuspid larger than fourth (fig. 13a);
    upper incisor lacking median tine (or minute lobe)
    when viewed anteriorly (fig. 13b); sagebrush and
    semiarid regions ....*merriami*

2b Third upper unicuspid smaller than fourth (fig. 17a);
    upper incisor with distinct median tine when viewed
    anteriorly (fig. 17b); montane forested regions.......
    ....*trowbridgii*

3a Third upper unicuspid smaller than fourth 4

3b Third upper unicuspid larger than fourth............ *lyelli*

4a Skull length greater than 19.0 mm................. 5

4b Skull length less than 19.0 mm; medial surface of
    upper incisor without inner ridge ....*sonomae*

5a First and second unicuspids equal or nearly equal; dorsum
    brown 6

5b First unicuspid smaller than second (fig. 16a); hind foot
    with distinct lateral fringe of stiff hairs; dorsum gray or
    black...*palustris*

6a Upper incisor without median tine or lobe; brown or
    reddish brown.................................... 7

6b Upper incisor with median tine or lobe; black ....
    ....*bendirii*

7a Pigment on anterior surface of upper incisor not extending
    dorsally to median tine or lobe; northern half of California,
    mostly in wet grasslands (fig. 14a).................*vagrans*

7b Pigment on anterior surface of upper incisor extending
    above median tine or lobe (fig. 14b) 8

8a Skull length 15.4 mm or more 9

8b Skull length 15.3 mm or less; arid regions south of
    Lake Tahoe ....*tenellus*

9a Cranium rather flat (fig. 15a); skull length 15.4–17.0 mm;
    third unicuspid conspicuously smaller than fourth 10

9b Cranium convex (fig. 15b); third unicuspid very slightly
    smaller than fourth; Sierra Nevada, San Gabriel, and San
    Bernardino Mountains.......................*monticolus*

10a Light brown; Central Valley to coast; Santa Catalina
    Island.....................................*ornatus*

10b Very dark brown or black; salt marshes at north end of
    San Pablo Bay, Suisun Bay, and Grizzly Island........
    ....*sinuosus*

MARSH SHREW
Sorex bendirii

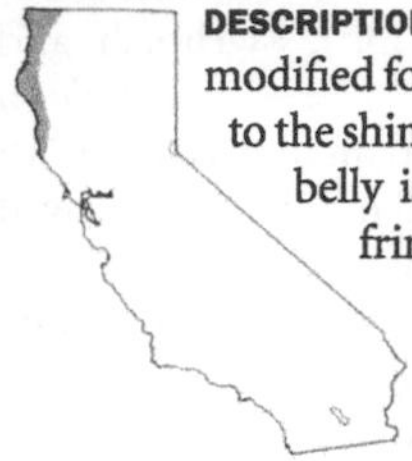

DESCRIPTION: A large, dark brown shrew, slightly modified for an aquatic life. Its pelage is dull, in contrast to the shiny pelage of the Water Shrew (*S. palustris*). Its belly is nearly as dark as its back. It has a small fringe of stiff hairs on the hind foot, which is less developed than in the Water Shrew. The first upper incisor has a median tine, and the third unicuspid is smaller than the fourth. TL 145–170 mm, T 60–80 mm, HF 18–21 mm, E 7–9 mm. Average weight: 16 g.

DISTRIBUTION: Found in the northwestern corner of California, south to Gualala. It prefers to be near standing water but is sometimes found along creeks. Its range extends north to British Columbia, along the coast.

FOOD: This shrew has not been well studied but presumably feeds on aquatic insects that occur in lowland marshes.

REPRODUCTION: At least one litter of four to six young (usually five) is born from March to June.

MT. LYELL SHREW
Sorex lyelli

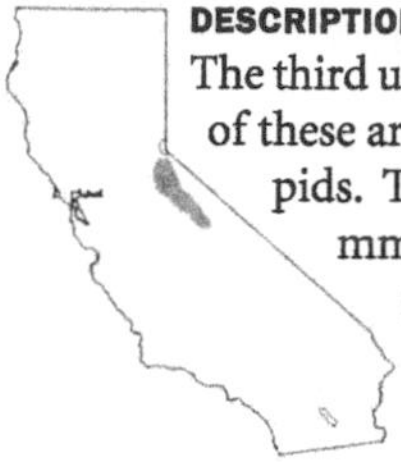

DESCRIPTION: An olive brown shrew with a gray belly. The third unicuspid is larger than the fourth, and both of these are smaller than the first and second unicuspids. TL 88–108 mm, T 36–43 mm, HF 11–12 mm, E 6 mm.

DISTRIBUTION: Known only at high elevations in the southern Sierra Nevada.

REPRODUCTION: One record indicated four embryos in June.

STATUS: This is a California species of special concern.

COMMENTS: This rare shrew has been collected only a few times, and very little is known about it.

MERRIAM'S SHREW
Sorex merriami
Fig. 13

DESCRIPTION: A small species with a conspicuously bicolored tail. The dorsum is gray brown, and the underparts and feet are white. The teeth are dark red and lack tines. The second unicuspid is the largest, and the fourth is smaller than the third (fig. 13a). TL 99–107 mm, T 33–42 mm, HF 11–13 mm, E 8–9 mm. Weight: 4.4–6.5 g.

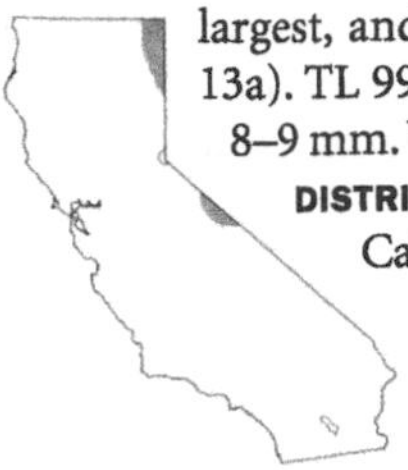

DISTRIBUTION: Found along the eastern border of California in high-elevation sagebrush and piñon-juniper communities. This is essentially a Great Basin species, occurring from eastern Washington to western North Dakota and Nebraska.

FOOD: This shrew eats small spiders, crickets, grasshoppers, and soil-dwelling insects such as beetle larvae and cutworms.

REPRODUCTION: Pregnancies are known to occur from late March to June. A litter has five to seven young, and females probably have only one litter a year.

MONTANE SHREW   *Sorex monticolus*
Figs. 14, 15

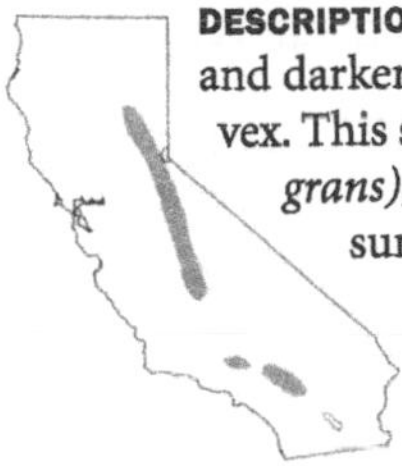

DESCRIPTION: A rather small shrew, brown in summer and darker and grayer in winter. The cranium is convex. This species resembles the Vagrant Shrew (*S. vagrans*), except that the pigment on the anterior surface of the upper incisor extends above the median tine (fig. 14b). TL 111–120 mm, T 46–55 mm, HF 13–15 mm, E 6–7 mm. Weight: 2.8–5.2 g.

DISTRIBUTION: Found in the Sierra Nevada and the San Bernardino Mountains in California, and from central Mexico through the Rocky Mountains north to Alaska. This shrew is well named, for it is truly a montane species.

FOOD: Many small insects, especially soil-dwelling larvae, and little vegetable food make up this shrew's diet.

REPRODUCTION: Pregnancies occur from April to May. Four to eight young are born in a litter, and it is possible that a female has only one litter a year.

ORNATE SHREW   *Sorex ornatus*
Fig. 15

DESCRIPTION: A rather small, dull brown shrew with a faintly bicolored tail. Members of some populations around the north shore of San Pablo Bay are very dark brown. This shrew is most similar to the Vagrant Shrew *(S. vagrans)* and Montane Shrew *(S. monticolus)* but is separable on the basis of geographic distribution and the characters in the key (fig. 15a). TL 89–108 mm, T 32–44 mm, HF 12–13 mm, E 6–7 mm. Weight: 3–7 g.

DISTRIBUTION: Occurs in the Central Valley and Coast Ranges in California, south to Baja California. It is typically found in rather open areas, including the salt marshes along the eastern shore of San Francisco Bay.

FOOD: This shrew feeds on small soil-dwelling insects.

REPRODUCTION: Breeding takes place from February until the adults die in early summer. From three to five young are born in a litter.

STATUS: The Buena Vista Lake Shrew *(S. o. relictus)*, which occurs in parts of Kings County and Kern County, is federally listed as endangered and is a California subspecies of special concern. Three other subspecies (the Monterey Shrew or Salinas Ornate Shrew *(S. o. salarius)*, the Southern California Saltmarsh Shrew *(S. o. salicornicus)*, and the Santa Catalina Shrew *(S. o. willetti)* are also California species of special concern.

COMMENTS: In the Central Valley, where this shrew is common, other shrew species are not likely to occur with it. Along lower elevations of the western slope of the southern Sierra Nevada, the Montane Shrew *(S. monticolus)* may occur with it.

WATER SHREW   *Sorex palustris*
Pl. 1, Fig. 16

DESCRIPTION: A large, black or dark brown shrew with distinctive dense, velvety, water-repellent fur and a fringe of rather stiff whitish hairs on the sides of its hind feet. Its upper incisor lacks a median tine, and the third unicuspid is smaller than the fourth (fig. 16a). TL 144–158 mm, T 73–78 mm, HF 18–21 mm, E 6 mm. Weight: 8–14 g.

DISTRIBUTION: Occurs throughout the Sierra Nevada, Cascade Range, and North Coast Ranges in California, at and above about 1,300 m (4,300 ft). It also inhabits much of the forested area north to Alaska, including the mountains of central Oregon and appropriate habitat in Washington. It is rarely found more than a meter from water and is common along some small mountain streams.

FOOD: This shrew mostly eats nymphal and larval stages of aquatic insects, especially mayflies, stoneflies, and caddisflies. It also commonly eats some small crustaceans and occasionally takes small fish and frogs.

REPRODUCTION: A litter comprises four to seven young, and females may possibly have two litters a year.

COMMENTS: This is the most aquatic North American shrew. Its water-repellent fur retains a bubble of air about its body and enables it to float like a duck. People who have been fortunate enough to observe this animal report that it swims and dives with the agility of an otter.

SUISUN SHREW   *Sorex sinuosus*

DESCRIPTION: A small, dark brown or blackish shrew similar to the Ornate Shrew (*S. ornatus*) but distinguishable on the basis of color and geographic occurrence. TL 98–106 mm, T 35–44 mm, HF 11–13 mm, E 7–8 mm. Weight: 4.5–6.8 g.

DISTRIBUTION: Restricted to salt marshes about San Pablo Bay, Suisun Bay, and Grizzly Island, California.

FOOD: This shrew subsists on insects and small crustaceans.

REPRODUCTION: Breeding occurs from late February to early June; lactating females have been found in July. Embryo counts indicate that a litter contains two to nine young. The female has one or two litters a year.

STATUS: This is a California species of special concern.

COMMENTS: It has been suggested that this shrew represents a very local subspecies of the Ornate Shrew (*S. ornatus*).

FOG SHREW   *Sorex sonomae*
Pl. 1

DESCRIPTION: A large shrew distinguished by its reddish, light brown color and by the absence of a median tine on the upper incisor. The tail is almost uniformly colored. The first and second unicuspids are equal or nearly so; the third unicuspid is smaller than the fourth. TL 120–158 mm, T 45–65 mm, HF 13–18 mm, E 7–8 mm. Weight: 14–18 g.

DISTRIBUTION: Inhabits coniferous coastal

forest and brushlands from Marin County north to the central Oregon coast. It is mostly found near creekside thickets.

FOOD: This shrew is known to feed heavily on centipedes and spiders. It also eats slugs and snails.

REPRODUCTION: There are three to five young in a litter, and the female probably has more than one litter a year.

COMMENTS: Although this large, brightly colored shrew is rather common within its range and habitat, it is not very well known.

INYO SHREW   *Sorex tenellus*

DESCRIPTION: A small, brownish shrew with a faintly bicolored tail. It resembles the Vagrant Shrew (*S. vagrans*) but has pigment on the upper incisor extending above the median tine or lobe. The third unicuspid is smaller than the fourth. The skull is 15.3 mm long or less. TL 85–106 mm, T 32–42 mm, HF 10–12 mm, E 6–7 mm. Weight: 4–8 g.

DISTRIBUTION: Occurs in the high arid region immediately east of the southern Sierra Nevada, California, southeast to southern Nevada.

FOOD: Unknown.

REPRODUCTION: Unknown.

TROWBRIDGE'S SHREW   *Sorex trowbridgii*
Pl. 1, Fig. 17

DESCRIPTION: A shrew that is brown in summer and gray in winter, with a bicolored tail. The upper incisor has a distinct median tine; the third unicuspid is smaller than the fourth (fig. 17b). TL 95–132 mm, T 40–62 mm, HF 12–15 mm, E 6–8 mm. Weight: 4.8 g.

DISTRIBUTION: Found in coniferous and mixed forests in the northern two-thirds of California. It is absent from the Central Valley. Its range extends north through western Oregon and western Washington to southern British Columbia.

FOOD: This generalized feeder takes many kinds of beetles, moths, heteropterans, spiders, and centipedes. It also eats earthworms in fall and winter, when soil moisture brings these invertebrates to the surface. At times it feeds heavily on conifer seeds, especially those of the Douglas-fir *(Pseudotsuga menziesii)*, but generally it does not take much vegetable food.

REPRODUCTION: Sexual activity starts in February (in the northern Sierra Nevada) and continues until the latter half of May or until the adults die. Pregnancies can occur as early as February, and some females have at least two broods. There are three to six young in a brood (most commonly five).

COMMENTS: Like most shrews, this is an annual species, with a complete turnover of generations every year. The adults die rather rapidly at about one year of age, and by the end of the summer almost the entire population consists of young born that spring.

VAGRANT SHREW   *Sorex vagrans*

Pl. 1, Fig. 14

DESCRIPTION: A rather variable species that differs in appearance from place to place. It is brown in summer and gray in winter, with a bicolored tail. The anterior surface of the upper incisor has pigment that does not extend to the median tine or lobe (fig. 14a); the third unicuspid is smaller than the fourth. TL 90–120 mm, T 34–42 mm, HF 11–12 mm, E 6–8 mm. Weight: 5–7 g.

DISTRIBUTION: Occurs in Coast Ranges and Sierra Nevada, usually in grassy meadows and other moist open areas. Its range extends east to Idaho and north to British Columbia.

FOOD: This shrew is an opportunistic feeder, taking small arthropods, earthworms, and slugs.

REPRODUCTION: Breeding takes place as early as January. There are three to eight in a litter, and the female may possibly have two or more litters.

STATUS: The Salt-marsh Wandering Shrew *(S. v. halicoetes)* is a California subspecies of special concern.

COMMENTS: Like most other shrews, this is an annual species; all adults die by the end of their second summer.

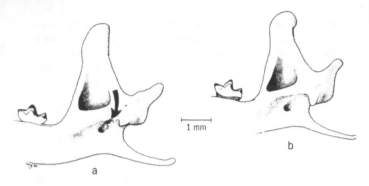

Figure 11. Lower jaws of long-tailed shrews (genus *Sorex*). (a) Subgenus *Sorex*, showing the characteristic postmandibular foramen (arrow); (b) subgenus *Otisorex*, which lacks a postmandibular foramen. From Junge and Hoffman (1981).

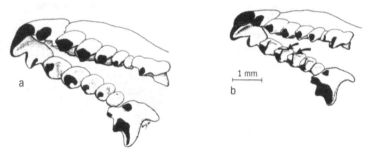

Figure 12. Inner surfaces of upper tooth rows of long-tailed shrews (genus *Sorex*). (a) Subgenus *Sorex*, with unicuspid teeth lacking pigmented ridges; (b) subgenus *Otisorex*, with pigmented ridges (arrows) on unicuspid teeth. From Junge and Hoffman (1981).

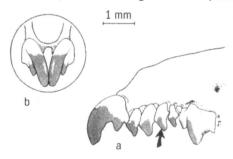

Figure 13. Teeth of Merriam's Shrew *(Sorex merriami)*. (a) Lateral view, showing the large third unicuspid (arrow); (b) front view of upper incisors, showing lack of median tine (compare with fig. 13b). From Junge and Hoffmann (1981).

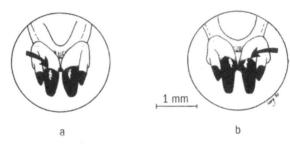

Figure 14. Front views of upper incisors. (a) Vagrant Shrew *(Sorex vagrans);* (b) Montane Shrew *(Sorex monticolus).* Arrows show extent of pigment on incisors. From Junge and Hoffman (1981).

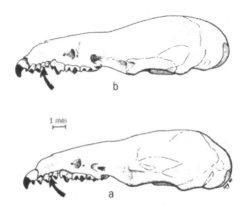

Figure 15. Lateral views of skulls. (a) Ornate Shrew *(Sorex ornatus),* with the third unicuspid (arrow) conspicuously smaller than the fourth (note the flatter skull); (b) Montane Shrew *(Sorex monticolus);* with the third unicuspid (arrow) only slightly smaller than the fourth (note the convex skull). From Junge and Hoffman (1981).

Figure 16. Skull of the Water Shrew *(Sorex palustris).* (a) The first unicuspid is smaller than the second, and the third unicuspid (arrow) is smaller than the fourth; (b) Front view of upper incisors. From Junge and Hoffmann (1981).

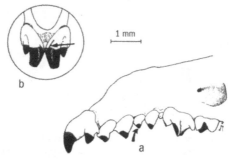

Figure 17. Teeth of Trowbridge's Shrew *(Sorex trowbridgii)*. (a) Lateral view, showing the small third unicuspid (arrow); (b) front view of upper incisors, with median tine (arrow). From Junge and Hoffmann (1981).

Moles (Talpidae)

These insectivores are fossorial (burrowing) mammals. They are most likely to be found in light, sandy soils and seem to avoid ground with a heavy clay content. Most species have greatly broadened forefeet for digging through the soil, although this specialization is less well marked in the Shrew-mole *(Neurotrichus gibbsii),* which tends to be shrewlike in habits. Moles have no external ear. They have three upper incisors per side, and all their teeth are white. The skull has a zygomatic arch (fig. 18b).

Key to Species of Moles (Talpidae) in California

1a   Forehand as wide as or wider than long; tail less than 25 percent of total length .... (2)

1b   Forehand longer than wide; tail more than 25 percent of total length .... *Neurotrichus gibbsii*

     2a   Unicuspid teeth evenly spaced (figs. 19b, 20b); usually black; found on coast north of San Francisco Bay 3

     2b   Unicuspid teeth unevenly spaced (fig. 18b); usually light gray or brownish; found throughout Pacific states except in arid regions ...... *Scapanus latimanus*

3a   Total length usually more than 200 mm; skull with distinct

ridge (sublacrymal maxillary ridge) above maxillary region
(fig. 20b) .... *Scapanus townsendii*

3b Total length less than 180 mm; skull lacking distinct ridge
(fig. 19)............................... *Scapanus orarius*

SHREW-MOLE   *Neurotrichus gibbsii*
Pl. 1

DESCRIPTION: A small, dark or black mole with only moderately expanded forefeet. The tail is distinctive: it is about one-third the total length of the animal and constricted at the base. As in shrews but not other moles, the pelage is directed to the rear. TL 110–125 mm, T 33–45 mm, HF 15–18 mm. Weight: 10–13 g.

DISTRIBUTION: Found in many humid habitats; unlike moles of the genus *Scapanus*, not restricted to light and sandy soils. It occurs in coastal redwood forests and along creek sides in coastal brushlands from Monterey Bay north to southern British Columbia. It is also found in the northern Sierra Nevada, south at least to Plumas County.

FOOD: Like most insectivores, the Shrew-mole seems to feed indiscriminately on a broad spectrum of soil-dwelling insects, pillbugs, and centipedes. At times it feeds heavily on earthworms, which may be its preferred food.

REPRODUCTION: The Shrew-mole appears to breed in late winter, although little is known about its reproductive patterns. There may be one to four young in a litter, born in spring.

COMMENTS: The Shrew-mole makes a shallow burrow. Unlike other moles, it is frequently active aboveground and may forage either by day or by night.

BROAD-FOOTED MOLE   *Scapanus latimanus*
Pl. 1, Fig. 18

DESCRIPTION: A light gray or black mole of medium size, with a forehand as wide as or wider than it is long, and a tail less than 25 percent of its total length. Its unicuspid teeth are unevenly spaced (fig. 18b). TL 135–190 mm, T 20–45 mm, HF 18–25 mm.

DISTRIBUTION: Widely distributed in California up to at least 2,000 m. It favors light, sandy soils but is absent from heavily cultivated areas. It is especially numerous on floodplains with high soil moisture and a strong growth of forbs and soil invertebrates. Its range extends from Oregon south to Baja California and east to extreme western Nevada.

FOOD: This mole feeds on soil invertebrates, especially earthworms, and underground parts of plants.

REPRODUCTION: A single litter of two to four young is born in late winter.

STATUS: The Alameda Island Mole *(S. l. parvus)* is a California subspecies of special concern.

COMMENTS: This is the most widely distributed and commonly encountered mole in the state.

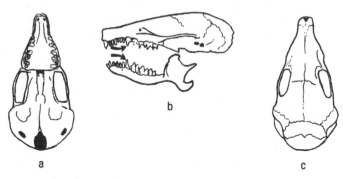

Figure 18. Skull of the Broad-footed Mole *(Scapanus latimanus)*. (a) Ventral view; (b) lateral view (arrows to unicuspid teeth); (c) dorsal view.

COAST MOLE   *Scapanus orarlus*
Fig. 19

DESCRIPTION: The smallest and most delicately built of the Pacific Coast moles, grayer than Townsend's Mole and sometimes brownish. The skull is delicate. The last upper unicuspid is larger than that immediately in front of it; the unicuspids are evenly spaced (fig. 19b). TL 165–175 mm, T 25–36 mm, HF 20–23 mm. Weight: 54–62 g.

DISTRIBUTION: Inhabits coastal California north of San Francisco Bay. Its range extends north to British Columbia and east to Idaho.

FOOD: This mole preys heavily on earthworms but probably takes most soil invertebrates.

REPRODUCTION: A single litter of two to four young is born in late winter.

COMMENTS: This mole may occur in gardens, but it is not the pest that Townsend's Mole can be.

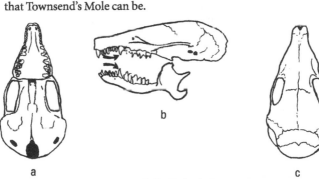

a   b   c

Figure 19. Skull of the Coast Mole *(Scapanus orarius).* (a) Ventral view; (b) lateral view (arrow to unicuspid tooth); (c) dorsal view.

TOWNSEND'S MOLE   *Scapanus townsendii*
Pl. 1, Fig. 20

DESCRIPTION: A large, black or dark gray mole whose size, color, and geographic distribution distinguish it from other California moles. The last upper unicuspid is about equal to that immediately in front of it; the unicuspids are evenly spaced (fig. 20b). TL 195–240 mm, T 33–52 mm, HF 24–28 mm. Weight: 115–150 g.

DISTRIBUTION: Occurs in the extreme northwestern corner of California. It prefers rich, moist soil and is common in meadows and river valleys at low elevations. It is also found in coastal Oregon and Washington in open valleys.

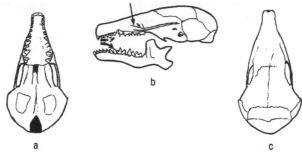

Figure 20. Skull of Townsend's Mole *(Scapanus townsendii).* (a) Ventral view; (b) lateral view (arrows to sublacrymal maxillary ridge and unicuspid teeth); (c) dorsal view.

FOOD: This mole eats soil invertebrates and sometimes underground parts of plants. It has been known to become a pest in commercial bulb plantings.

REPRODUCTION: A single litter of two to four young is born in midwinter.

CHIROPTERA

Bats have been flying since the Eocene and are highly specialized in both structure and behavior. The fingers are extremely long and slender, and they are connected by a thin, delicate membrane. The hind legs, too, are connected by a membrane in most species (including all of those on the Pacific Coast). This interfemooral membrane (or uropatagium) usually encloses the tail. Typically a small bone, the calcar, projects from the hind foot and extends the interfemoral membrane when the bat is in flight (see fig. 21). The eyes are frequently greatly reduced, but there is evidence that bats see well. The ears are large in many species; so-called gleaner species, which often pick up insects from the surface of leaves, frequently have rather long ears. In some species the ears are highly mobile and can be directed independently of each other. The inner ear is sensitive to extremely high frequencies.

The pectoral muscles are large and strong, and in most species the sternum is keeled, like the "breastbone" of a fowl, providing attachment for the pectoral muscles. Different kinds of bats have different specializations for flight, and these are reflected in the wing structure. For this reason the length of the forearm (labeled "F" in the species accounts) is of importance in identification.

The history of bats is extremely interesting, for the earliest fossils are clearly bats, in some respects very much like modern bats, and bats do not appear to be close to any other order. Relationships among bats are suggested by chemical evidence. Although fossil bats date from the Eocene, many of their features indicate a much earlier origin. Moreover, noctuid moths, which are heavily preyed upon by some bats, have well-developed hearing mechanisms that may have evolved to detect the high-frequency emissions of echolocating bats; and noctuid moths are known from fossils 70 million years old. Bats may have become distinct from other mammals in the Cretaceous. Some students have suggested that bats may be related to colugos (Dermoptera) of southeast Asia, or perhaps to insectivores or primates.

Perhaps because bats were flying mammals by the Eocene and because their skeleton is very delicate, there are not many known fossils of bats. The genus *Myotis* dates from the Oligocene in Europe, and free-tailed bats, *Tadarida,* are known

Figure 21. Calcars of *Myotis* spp.: (a) with a keel (arrow); (b) without a keel (arrow). The presence of a keel is usually apparent in fresh specimens but may be difficult to see after the skin is dry.

from Oligocene fossil beds in Europe and South America. The genus *Lasiurus* is known from Pliocene fossils from North America.

The oldest fossils have a keeled sternum, and the ears indicate that echolocation had developed by the Eocene. Presumably bats developed echolocation after they began flying, possibly in order to avoid obstacles and also to detect flying insects, two very different auditory capabilities. Because many insects are active at night, they were a rich food supply not exploited by birds.

Much of our knowledge of the biology of bats, and especially of their migrations, is because of their danger as vectors and reservoirs of rabies. Because bats can transmit rabies and also migrate vast distances, they have great potential to disseminate this fearful disease.

Bat Conservation International maintains a Web site (www.batcon.org) that offers current information about bats, bat-related products (such as bat houses), and links to resources for educators and others interested in the conservation of these beneficial mammals, which eat millions of tons of insects.

Flight

Bats are powerful fliers with a much larger wing membrane than gliding mammals. The wings, mostly devoid of fur, contain blood vessels and allow dissipation of heat and water. The bones are solid, in contrast to the hollow bones of birds, but as in birds, most of the muscles lie within the body rather than in the wing itself.

In cross section, the wing is curved, concave on the lower side and convex on the upper. In flight, air passing over the upper and lower surfaces must reach the trailing edge of the wing at the same time. Therefore, because of the shape of the wing, air must move along the dorsal surface more rapidly. For this reason, air pressure on the dorsal surface is less than that on the ventral surface (the chord). This differential produces lift, which is roughly proportional to the area of wing surface and increases with the square of the bat's flight speed.

Wing loading, the amount of body weight supported by the wing surface (that is, the body mass divided by the wing surface area), influences the efficiency of flight, as does flight speed. Some bats may forage at some distance from their diurnal roost; body mass and wing loading are increased by the mass of the insects consumed. Also, most California bats either hibernate or migrate, and both of these activities require deposits of fat, increasing wing loading. Pregnancy, too, increases wing loading.

At the trailing edge of the wing, the turbulence caused by the meeting of the dorsal and ventral air creates drag, which tends to impede forward movement. Drag is modified by the aspect ratio, or wing span^2/wing area: as the aspect ratio increases, drag decreases.

The features described above all relate to flight ability when the bat is either gliding or flapping. In addition, flapping, which distinguishes bats from gliding mammals, produces thrust and lift. The wingbeat increases the differential in pressure above and below the wing surface. It also increases drag, which is overcome by the forward thrust of a flying bat. In flight the wing is rotated between upstroke and downstroke.

Wing shape, wing loading, and speed of flight are all adapted to the habitat and foraging habits of the bat. High wing loading characterizes bats with narrow wings and greater flight speed. A low aspect ratio is typical of woodland bats. Typically they have a rather fluttering flight, which facilitates foraging in the leafy canopy. Bats with reduced wing loading are more maneuverable; these slower-flying species may be gleaners. High aspect ratios characterize species that forage in open areas and at greater speeds.

Echolocation

Bats are very sensitive to the position and size of a flying insect. However, they advertise their presence when they produce sound, which some insects can detect. Moths of the family Noctuidae, for example, can detect bats' ultrasonic emissions, even though the moths do not communicate among themselves by sound. A moth can hear echolocation emissions of up to 110 kHz at greater distances than those at which the bat can detect the echo. To counter this advantage of the moth, a bat may use low intensity and high frequencies. Echolocation is discussed at length under Senses.

Reproduction

Reproduction in the Chiroptera is characterized by several unusual, but not unique, features. In most California species, males experience testicular enlargement in fall and females ovulate in spring. Thus, mating occurs in autumn, but the spermatozoa then lie dormant in the uterine tract until ovulation and fertilization occur in spring. Birth may take place five or six months after mating, with actual embryonic development occupying approximately two months. The gestation is remarkably long for such a small mammal. Most bats typically bear a single young at one time, but species of *Lasiurus*, such as the Hoary Bat (*Lasiurus cinereus*), produce from two to four, and rarely five. Most female bats possess a single pair of mammary glands, thoracic in position, but in *Lasiurus* there are two pairs of mammary glands and nipples, providing for the nursing of two or more young. There are two ovaries, but ovarian activity is often confined to one side. The penis is pendant, as in primates and colugos. Testes descend into a scrotum seasonally.

Perhaps because many bats remain in caves, mine shafts, or hollow trees during the day and such environments are localized, bats tend to be social creatures and sometimes assemble in large numbers. In spring, when young are born, females may form nursery colonies, in which the young may suckle from any female, randomly, whereas other females forage. The extent to which this occurs and varies among species is not well understood. Young bats are precocial, fully furred, and capable of clinging to the mother while she is in flight.

Key to Families of Bats (Chiroptera) of the Pacific Coast States

1a Tail entirely (or almost entirely) contained within interfemoral membrane; if tail tip projects beyond margin of interfemoral membrane, face has leaflike appendage 2
1b Tail projecting conspicuously beyond margin of interfemoral membrane; face without leaflike appendage . . . . . .
. . . . Molossidae
   2a Face with flat leaflike appendage . . . . . Phyllostomidae
   2b Face without leaflike appendage . . . .
. . . . Vespertilionidae

Leaf-nosed Bats (Phyllostomidae)

This diverse group of bats is found mostly in Central and South America. Its members are distinguished by peculiarities of the skull and, in most species, by a distinctive nose leaf on the tip of the snout. The species differ from one another in wing structure, flight pattern, and diet. Some eat fruit, and others eat nectar, pollen, or insects. California is the only Pacific Coast state in which leaf-nosed bats occur.

Key to Leaf-nosed Bats (Phyllostomidae) in California

1a Snout very long. 2
1b Snout not elongate; upper molars with distinct W pattern (fig. 23a); tail long, projecting slightly beyond interfemoral membrane; lower incisors present . . . *Macrotus californicus*
   2a Tail short, extending less than halfway to margin of interfemoral membrane; lower incisors absent. . . . .
. . . . *Choeronycteris mexicana*
   2b Tail minute, not extending into interfemoral membrane; lower incisors present . . . .
. . . . *Leptonycteris curasoae*

HOG-NOSED BAT or LONG-TONGUED BAT

Choeronycteris mexicana

Pl. 2, Fig. 22

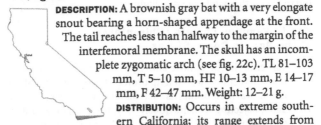

DESCRIPTION: A brownish gray bat with a very elongate snout bearing a horn-shaped appendage at the front. The tail reaches less than halfway to the margin of the interfemoral membrane. The skull has an incomplete zygomatic arch (see fig. 22c). TL 81–103 mm, T 5–10 mm, HF 10–13 mm, E 14–17 mm, F 42–47 mm. Weight: 12–21 g.

DISTRIBUTION: Occurs in extreme southern California; its range extends from southern Nevada and Arizona south through Mexico, including Baja California, to Guatemala.

FOOD: The extremely long snout and extrudable tongue are adaptations for feeding on the pollen and nectar of night-blooming flowers.

REPRODUCTION: A single young is born in spring. Until the infant is rather large, the mother carries it when she forages.

STATUS: This is a California species of special concern.

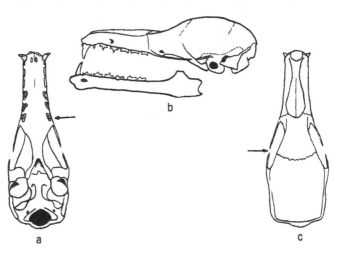

Figure 22. Skull of the Hog-nosed Bat *(Choeronycteris mexicana);* (a) ventral view (arrow to upper molar); (b) lateral view; (c) dorsal view (arrow to incomplete zygomatic arch).

COMMENTS: This bat roosts in the darkest regions of caves and mines.

CALIFORNIA LEAF-NOSED BAT   *Macrotus californicus*
Pl. 2, Fig. 23

DESCRIPTION: A light chocolate brown bat; the ends of the hairs are conspicuously darker than the underfur, and the belly is pale. The snout has a simple leaflike appendage at the tip. The ears are very large (more than 20 mm from crown to tip). The tail extends slightly beyond the interfemoral membrane. The upper molars have a pigment pattern forming a W (fig. 23a). TL 77–108 mm, T 25–42 mm, E 24–29 mm, F 44–58 mm. Weight: 10–14 g. (See fig. 23.)

DISTRIBUTION: Found in the arid extreme southern regions of California. Its range extends from southern Nevada and Arizona south through most of Mexico, including Baja California, and the West Indies.

FOOD: It eats mostly large, heavy-bodied insects, such as noctuid moths, crickets, grasshoppers, and scarabid and carabid beetles.

REPRODUCTION: A very unusual reproductive pattern distinguishes this nonhibernating bat from other North American bats. Both mating and fertilization occur in autumn. Embryonic development is slow for the first five months, then accelerates in

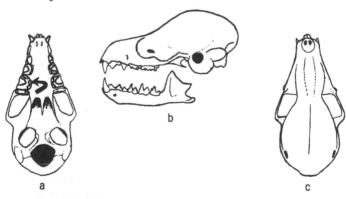

Figure 23. Skull of the California Leaf-nosed Bat *(Macrotus californicus):* (a) ventral view (arrow to molar); (b) lateral view; (c) dorsal view.

spring, and birth follows mating by approximately eight months. Ovulation occurs only in the right ovary, implantation only in the right uterine horn.

STATUS: This is a California species of special concern.

COMMENTS: This bat roosts in buildings, mines, and caves.

SOUTHERN LONG-NOSED BAT   *Leptonycteris curasoae*

DESCRIPTION: A reddish brown or cinnamon bat with short fur and an elongate snout and tongue. The very short tail does not extend into the abbreviated interfemoral membrane. Lower incisors are present. TL 69–84 mm.

DISTRIBUTION: Found in the city of Oceanside, San Diego County, and in San Bernardino County. Its range extends from northern South America north to the southern United States.

FOOD: This bat feeds on nectar, fruits, and pollen.

REPRODUCTION: Mating takes place in autumn. A single young is born in spring.

STATUS: The Lesser Long-nosed Bat (*L. c. yerbabuenae*) is federally listed as endangered.

Vesper Bats (Vespertilionidae)

This is the predominant family of bats in the Northern Hemisphere. It is structurally diverse, and its most important characteristic features are skeletal details. The ears are large in some species and small in others, and may be either joined or separate at the top of the head. The snout never has a leaflike appendage at the tip. The tail extends to, but never beyond, the margin of the interfemoral membrane. The anterior incisors are separated from each other by a toothless space medially, so there is an anterior concavity on the roof of the mouth when viewed ventrally.

Almost all species are insectivorous. Two species (not present in California) capture small fish at the surface of quiet waters. Vespertilionid species engage in sperm storage, with delayed fertilization occurring in spring, shortly after arousal from hibernation. In some species males hibernate separately from females. At least short migrations characterize most vespertilionid bats, and some kinds move hundreds of kilometers each year.

You are most likely to observe bats as they fly at dusk. Like many small mammals, bats are crepuscular, flying at dusk and again at dawn. During the night they rest in a nighttime roost and may pass scats from their evening meal. A nighttime roost may be under the eaves of a porch roof, beneath which there may be pieces of insect wings and legs that the bats have removed while feeding.

Key to Genera and Species of Vesper Bats (Vespertilionidae) in California

1a With two pairs of upper incisors 5
1b With one pair of upper incisors 2
    2a Ears very long; interfemoral membrane nude; snout blunt with ridge on side and top ....
        .... *Antrozous pallidus*
    2b Ears short; interfemoral membrane well furred dorsally; snout without ridge on sides and top 3
3a Color reddish or brownish gray, tips of hairs white; two upper premolars, usually with inner peglike upper premolar (fig. 27a) 4
3b Color yellow or yellowish brown; one upper premolar; tips of hairs not white.................... *Lasiurus xanthinus*
    4a Color brick red; forearm less than 45 mm ....
        .... *Lasiurus blossevillii*
    4b Color grayish brown; forearm more than 45 mm.....
        .... *Lasiurus cinereus*
5a Color brown or blackish and more or less uniform; lower canine large and pointed 6
5b Black with three large white spots on dorsum; lower canine small and appearing bifid laterally......................
    .... *Euderma maculatum*
    6a With two lower premolars 9
    6b With three lower premolars........................ 7
7a Ears short; color black, hairs tipped with white; interfemoral membrane furred; inner upper incisor bicuspidate (fig. 26a, b) .... *Lasionycteris noctivagans*
7b Ears long; color brownish or olive; interfemoral membrane nude or scantily furred 8
    8a Muzzle with two large fleshy glands on sides.........
        .... *Plecotus townsendii*
    8b Muzzle without large glands..... *Idionycteris phyllotis*

9a   With two pairs of upper premolars 10
9b   With one pair of upper premolars (fig. 24b) . . . .
        . . . . *Eptesicus fuscus*
        10a Tragus about one-third ear length; upper incisors
              about equal. . . . .
              . . . . *Pipistrellus hesperus*
        10b Tragus more than one-half ear length; outer upper in-
              cisor clearly larger than inner 11
11a Margin of interfemoral membrane with fringe of fine hairs
        . . . . *Myotis thysanodes*
11b Margin of interfemoral membrane with fine hairs absent or
        extremely few 12
        12a Ear short, not extending more than 2 mm beyond
              muzzle when laid forward 13
        12b Ear long, extending more than 6 mm beyond muzzle
              when laid forward. . . . . *Myotis evotis*
13a Calcar keeled (fig. 21a) 14
13b Calcar not keeled (fig. 21b) 16
        14a Underside of wing not furred to elbow; hind foot less
              than 8.5 mm; forearm usually less than 35 mm. . . . . . .
              15
        14b Underside of wing with fur to elbow; hind foot more
              than 8.5 mm; forearm usually more than 35 mm . . . . .
              . . . . *Myotis volans*
15a Ears black; pelage glossy; cranium low (fig. 30b) . . . .
        . . . . *Myotis ciliolabrum*
15b Ears dark brown; pelage dull; cranium high (fig. 29b) . . . . .
        . . . . *Myotis californicus*
        16a Forearm less than 40 mm; cranium without distinct
              sagittal crest (figs. 31b, 33b). 17
        16b Forearm more than 40 mm; cranium with sagittal
              crest (fig. 32b, c) . . . . *Myotis velifer*
17a Pelage dull; forearm 32–37 mm; braincase abruptly raised
        (see fig. 33b) . . . . *Myotis yumanensis*
17b Pelage glossy; forearm 35–40 mm; braincase gradually ele-
        vated (see fig. 31b) . . . . *Myotis lucifugus*

PALLID BAT

Antrozous pallidus

Pl. 2

DESCRIPTION: A medium-sized bat with buff or sandy-colored fur; the hairs are darker at the tips and lighter next to the skin. The ears are very long and clearly separated at the base. (In contrast, Townsend's Long-eared Bat *[Plecotus townsendii]* and Allen's Long-eared Bat *[Idionycteris phyllotis]* have ears joined at the base; the California Leaf-nosed Bat *[Macrotus californicus]* has a triangular leaflike appendage on the snout, and the tail extends beyond the interfemoral membrane.) There is only a single pair of upper incisors. TL 92–135 mm, T 43–49 mm, E 28–32 mm, F 48–61 mm. Weight: 16–19 g. Dentition: 1/2, 1/1, 1/2, 2/3.

DISTRIBUTION: Occurs in most of California; especially common in open, lowland areas, generally below 2,000 m. It is widely distributed in Oregon except in the high mountains and is also found along the Snake and Columbia River Valleys in Washington. Indeed, its range extends through much of western North America, from central Mexico to British Columbia.

FOOD: The Pallid Bat feeds largely on flightless arthropods, which it captures by foraging on the ground. Jerusalem crickets, scorpions, and June beetles figure largely in its diet.

REPRODUCTION: Mating occurs in autumn; fertilization is delayed, and birth occurs in late June. From one to three embryos have been found, but litters usually consist of two young (less commonly one). The female hangs upside-down during labor, and the young drop into a sack formed from the interfemoral membrane.

STATUS: This is a California species of special concern.

COMMENTS: This bat roosts in fissures in cliffs, abandoned buildings, and bird boxes, as well as under bridges. It frequently collects in small nursery colonies. The species makes local seasonal movements but apparently is not migratory.

BIG BROWN BAT

Eptesicus fuscus

Pl. 2, Fig. 24

DESCRIPTION: Larger than the Pacific Coast species of *Myotis,* this bat is distinguished by a combination of features. Its ears and

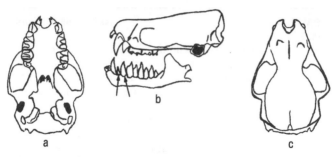

Figure 24. Skull of the Big Brown Bat *(Eptesicus fuscus):* (a) ventral view; (b) lateral view (arrows to single upper premolar and two lower premolars); (c) dorsal view.

wing membranes are nude and darkly pigmented; it is brownish and rather glossy; and it has two lower premolars and a single upper premolar (see fig. 24b). TL 105–120 mm, T 39–51 mm, E 13–19 mm, F 39–45 mm. Weight: 14–23 g.

DISTRIBUTION: This widespread and common species ranges from Colombia and Venezuela through the Greater Antilles north to Canada and east to the Atlantic coast.

FOOD: It captures beetles and flies.

REPRODUCTION: Mating occurs in August or September, ovulation and fertilization take place in April, and a single young is born in June.

COMMENTS: This is an early-flying species, usually appearing long before the sky is dark. The Big Brown Bat frequently enters buildings, sometimes invading attics and abandoned buildings in numbers. It also enters caves, especially in winter, though it tends to remain near the entrance. This bat hibernates in cold weather, but it is much hardier than most species and not infrequently forages on winter evenings. It is sometimes considered to be the same species as the Old World *Eptesicus serotinus.*

SPOTTED BAT   *Euderma maculatum*

Pl. 2

DESCRIPTION: A black bat with huge ears and three large white spots dorsally. The throat has a large, apparently nonglandular, nude area. This species is quite unlike any other California bat. TL 107–115 mm, T 47–50 mm, E 44–50 mm, F 48–51 mm. Weight: 13–18 g. Dentition: 2/2, 1/1, 2/2, 3/3.

DISTRIBUTION: Known from central Mexico north to southern British Columbia and east to Texas. It is found in both montane open coniferous forests and low deserts.

FOOD: This bat seems to favor noctuid moths; it is known also to take terrestrial insects.

REPRODUCTION: A single young is born in late May or early June; nursing continues into August.

STATUS: This is a California species of special concern.

COMMENTS: This species dwells primarily in caves.

ALLEN'S LONG-EARED BAT   *Idionycteris phyllotis*
Fig. 25

DESCRIPTION: A medium-sized bat, pale tawny in color, with the base of the hairs dark brown. The very long ears are joined at the top of the head; a pair of flaps from the ears project over the forehead and snout. The muzzle has only faintly developed dermal glands. The calçar has a strong keel (see fig. 21a). TL 103–108 mm, T 46–55 mm, E 38–43 mm, F 39–49 mm. Weight: 7–17 g. Dentition: 2/3, 1/1, 2/3, 3/3.

DISTRIBUTION: Not yet known to occur in California. However, it is found in montane forests of oaks and pines from central Mexico to southern Nevada, Arizona, and New Mexico, and it probably enters the adjoining area of California.

FOOD: This bat forages for flying insects.

REPRODUCTION: A single young is born in June or July.

COMMENTS: This bat emerges late in the evening and calls loudly while hunting. It roosts in caves and mineshafts. Females are known to form nursery colonies.

This species has long been placed in the genus *Plecotus*. Primarily on the

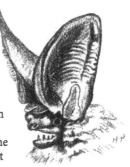

Figure 25. Allen's Long-eared Bat (*Idionycteris phyllotis*).

basis of chromosomal morphology, it is considered to be more closely related to *Euderma maculatum* than to those species currently placed in *Plecotus*. Its designation as the sole member of the genus *Idionycteris* is not universally accepted.

SILVER-HAIRED BAT   *Lasionycteris noctivagans*
Pl. 2, Fig. 26

DESCRIPTION: A bat whose conspicuously white-tipped hair gives its dark brown or black dorsal fur a frosted appearance. The ears are short and broad. The proximal half of the interfemoral membrane is well furred. The inner incisor of the upper jaw is strongly bicuspidate (see fig. 26b). TL 92–115 mm, T 35–45 mm, E 11–14 mm, F 38–44 mm. Weight: 6–12 g. Dentition: 2/3, 1/1, 2/3, 3/3.

DISTRIBUTION: Common in forested areas in most of California,

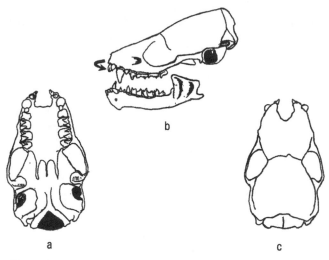

a   b   c

Figure 26. Skull of the Silver-haired Bat *(Lasionycteris noctivagans)*: (a) ventral view; (b) lateral view (arrow to bicuspid upper incisor); (c) dorsal view.

Oregon, and Washington. Its range extends from extreme north-eastern Mexico north to Alaska and east to the Atlantic coast.

FOOD: It eats flying insects of many orders but is particularly fond of moths.

REPRODUCTION: Mating occurs in September or October; ovulation takes place the following spring. After a gestation of 50 to 60 days, a litter of two young is born. Young are fledged in about three weeks and weaned shortly thereafter.

COMMENTS: This is a tree-dwelling species, but it sometimes enters buildings.

Genus *Lasiurus*

These colorful bats have sometimes been called hairy-tailed bats, for most or all of the dorsal surface of the interfemoral membrane is well furred. The undersides of the wings are also densely covered with hair along the bones and over muscles and tendons. The ears are short and rounded, generally with nude, darkly pigmented margins. Unlike most bats, these species have two pairs of mammae and have litters of two to four (occasionally five) young. Some species are highly migratory and make extensive flights over the oceans. They typically roost in the dense foliage of trees and are rarely found in concentrations. In parts of their winter range, tree bats may be active on rather cold nights, and they are much more tolerant of low temperatures than are species of *Myotis*. (Species of *Lasiurus* are not known to hibernate in nature, although it is possible that they do.) Dentition: 1/3, 1/1, 1–2/2, 2/2.

Key to Species of *Lasiurus* in the Pacific Coast States

1a Usually with small, peglike premolar next to inner margin of upper incisor (fig. 27a); dorsal pelage russet or brownish, but hair white tipped 2

1b Without small, peglike premolar; dorsal pelage even yellow brown, hairs not white tipped . . . . *Lasiurus xanthinus*

    2a Color reddish or russet; forearm 37–44 mm . . . . *Lasiurus blossevillii*

    2b Color brownish; forearm 50–56 mm . . . . *Lasiurus cinereus*

RED BAT   *Lasiurus blossevillii*
Pl. 2

DESCRIPTION: Basically bright russet dorsally, with white tips on the hairs. TL 98–110 mm, T 38–43 mm, E 8–11 mm, F 37–44 mm. Weight: 9–15 g.

DISTRIBUTION: Widespread in California, from the northern border south to Los Angeles, San Bernardino, and Inyo Counties. Its range extends north to Canada, south to Chile and Argentina, and east to the Atlantic coast, though it is absent from most of the Great Basin and Rocky Mountain region.

FOOD: It feeds frequently on moths and is also known to take terrestrial insects.

REPRODUCTION: Mating occurs in late summer (August to September) and ovulation the following spring. Usually three or four young are born in June. Young become independent at about four to five weeks of age.

COMMENTS: This bat flies early in the evening, well before dark, foraging at decreasing heights as the sky becomes darker. It is found in and near deciduous trees, frequently in orchards, and is most commonly encountered in August and September when migrating.

This is one of the most cold-tolerant bats in the United States. Its dense pelage reduces heat loss, and the furred interfemoral membrane can be drawn over the ventral surface of the body as additional protection from the cold. The species survives in temperatures well below freezing, compensating by increasing its heartbeat, and it does not hibernate.

HOARY BAT   *Lasiurus cinereus*
Pl. 2, Fig. 27

DESCRIPTION: A dull chocolate brown bat with white-tipped dorsal hairs that give it a frosted appearance. It is quite unlike any other bat in our state and is easily distinguished from the Red Bat by its color and larger size. It has a peglike upper premolar (see fig. 27). TL 126–143 mm, T 48–63 mm, E 9–14 mm, F 50–56 mm. Weight: 23–27 g.

DISTRIBUTION: Generally distributed in wooded areas of California, Oregon, and Washington. It occurs throughout the United States and Canada as well as south to Chile and Argentina; there are also populations on the Hawaiian and Galápagos Islands. This species

 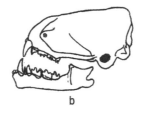 

Figure 27. Skull of the Hoary Bat *(Lasiurus cinereus):* (a) ventral view (arrow to minute, inner premolar); (b) lateral view; (c) dorsal view.

can be found in California in winter, migrates north to Hudson Bay in summer, and visits the Farallon Islands on migration.

FOOD: This bat captures many moths and beetles.

REPRODUCTION: Usually two (occasionally three or four) young are born in June or July. Newborn young may cling to a twig while the mother forages in the evening, but she is capable of carrying them both when moving to a new roosting site.

COMMENTS: This bat begins to forage early in the evening, long before dark, and is sometimes seen flying in the daytime in September. In winter, it sometimes roosts on the vertical trunks of trees, even 1 m from the ground, and then closely resembles a protruding part of the tree. It is relatively common but does not occur in aggregations.

When picked up alive, individuals of this species produce a most startling rattling hiss accompanied by an impressive show of teeth.

WESTERN YELLOW BAT   *Lasiurus xanthinus*
Fig. 28

DESCRIPTION: A medium-sized bat with light yellow fur. The anterior half of the dorsal surface of the interfemoral membrane is furred. The ears are short. TL 109–124 mm, T 45–50 mm, HF 8–11 mm, E 11–13 mm, F 43–52 mm. Weight: 12–19 g.

DISTRIBUTION: Found in the extreme southwestern deserts of California, north to Los Angeles and Riverside Counties and east to Arizona. Its range extends south to Argentina and Uruguay.

FOOD: This species feeds on flying insects.

REPRODUCTION: Two or three young are born in late June.

COMMENTS: This bat emerges late in the evening. It is sometimes found roosting in dense palm foliage. The species has also been known as *Lasiurus ega.*

Figure 28. Western Yellow Bat (*Lasiurus xanthinus*).

Genus *Myotis*

This is one of the more difficult groups of small mammals to identify, as the species are remarkably similar in general color and form. Their distinguishing features are most significant in combination; color and form must be considered together with cranial features. In most species the tail vertebrae extend to the margin of the interfemoral membrane, but in some they may project slightly beyond it. In some species the calcar is clearly keeled, whereas in others it is rounded in cross section (see fig. 21). Like many other bats, these species have a small projection at the base of the ear, the tragus, and this may be an important diagnostic feature.

No single common name exists for bats of this genus, and they are sometimes referred to by their generic name; for example, *Myotis volans* has been called the Long-legged Myotis. For many years, *Myotis lucifugus* has been called the Little Brown Bat; although this name is familiar to generations of American mammalogists, the genus consists of many species of "little brown bats." We have tried to adopt the most familiar and descriptive common name for each species, but the realistic solution to this problem is to learn the scientific names.

CALIFORNIA BAT   *Myotis californicus*

Pl. 2, Fig. 29

DESCRIPTION: A small, buff-colored bat with darkly pigmented ears and face. The medium-sized ears barely extend beyond the

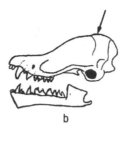

Figure 29. Skull of the California Bat *(Myotis californicus):* (a) ventral view; (b) lateral view (arrow to high cranium); (c) dorsal view.

tip of the snout when laid forward. This species is definitely distinguishable from the Small footed Bat *(M. ciliolabrum)* only on the basis of the outline of the skull (see fig. 29b, 30b) and the small tip of the tail, which protrudes slightly from the interfemoral membrane in the Small-footed Bat but not in the California Bat. TL 78–87 mm, T 35–40 mm, E 11–15 mm, F 29–36 mm. Weight: 3–5 g.

DISTRIBUTION: Occurs widely in the Pacific Coast states. It is associated with oak woodlands and also juniper-piñon communities of the desert; it prefers lower elevations, not commonly inhabiting mountains or coniferous forests. Its range extends from central Mexico north to British Columbia and east to the Rocky Mountains.

FOOD: This species forages for flying insects.

REPRODUCTION: A single young is born late May to early July.

COMMENTS: This bat forages early in the evening, often hunting only 2 or 3 m above the ground. It roosts in buildings, caves, and mine shafts. Recovery of banded individuals reveals that this species may live up to 15 years in the wild.

SMALL-FOOTED BAT   *Myotis ciliolabrum*
Fig. 30

DESCRIPTION: This bat is light or golden brown dorsally; the tips of the hairs tend to be glossy. The ears and face are frequently black;

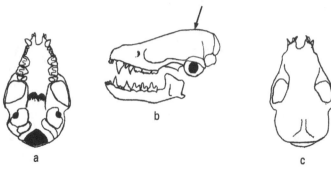

Figure 30. Skull of the Small-footed Bat *(Myotis ciliolabrum):* (a) ventral view; (b) lateral view (arrow to low cranium); (c) dorsal view.

the ears scarcely extend beyond the tip of the snout when laid forward. The calcar is keeled (see fig. 21a). This species is similar to the California Bat *(M. californicus)* but distinguishable based on the outline of the skull (see figs. 29, 30). Also, in the Small-footed Bat the tail projects slightly beyond the interfemoral membrane, whereas in the California Bat it does not. TL 60–90 mm, T 32–45 mm, E 12–15 mm, F 28–36 mm. Weight: 3–6 g.

DISTRIBUTION: Found in most of California except the coastal redwood region. Its range extends from central Mexico north to southern Canada and east to the Rocky Mountains, including eastern Oregon and Washington north to British Columbia; it is also found in the eastern United States.

FOOD: This bat captures small moths, beetles, and other flying insects.

REPRODUCTION: A single young is born in June or July.

COMMENTS: This bat emerges rather early in the evening. It is usually solitary and does not appear to form nursery or other colonies.

LONG-EARED BAT   *Myotis evotis*
Pl. 2

DESCRIPTION: This bat has a pale gold dorsum; the ears tend to be darkly pigmented. The interfemoral membrane sometimes has a few marginal hairs, but these are much less conspicuous than in the Fringed Bat. The ears are long, extending 7 mm beyond the tip of the snout when laid forward. The sagittal crest is weak or absent. TL 84–95 mm, T 38–41 mm, E 18–23 mm, F 35–42 mm. Weight: 4–8 g.

DISTRIBUTION: Found over most of California, Oregon, and Washington, in montane forests, though seldom in large numbers. Its range extends through western North America from Mexico to Canada.

FOOD: This bat gleans insects resting on leaves.

REPRODUCTION: A single young is born in June.

COMMENTS: This is a late-flying species and a hovering one. Emerging after dark, it forages low, from 1 to 2 m above the ground. It apparently does not form nursery colonies.

LITTLE BROWN BAT   *Myotis lucifugus*
Fig. 31

DESCRIPTION: This is perhaps the best-known member of the genus in North America and may serve as a basis of comparison in identifying other species. The rather glossy fur is dark brown or blackish. The ears are not large, extending about to the tip of the snout when laid forward. The calcar is rounded and without a keel (see fig. 21b). The skull lacks a sagittal crest (see fig. 31b). TL 80–93 mm, T 31–39 mm, HF 8–10 mm, F 35–41 mm. Weight: 7–10 g.

DISTRIBUTION: Found throughout much of California, Oregon, and Washington, but absent from southwestern California. Its

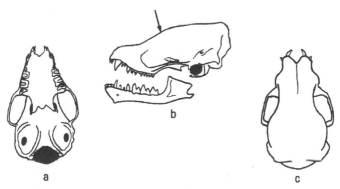

Figure 31. Skull of the Little Brown Bat *(Myotis lucifugus):* (a) ventral view; (b) lateral view (arrow to gradual rise in braincase); (c) dorsal view.

range extends over much of North America, from northern Mexico far into Alaska and east to Newfoundland.

FOOD: As in many species of the genus, the diet depends largely on availability, and the Little Brown Bat feeds on a variety of flying insects. Some workers report that it has a penchant for adult forms of flying insects, especially aquatic species, such as midges.

REPRODUCTION: This bat, like virtually all local vespertilionid species, mates in autumn, usually September or October. Sperm are stored in the upper region of the uterine tract, if not in the oviduct itself; ovulation and fertilization occur the following spring, and the single young is born about 60 days later. Bats normally roost head downward; during the birth process this position is reversed, and the infant is dropped into a basket the mother forms with her interfemoral membrane. When very small the young bat clings to its mother while she forages at night. It is capable of flight at about three weeks of age and is weaned shortly thereafter.

STATUS: The Occult Little Brown Bat *(M. l. occultus)* is a California subspecies of special concern.

COMMENTS: This bat is frequently abundant in buildings. Much of the research on bats has been conducted on this species. Movements and mortality have been studied by placing number bands on individuals. By this technique individuals have been found to live up to 31 years.

FRINGED BAT   *Myotis thysanodes*

DESCRIPTION: A brown-backed bat with a distinct cinnamon aspect. The interfemoral membrane has a delicate fringe of fine hairs. The ears are of medium length, extending some 3–5 mm beyond the tip of the snout when laid forward. The calcar lacks a keel (see fig. 21b). The skull has a sagittal crest. TL 80–91 mm, T 34–42 mm, E 16–19 mm, F 40–66 mm. Weight: 3–6 g.

DISTRIBUTION: Occurs throughout California, Oregon, and Washington; most frequently in coastal and montane forests and about mountain meadows. Its range extends from British Columbia east to the Rocky Mountain states and south to Mexico.

FOOD: This bat presumably consumes a broad variety of flying insects. It seems to favor beetles and moths.

REPRODUCTION: A single young is born in late June to early July.

COMMENTS: This species forms nursery colonies in caves or old buildings.

CAVE BAT   *Myotis velifer*
Fig. 32

DESCRIPTION: A large *Myotis* with a brown or black and dull (not glossy) dorsum and small ears. The forearm is unusually long. The calcar is not keeled (see fig. 21b). The mature skull has a distinct sagittal crest (see fig. 32b, c). TL 90–104 mm, T 39–47 mm, HF 9–10 mm, E 13–16 mm, F 35–47 mm. Weight: 9–13 g.

DISTRIBUTION: Found along the Colorado River, California. Its range extends east to Oklahoma and Kansas and south to Honduras.

FOOD: This bat seems to favor beetles and small moths.

REPRODUCTION: As in other species of the genus, mating occurs in

a   b   c

Figure 32. Skull of the Cave Bat *(Myotis velifer):* (a) ventral view; (b) lateral view; (c) dorsal view (arrows in parts b and c to sagittal crest). The sagittal crest may not be developed in young individuals.

late summer or fall and ovulation and fertilization the following spring. A single young is born in mid-June or July. It is capable of flight at about three weeks of age; weaning follows a few days later.

STATUS: This is a California species of special concern.

COMMENTS: This bat emerges rather late in the evening, commonly after dark, and frequently feeds over streams and ponds. It is found in caves and buildings, occasionally in large colonies.

LONG-LEGGED BAT   *Myotis volans*

DESCRIPTION: A rather long-haired bat, with tawny to dark brown dorsal fur. The wing membrane and the base of the interfemoral membrane are furred. The ears are short, reaching about to the tip of the snout when laid forward. The calcar is keeled (see fig. 21a). The cranium is rather highly elevated with a poorly developed sagittal crest. TL 89–98 mm, T 34–45 mm, E 10–14 mm, F 35–42 mm. Weight: 5–7 g.

DISTRIBUTION: Found throughout California, Oregon, and Washington, in both forested regions and brushy areas, up to 2,500 m in deserts and 3,300 m in the mountains. Its range extends throughout the western United States from northern British Columbia south to Central Mexico.

FOOD: This bat emerges early, long before the sky is dark, and presumably eats early-flying nocturnal insects.

REPRODUCTION: A single young is born in very late spring or early summer.

COMMENTS: This bat roosts in buildings, trees, and crevices in cliffs.

YUMA BAT   *Myotis yumanensis*
Fig. 33

DESCRIPTION: A bat with a dull brown or buff dorsum. This species is similar to the Little Brown Bat (*M. lucifugus*), but smaller. The calcar lacks a keel (see fig. 21b). The skull has a conspicuous rise but no sagittal crest (see fig. 26b). TL 82–88 mm, T 32–38 mm, E 11–15 mm, F 32–38 mm. Weight: 6–8 g.

DISTRIBUTION: Occurs throughout California, Oregon, and Washington, especially along wooded canyon bottoms. It is common in the deserts of southeastern California. Its range extends from

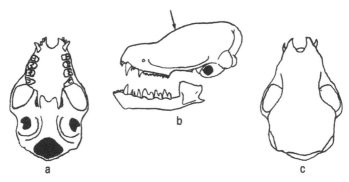

Figure 33. Skull of the Yuma Bat *(Myotis yumanensis)*: (a) ventral view; (b) lateral view (arrow to abrupt rise in braincase); (c) dorsal view.

Central Mexico (including Baja California) north to British Columbia and east to Colorado and Texas.

FOOD: This bat subsists on flying insects, especially small moths, beetles, and midges.

REPRODUCTION: A single young is born in spring or early summer.

COMMENTS: This is a colonial species, roosting in caves and old buildings in aggregations of up to 2,000 individuals. Prior to birth of the young, females segregate into nursery colonies, occasionally of large size.

WESTERN PIPISTRELLE *Pipistrellus hesperus*
Pl. 2

DESCRIPTION: A very small bat with yellowish or dull pelage. The base of the hairs is black, the ears are short and darkly pigmented, and the face is black. The tragus is blunt, and the calcar is keeled (fig. 21a). TL 69–81 mm, T 25–31 mm, E 10–11 mm, F 29–32 mm. Weight: 3–5 g. Dentition: 2/3, 1/1, 2/2, 3/3.

DISTRIBUTION: Found throughout most of California, often in open arid areas at lower elevations. It also occurs at lower elevations in eastern Oregon and in the Snake and Columbia River Valleys in Washington. Its range extends from Central Mexico north to Washington and east to Oklahoma.

FOOD: This species takes small moths, beetles, and dipterans.

REPRODUCTION: Usually two young are born in late June.

COMMENTS: This bat emerges early in the evening, long before the sky is dark, and thus is commonly seen. It also flies at dawn. It forages close to the ground and can be recognized by its rather erratic flight. It lives alone or in small colonies; nursery colonies have been found. It hibernates in caves but may emerge sporadically in winter.

TOWNSEND'S LONG-EARED BAT *Plecotus townsendii*
Pl. 2

DESCRIPTION: A medium-sized, light brown bat with a brown belly. The ears are very long and joined at the top of the head. Large glandular swellings are present on the sides of the snout. (Compare to the Pallid Bat [*Antrozous pallidus*].) TL 89–112 mm, T 35–54 mm, E 30–41 mm, F 39–47 mm. Weight: 7–12 g. Dentition: 2/3, 1/1, 2/3, 3/3.

DISTRIBUTION: Found all over the state, including the Channel Islands. It frequents desert scrub and piñon-juniper habitats. Its range extends from Mexico to British Columbia and the Rocky Mountain states; it also occurs in some parts of the central Appalachians.

FOOD: This bat picks off insects that are sitting on leaves. It favors small moths.

REPRODUCTION: Mating takes place in fall, and a single young is born in May or June in California. This species is well known for sperm storage in females.

STATUS: This is a California species of special concern.

COMMENTS: This bat emerges late in the evening. Like other long-eared bats, it is a hovering species. It lives in caves but is also found in buildings. Females often form small groups, but males tend to be solitary. Nursery colonies have been found in caves, mine shafts, and buildings.

This species is sometimes placed in *Corynorhinus*, which many zoologists consider to be a subgenus of *Plecotus*.

Free-tailed Bats (Molossidae)

This family is mostly tropical, but several species enter temperate regions. Those in California have short, dense fur of a dark chocolate color. The interfemoral membrane is naked (or only thinly provided with hair) and distinctly thick and leathery. The tail lies within the interfemoral membrane, but the tip projects beyond the margin of the membrane—hence the name "free-tailed."

Key to Species of Free-tailed Bats (Molossidae) in California

1a Size not large (total length less than 140 mm); anterior margin of bony palate with marked indentation between upper incisors 2
1a Size large (total length greater than 140 mm); upper incisors virtually in contact with each other . . . . *Eumops perotis*
    2a Ears long, extending conspicuously beyond muzzle when laid forward; inner bases of ears connected to each other 3
    2a Ears shorter, not extending much beyond muzzle when laid forward; inner bases of ears not connected to each other . . . . *Tadarida brasiliensis*
3a Forearm longer than 57 mm . . . . *Nyctinomops macrotis*
3a Forearm shorter than 50 mm . . . . *Nyctinomops femorosacca*

WESTERN MASTIFF BAT   *Eumops perotis*
Pl. 2

DESCRIPTION: A large bat with short, dull fur, gray or dark brown in color, with the hairs white or very light at the base. The broad, truncate ears are joined across the top of the head. TL 155–185 mm, T 35–45 mm, E 27–32 mm, F 73–82 mm. Dentition: 1/2, 1/1, 2/2, 3/3. Weight: 80–100 g.

DISTRIBUTION: Found mostly in the southern half of California, but ranges north to Butte County. It prefers open, arid areas with high cliffs. Its range extends through southern Arizona and Texas south into Argentina and Cuba, but its distribution is discontinuous.

FOOD: This bat catches strong flying insects such as dragonflies, moths, beetles, and hymenopterans.

REPRODUCTION: Mating takes place in early spring. From one to two young are born any time between late June and September.

STATUS: This is a California species of special concern.

COMMENTS: Emerging when the evening light has nearly disappeared, this species forages very high (600–700 m), usually over mesquite. This is a very vocal bat. Its hunting calls are audible to the human ear.

It roosts in crevices in small colonies and can also be found in caves and buildings. It differs from many other bats in that the sexes remain together throughout the year, including the period of birth of the young. It makes local seasonal movements but remains in California throughout the year.

POCKETED FREE-TAILED BAT *Nyctinomops femorosaccus*
Fig. 34

DESCRIPTION: A medium-sized free-tailed bat, dark brown or gray in color, with the base of the hairs distinctly lighter than the tips. The ears are joined at the base. TL 100–108 mm, T 38–43 mm, E 21–24 mm, F 45–49 mm. Dentition: 1/2, 1/1, 2/2, 3/3. Weight: 9–13 g.

DISTRIBUTION: Occurs in the southern part of California in desert areas of Riverside, San Diego, Imperial, and San Bernardino Counties. Its range extends east through southern Arizona to Texas and south to central Mexico, including Baja California.

FOOD: This bat captures flying insects, especially moths and beetles.

REPRODUCTION: A single young is born in late June or July. Nursing may continue into September.

STATUS: This is a California species of special concern.

COMMENTS: This species flies

Figure 34. Pocketed Free-tailed Bat *(Nyctinomops femorosaccus).*

late in the evening. It roosts in buildings and also in crevices in high, vertical rock outcrops.

BIG FREE-TAILED BAT *Nyctinomops macrotis*
Fig. 35

DESCRIPTION: A glossy, reddish brown to black bat, with hairs that are distinctly lighter at the base. This species can be recognized by its large ears, which extend far beyond the tip of the snout when laid forward and are united at the base. TL 108–141 mm, T 48–61 mm, E 26–29 mm, F 57–62 mm. Dentition: 1/2, 1/1, 2/2, 3/3. Weight: 12–19 g.

DISTRIBUTION: Resides in piñon-juniper regions of the arid parts of California, though it is apparently uncommon there. Its range extends from southern California north to San Mateo County, east to Kansas, and south to Uruguay. It is known to move north to British Columbia in late summer.

FOOD: This species captures moths and beetles.

REPRODUCTION: A single young is born in June or July; nursing may continue to August.

STATUS: This is a California species of special concern.

COMMENTS: This species is known to be very vocal in flight. It is associated with high cliffs and rocky outcrops, where it roosts in crevices. Females aggregate in nursing colonies.

Figure 35. Big Free-tailed Bat (*Nyctinomops macrotis*).

GUANO BAT
Pl. 2, Fig. 36
Tadarida brasiliensis

DESCRIPTION: A small, chocolate brown free-tailed bat. Its fur is dull, and the hairs are not usually much (if any) lighter at the base. The ears are close together at the base but definitely not joined. They extend no more than slightly beyond the tip of the snout when laid forward. The anterior margin of the bony palate is concave (fig. 36a). TL 88–112 mm, T 31–41 mm, E 14–20 mm, F 36–46 mm. Dentition: 1/3, 1/1, 2/2, 3/3. Weight: 8–14 g.

DISTRIBUTION: This is the most common free-tailed bat in California, with some numbers present throughout the year. Its range extends into southwestern Oregon, across the southern United States to the Atlantic coast, and south through northern South America, including the Greater Antilles. It survives temperatures as low as 5 degrees C.

FOOD: This bat feeds largely on small moths and is said to consume 1 g per night.

REPRODUCTION: Breeding occurs in late winter, and gestation takes about 100 days. Implantation occurs almost exclusively in the right uterine horn. A single young is born from late June to early July. Very large nursery colonies are formed, and the mother-young relationship seems not to be firm; a lactating female allows any of the young to take milk and seems unable to distinguish her own young from others.

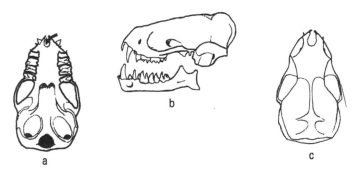

Figure 36. Skull of the Guano Bat *(Tadarida brasiliensis):* (a) ventral view (arrow to indentation of bony palate); (b) lateral view; (c) dorsal view.

COMMENTS: This bat emerges early in the evening and is a very rapid flyer.

It regularly occupies caves, sometimes together with bats of other species; and also commonly lives under roofing tiles. On occasion, it assembles in large numbers in buildings, caves, or mine shafts, from sea level to 1,500 m or more. The Guano Bat is so named because it created deposits of droppings up to 30 m in depth in the Carlsbad Caverns of New Mexico. These deposits sustained a commercial guano mine for the first 20 years of the twentieth century. In recent years, insecticides have caused drastic declines in the maternity colonies in these caves.

Professor E. Lendell Cockrum of the University of Arizona has documented this species' northerly migration in spring. California populations make only short migrations, but Mexican groups may move 1,600 km (see the section "Migration and Movements").

This species is also known as the Brazilian Free-tailed Bat. When its specific name was *Tadarida mexicana*, it was called the Mexican Free-tailed Bat. Such name changes serve no useful purpose.

CARNIVORA

Two major groups of meat-eating mammals have existed: the Creodonta and the Carnivora. Members of the Carnivora are conventionally referred to as "carnivorans" to separate them from the creodonts, which died out in the Miocene. Most carnivorans are clearly modified for a diet of meat and the capture of living prey. They feed mostly on birds and on other mammals, but some, such as the pinnipeds, feed largely on fish and marine invertebrates. Bears and raccoons are generalists, eating fruits when available. Carnivoran dentition is distinctive; the incisors are rather small and used for grooming and pulling, whereas the canines are conical and long, suitable for holding and tearing meat (fig. 4). The molars are variously shaped and often serve to shear or crush bones. The cheek teeth are sometimes modified for cutting flesh; the inner side of the fourth upper premolar and the outer side of the first lower molar fit tightly together and slide past each other like the blades of scissors. These two teeth are referred to as carnassials and are characteristic of those carnivorans that are primarily flesh eaters, such as cats. Like meat eaters in general, carnivorans have a rather short, simple gastrointestinal tract.

The first Carnivora may have arisen from the miacoids, carnivorans that existed from the Late Cretaceous to the early Paleocene. Carnivorans first appeared in the Paleocene but did not begin to diverge until the late Eocene and early Oligocene. The earliest carnivorans existed together with the creodonts, which survived to the Oligocene in North America. The carnivorans replaced the creodonts in the Oligocene, but there is no evidence to support a close relationship between the two orders.

The earliest relatives of dogs (Canidae) diversified into running types in North America, where they were the dominant carnivorans for the first part of the Cenozoic. Bears (Ursidae) are derived from an ancient group, the Arctoids, and underwent evolutionary changes in Europe. The family Ursidae first appeared in North America in the late Miocene. Early ursids did not resemble modern bears, for they walked on their toes (digitigrade) and were adapted for running (cursorial). The modern bears with which we are familiar date from the Oligocene and entered North America in the Pliocene. Procyonids, which include the Ringtail *(Bassariscus astutus)* and Raccoon *(Procyon lotor)* evolved in North America. The weasels and their friends (Mustelidae) are probably a loosely related assemblage, for skunks and otters are not really close to weasels *(Mustela* spp.) and martens *(Martes* spp.). Presumably they radiated in Asia, where fossils date from the early Oligocene, and entered North America in the early Miocene.

Major groups of carnivorans are classified on the basis of the structure of the foot. Bears and Raccoons, for example, walk on the palms or plantar surface of the feet (plantigrade). In contrast, dogs and cats are digitigrade. Weasels are intermediate, with some species plantigrade and others digitigrade. Seals (Phocidae), sea lions (Otariidae), and walruses *(Odobenus)* are appropriately designated pinnipeds—literally, "fin feet." Although they differ in some features, they constitute a natural group. Legs and body shape are correlated with habits: martens, with retractile claws, are climbers; plantigrades, such as bears, are mostly terrestrial; pinnipeds, with flippers, are swimmers and awkward on land.

Most carnivorans other than dogs tend to be solitary except during the mating season.

Dogs, Foxes, and Allies (Canidae)

Dogs and their relatives (wolves, foxes, and the like) are somewhat generalized carnivorans that are adapted for running. Light bodied, with rather slender legs, they are digitigrade and have nonretractile claws. The long tail is frequently bushy. The grooved baculum, or penis bone, characterizes the family. The skull is rather long, with an elongate snout and strong zygomatic arches. The eyes are rather large, lateral in position but directed forward somewhat, and vision is binocular. The bullae are rounded and somewhat inflated, and the ears are erect. The dental formula of California species is 3/3, 1/1, 4/4, and 2/3. The molars are formed for both shearing and crushing bones, indicating a dietary position somewhat between cats and bears (figs. 37–39).

Numerous genera of canids exist today, with the greatest diversity in South America and Africa. Canids first appeared in North America in the middle Eocene, in the absence of the bear-like arctoids, and invaded Asia in the early Pliocene. There was a great diversity of canids in North America from the late Eocene until the late Miocene. The arid climate of the early Miocene favored an increase of grasslands and diversification of herbivores, such as horses (*Equus* spp.) and the Pronghorns (*Antilocapra americana*), which perhaps accounted for the differentiation of canids into doglike predators specialized for running. With the development of grasslands, the woodlands became restricted.

Dogs and their allies are rather sociable, forming semipermanent groups, and pairing (monogamy) is typical of many species. Both parents tend to care for the growing young, with the male commonly providing food when the pups are small. Even when the young mature, there is a strong bond in a dog family, which sometimes continues as a hunting group. Young of the previous year may help care for their growing siblings.

COYOTE   *Canis latrans*
Pl. 4, Fig. 37

DESCRIPTION: A medium-sized, rangy, doglike carnivoran, approximately the shape of a German shepherd. It is gray, sandy, or brown in color. Most Coyotes weigh from 8 to 20 kg. Some extremely large individuals, weighing from 25 to 33 kg, could be mistaken for Wolves (*C. lupus*). TL 1.0–1.3 m, T 300–400 mm. (See fig. 37.)

DISTRIBUTION: Found throughout the western states, typically in open country. The Coyote can occur almost anywhere, even in parts of Los Angeles. It is common in the sagebrush plains of the Great Basin. Though it is most frequently seen west of the Mississippi River, in the last 40 years—perhaps because of the disappearance of the Wolf—it has spread to many regions to the east, even reaching the Atlantic coast in some areas. Its range extends from Alaska and central Canada south to Panama.

FOOD: The Coyote preys extensively on jackrabbits (*Lepus* spp.), cottontails (*Sylvilagus* spp.), and ground squirrels (*Spermophilus* and *Ammospermophilus* spp.), supplementing this fare with small mice, fruits, berries, insects, carrion, and domestic sheep.

REPRODUCTION: Mating takes place in February in California; from five to 10 pups are born some two months later. Both parents remain with the young until fall. You may encounter pairs or even groups of three or five animals, perhaps comprising parents

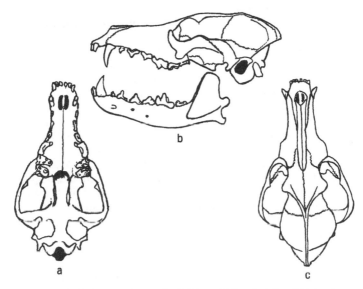

Figure 37. Skull of the Coyote *(Canis latrans):* (a) ventral view; (b) lateral view; (c) dorsal view.

and grown young. Family groups tend to disperse in early winter, however, when food becomes scarce.

COMMENTS: Some montane populations tend to move downhill in autumn.

Coyotes are sometimes seen mousing, stalking and pouncing on mice, often quite oblivious to human observers. They are also seen on occasion accompanying foraging Badgers *(Taxidea taxus),* capturing rodents dislodged by the digging of these predators.

Versatility is the key to the Coyote's success. Its major conflict with human society stems from the penchant some individuals develop for domestic stock, especially sheep. Once such a habit is formed, it is not lost, and such individuals do economic damage. However, the Coyote is very prolific and has sustained its numbers in the face of many decades of persistent hunting and poisoning.

This species has been known to mate with domestic dogs; the hybrid, called a coydog, is sometimes found in the wild.

GRAY FOX   *Urocyon cinereoargenteus*
Pl. 3, Fig. 38

DESCRIPTION: A silvery gray fox with conspicuous patches of yellow, russet, brown, or white on the throat and belly and a black-tipped tail, with a middorsal crest of stiff black hairs, which is not present on other foxes (except the Island Fox *[U. littoralis]*). The Gray Fox has rather short legs, perhaps an adaptation for climbing trees. The skull has two distinct sagittal crests, parallel ridges that end in a lyre-shaped flare posteriorly (see fig. 38b, c). TL .8–1.1 m, T 275–443 mm, HF 100–150 mm. Weight: 3–5 kg, occasionally to 7 kg.

DISTRIBUTION: By far the most common and widespread fox in the Pacific states, the Gray Fox survives well in cultivated land, chaparral, and forested areas. It ranges through most of the United States, except the Rocky Mountains, and south through most of South America.

FOOD: This fox eats small rodents, birds, and berries, as well as insects and fungi. Its climbing ability allows it to obtain a greater variety of foods than other foxes.

REPRODUCTION: Mating in late winter is followed by birth of three to five young in April or May. Larger litters may represent combined broods of two females that have denned together. This fox sometimes dens under farm outbuildings or even under suburban homes. The male remains with the female while the young are dependent.

COMMENTS: This is a very beautiful fox that has a rather good quality fur. Because of its abundance and wide distribution, it is

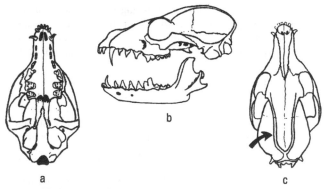

Figure 38. Skull of the Gray Fox *(Urocyon cinereoargenteus):* (a) ventral view; (b) lateral view; (c) dorsal view (arrow to sagittal crest).

an important predator. Like others in the dog family, it is playful at times. One was observed tossing a dried cow pie into the air, chasing after it, and flinging it up again, whether by mouth or paw was not clear.

ISLAND FOX   *Urocyon littoralis*

DESCRIPTION: A diminutive replica of the Gray Fox *(U. cinereoargenteus)*, which it closely resembles in every way but size. TL 590–780 mm, T 110–290 mm, HF 98–157 mm. Weight: 2.0–2.2 kg.

DISTRIBUTION: Found on six of the Channel Islands, where it has been generally abundant but has recently declined.

FOOD: This fox is omnivorous, taking more plant and insect food than does the Gray Fox *(U. cinereoargenteus)*. It feeds extensively on berries such as manzanita, toyon, saltbush, prickly pear, and ice plant, and also eats mice.

REPRODUCTION: Mating takes place in February and March; a litter, usually of two kits, is born in late April or May. The Island Fox dens in holes or hollow trees. Both parents care for the young, and the family forages as a group.

STATUS: This species is listed by the state as threatened. Four subspecies (the San Miguel Island Fox *[U. l. littoralis]*; the Santa Rosa Island Fox *[U. l. santarosa]*; the Santa Cruz Island Fox *[U. l. santacruzae]*; and the Santa Catalina Island Fox *[U. l. catalinae]*) have been proposed for federal listing as endangered.

COMMENTS: Although presumably nocturnal, the Island Fox is frequently seen in the daytime.

This little fox does not conflict with human activity and apparently does not suffer much from the presence of humans. Its main competitor is the domestic house cat *(Felis sylvestris)*. Golden Eagles, which only recently arrived on three of the northern islands of this fox's habitat, prey on it and are being removed.

KIT FOX   *Vulpes macrotis*
Pl. 3

DESCRIPTION: A small gray fox with exceptionally large ears. Its color is rather uniform except for a black tip to the tail. Of the similar species, the Red Fox *(Vulpes vulpes)* is appreciably larger, has smaller ears, and is reddish with a white-tipped tail, whereas the Gray Fox has much shorter legs and is usually a mixture of red and gray. The Kit Fox skull also lacks the sagittal crests of the Red

Fox and Gray Fox skulls. TL 730–840 mm, T 260–325 mm, HF 113–137 mm, E 78–94 mm. Weight: 1.7–2.5 kg.

DISTRIBUTION: Found in open, arid regions of the southern part of the Great Basin and the interior valleys of California and southeastern Oregon. One or more relict populations persist in Contra Costa County, California. The range of this species extends from southeastern Oregon east to west Texas and New Mexico and south to central Mexico, including Baja California.

FOOD: This fox eats various small rodents, especially kangaroo rats, as well as mice and small squirrels, lizards, insects, and berries of wild shrubs. Brush Rabbits *(Sylvilagus bachmani)* sometimes form an important part of its diet.

REPRODUCTION: Mating takes place in winter; three to five young are born in February or March. The female spends much of her time in the den when the young are small. Some family groups consist of one male with two females and their offspring. The young disperse in fall.

STATUS: The San Joaquin Kit Fox *(V. m. mutica)* is federally listed as endangered and listed by the state as threatened.

COMMENTS: The Kit Fox's range has narrowed greatly in recent years. This delicate little fox does not conflict with human activities, but it has declined with the expansion of agriculture and intensification of predator control. Intensive agricultural use of much of the San Joaquin Valley renders the habitat unsuitable for it. Moreover, the Kit Fox picks up poisoned baits left out for Coyotes. It is believed to increase in areas where the Coyote is reduced in numbers.

The use of the specific name *macrotis* is a matter of controversy; some authorities use *velox*. We have followed Wilson and Reeder (1993).

RED FOX   *Vulpes vulpes*
Pl. 3, Fig. 39

DESCRIPTION: A rather bright reddish or yellowish fox with much black on the legs and feet and a white tip to the tail. Color variations include an all black or melanistic silver fox and a brown and gray cross fox, but all of these are nevertheless Red Foxes. The skull is distinctive in having sagittal crests that come to a point posteriorly (see fig. 39c), in contrast to the lyre-shaped sagittal crests of the Gray Fox (see fig. 38c). TL .88–1.0 m, T 340–390 mm, HF 140–165 mm. Weight: 5–8 kg.

DISTRIBUTION: Found in two widely separated parts of California. One population lives at the higher elevations (1,500 m and above) of the Sierra Nevada, north through the Cascade Range into British Columbia. The other lives in coastal regions from the northern Sacramento Valley south to the Los Angeles area. This species is found throughout North America and also Eurasia and northern Africa.

FOOD: Small rodents, birds, berries, and insects form the bulk of the Red Fox's diet. It can capture birds up to the size of a Pheasant *(Phasianus colchicus)* or Mallard *(Anas platyrhynchos)*. In recent years the Red Fox has invaded salt marshes around San Francisco Bay, where it preys on the Clapper Rail *(Rallus longirostris)*.

REPRODUCTION: Little is known of the breeding of the Red Fox in the Pacific states. A litter of five to 10 kits is born in early spring. As in other members of the dog family, the male helps the female provide food for the young.

STATUS: The Sierra Nevada Red Fox *(V. v. necator)* is listed by the state as threatened.

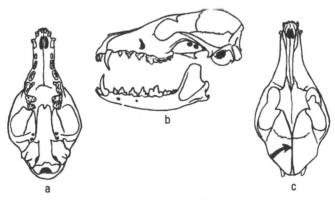

Figure 39. Skull of the Red Fox *(Vulpes vulpes)*: (a) ventral view; (b) lateral view; (c) dorsal view (arrow to sagittal crest).

COMMENTS: The Sierra Nevada population is native to California. The coastal population stems from a population of the eastern Red Fox introduced into the lowlands of the state in the nineteenth century, presumably by release or escape from fur farms, and its members most closely resemble Red Foxes from the northern Central Plains states.

GRAY WOLF   *Canis lupus*
Fig. 40

DESCRIPTION: A very large dog, with the general aspect of a Coyote (*C. latrans*) or a German shepherd. It is usually gray with varying amounts of black on the back, but it is sometimes white or all black. The face is often sharply bicolored, with whitish lower cheeks contrasting with dark fur surrounding the eyes and covering the forehead. The legs are long, and the chest is deep. The auditory bullae are well rounded and rather smooth, in contrast to domestic dogs, in which the bullae are relatively flat and rugose (wrinkled). The canines are well developed. Weight: 30–75 kg (adult males).

DISTRIBUTION: Once inhabited the eastern margin of California, west to Sacramento. The Gray Wolf is characteristically a forest- and plains-dweller, avoiding deserts. Its range extends throughout the Northern Hemisphere, including North America from the Arctic to central Mexico. In the contiguous states, it is now known in Minnesota, Montana, Idaho, Wyoming, Washington, Oregon, Utah, and Nevada.

FOOD: The Gray Wolf is primarily a predator of hoofed mammals, such as sheep, deer (Cervidae), and Caribou (*Rangifer tarandus*); it is known to kill young bison and moose. In addition, it takes many Beavers (*Castor canadensis*) and other rodents, as well as rabbits (*Lepus* spp.).

REPRODUCTION: Mating takes place in middle or late winter. After a gestation of about two months, from five to nine pups are born in a den, either in a hole in the ground or among rocks. Both parents, and sometimes young of the previous year, provide food for the pups. The den is clean, for the mother eats the scats of her young until weaning. In fall, family groups remain as a unit, and several families hunt together until mating. Young are sexually mature at two to three years.

STATUS: This species is federally listed as endangered in most of the contiguous states.

Figure 40. Differences between Coyote and Gray Wolf. Coyote (top): medium size, slender snout, large ears, face fox-like, small feet, tail usually carried low when running. Wolf (bottom): Large size, heavy snout, medium ears, face dog-like, large feet, tail usually carried high when running.

COMMENTS: The wolf has long had a reputation as a formidable predator and for centuries has been feared in North America and Eurasia. In early United States history, the Gray Wolf preyed on livestock, which led to persistent trapping and poisoning that eliminated the species from almost all of its original range in the contiguous states. Large-scale poisoning of wolves in the early twentieth century, under the auspices of state and federal agencies, also resulted in the loss of hundreds of thousands of individuals of nontarget species, including, but not limited to, Black-footed Ferrets (*Mustela nigripes*), skunks, and other innocuous mammals.

In 1995 a small population of Gray Wolves was introduced into Yellowstone National Park, and releases in other states followed. The populations are gradually spreading, resulting in the return of wolves to the western states, where they had been extirpated. Last known in California in 1924, the Gray Wolf is expected to return to our state in the near future.

Many people would welcome the reappearance of the Gray Wolf in California. Although ranchers have reason to be concerned for their livestock, in other states stockmen are reimbursed for losses due to wolf predation with funds provided by conservation groups and governments.

In *The Wolves of North America*, Stanley Young and Edward Goldman (1944) reviewed the encounters of wolves and humans in North America. Although these accounts and similar ones from Eurasia are anecdotal and may be sensationalized, there can be no doubt that, at times, wolves will attack, kill, and eat humans. Reports from India also attest to this (Prater 1971). Nevertheless, with a population of more than 200 wolves and thousands of hikers in the greater Yellowstone area, no attacks have been reported as of 2002, and biologists studying this species have remarked on its fear of humans.

Long-term studies have shown that wolf predation removes chiefly ill, old, and very young animals, thus improving the health and preventing excessive growth of the prey population. Because it hunts in family groups or larger aggregations, a wolf is able to capture animals many times its size. Wolves are intolerant of Coyotes and suppress that species where their ranges overlap.

Bears (Ursidae)

Bears are large, heavy bodied, almost tail-less carnivorans. They have stout limbs and heavy, blunt, nonretractable claws, and they are plantigrade. Reflecting their omnivorous diet, bears' molars are flat crowned—shaped for crushing, like those of swine and humans, not for cutting, like those of cats (fig. 42). Dentition 3/3, 1/1, 4/4, and 2/3.

Most of the evolutionary history of bears took place in the Old World, starting in the Miocene. The genus *Ursus* dates from the Pliocene in North America and Eurasia.

Three species of bears are found in North America. The Polar Bear *(Ursus maritimus)* is an arctic species, and the Grizzly Bear, or Brown Bear, *(U. arctos)* no longer exists in the three coastal states (see the section "Mammals and California Society"). The Black Bear *(U. americanus)*, however, is common in many parts.

BLACK BEAR *Ursus americanus*

Pl. 4, Figs. 41, 42

DESCRIPTION: A large, stout bear with coarse black, brown, or cinnamon fur, a white or pale patch on the throat or belly, and a minute tail. It sometimes weighs more than 190 kg, but this is unusual. Its claws are about the same length on the forefeet and hind

Figure 41. Black Bear
(Ursus americanus).

feet; in comparison, the Grizzly Bear *(U. arctos)* has much longer claws on the forefeet. (See fig. 42.)

DISTRIBUTION: Found in most forested regions; in the Cascade Range and Sierra Nevada it occurs from the upper edge of the forested elevations down to about 1,000 m or less. In the northwest coastal forests (Marin County and Yolo County) it may occur at sea level and even venture out on the beaches. The Black Bear has moved into regions formerly occupied by the Grizzly Bear, especially in southwestern California, including the Tehachapi Mountains and southern coastal regions. Its range extends from the Canadian coniferous forests to Mexico.

FOOD: The Black Bear is a true omnivore, finding nutrition in almost any organic food. It is fond of berries, nuts, and other vegetable foods, and in fall it often subsists on manzanita berries and acorns. Like other forest dwellers, it is also fond of underground fungi, or truffles. Most of its animal food consists of insects, espe-

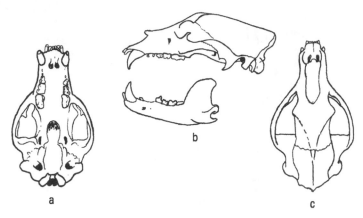

Figure 42. Skull of the Black Bear (Ursus americanus): (a) ventral view; (b) lateral view; (c) dorsal view.

cially ant larvae and beetle larvae, but it also eats mice, ground squirrels, and occasionally a ground-nesting bird.

REPRODUCTION: Mating takes place in summer, and implantation is delayed for several months. One or two small young are born in midwinter, when the female has retreated for a winter rest. The cubs nurse and grow during the winter, while the mother remains semitorpid, with reduced heart and breathing rates. Her fat stores, accumulated during the previous summer and fall, contribute to the manufacture of milk, which is the sole nourishment of the rapidly growing young until springtime, when they emerge from their winter home. This species breeds only every other year. Apparently this schedule is necessary because of the tremendous drain on the female's stored energy during her extended period of fasting.

COMMENTS: Adult male bears are known to kill younger bears. The removal of adult males, as by hunting, probably does not reduce bear populations by more than the number killed. The Black Bear frequently forages in garbage dumps in mountain communities and is a pest in some montane parks. It may invade apple orchards in fall and also cause great damage to beehives. Despite their bulk and short limbs, Black Bears climb well.

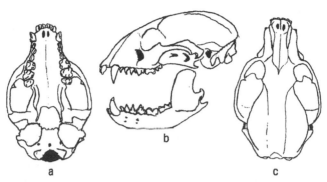

Figures 43 a–c. Skull of the Raccoon *(Procyon lotor):* (a) ventral view; (b) lateral view; (c) dorsal view.

Raccoons and Ringtails (Procyonidae)

Modern procyonids are diverse morphologically and ecologically, but they have certain common characteristics, including similar inner structures of the ear and a bilobed baculum. The Raccoon *(Procyon lotor)* and Ringtail *(Bassariscus astutus),* together with their Central American relatives, the Coati *(Nasua narica)* and Kinkajou *(Potus flavus),* are related to bears and pandas. Coatis occur as far north as Arizona and New Mexico and are getting close to California. Procyonids are plantigrade, walking on the bare or partly furred soles of the feet. Raccoons, like bears, have low-crowned, tuberculate molars shaped for crushing, whereas Ringtails, like cats, have teeth formed for shearing (fig. 43). Dentition in California species 3/3, 1/1, 4/4, and 2/2 or 2/3.

The Procyonidae date from the early Eocene in North America and are known from Oligocene fossil beds in Europe.

Figure 43d. Skull of the Ringtail *(Bassariscus astutus):* lateral view.

RINGTAIL

Bassariscus astutus

Pl. 5, Figs. 43, 44

DESCRIPTION: Somewhat like a small Raccoon *(Procyon lotor)*, but with a slender build and an extremely long tail. The huge, dark eyes of the Ringtail replace the black facial mask of the Raccoon, but the relationship is still obvious. The soles are mostly or partly furred; the claws are partly retractable. TL 620–800 mm, T 315–504 mm, HF 55–75 mm, E 45–50 mm. Weight: .9–1.2 kg. Dentition: 3/3, 1/1, 4/4, 2/2.

DISTRIBUTION: Found in brushy and wooded areas, generally at lower and middle elevations. The Ringtail is especially common in foothill canyons and in some areas of the Sacramento Valley north to southern Oregon. It is less common in the high mountains but is known to live up to 2,600 m. A population is found in the Sutter Buttes. The Ringtail, like the Raccoon, seems to prefer to live along watercourses. Its range extends from Oregon and Colorado south into Central America.

FOOD: The studies of Professor Gene Trapp of Sacramento State University have revealed much about the habits of the Ringtail. It preys on mice and wood rats *(Neotoma* spp.), but also takes berries and soft fruits such as cherries, raspberries, and the fruits of the Pacific madrone *(Arbutus menziesii),* as well as arthropods and small birds. It does much of its foraging in trees. It does not forage in water or eat aquatic organisms.

REPRODUCTION: Little is known of the breeding pattern of this animal. Mating occurs in late winter, and three to four kits are born

Figure 44. Ringtail
(Bassariscus astutus).

in May or June. Dens are secreted among large boulders near canyon bottoms and in hollow trees.

STATUS: The Ringtail is fully protected in California.

COMMENTS: The Ringtail's hind foot is capable of a wide rotation, enabling the animal to descend from a tree headfirst.

Like the Raccoon, the Ringtail is nocturnal. It shuns urban regions but is known to enter cabins in the mountains. According to early stories, gold miners welcomed the Ringtail because it kept their cabins free of mice and wood rats . This little carnivoran is friendly and unafraid, and its presence should be encouraged.

RACCOON   *Procyon lotor*
Pl. 5, Figs. 43, 45

DESCRIPTION: Perhaps the most familiar carnivoran in North America. Its black mask and ringed tail distinguish it from all other carnivorans but the Ringtail (*Bassariscus astutus*). TL 780 930 mm, T 300–390 mm, HF 100–130 mm, E 50–60 mm. Weight: 4–8 kg; females tend to be smaller than males. Dentition: 3/3, 1/1, 4/4, 2/2. (See fig. 43.)

DISTRIBUTION: Found almost everywhere throughout North America, in forests, swamps, and marshes, though not in extremely dry regions. It may live in woodlands far from water or in cattail marshes far from trees. It is found about homes and is a common resident of San Francisco parks It persists along creeks that course through urban areas. In some cities it has learned to live in storm drains, presumably preying upon rats. Its range extends far into Central America.

FOOD: The Raccoon eats vertebrates, invertebrates, fruits, nuts, and berries. It commonly forages along watercourses for crayfish

Figure 45. Raccoon
(*Procyon lotor*).

and frogs. It also eats mice and small birds, including young and eggs from birds' nests.

REPRODUCTION: Mating takes place in late winter. A litter of three to six young is born in a hollow log or tree. Mother and offspring remain together until the end of summer, at which time the young are nearly full grown. Raccoons are sociable, and family groups may remain in a unit through the winter. Dispersal occurs at about one year of age.

COMMENTS: The Raccoon survives where other carnivorans disappear. Its persistence is doubtless due in part to its adaptability and perhaps also to its tree-climbing ability and its extreme ferocity when encountering dogs.

The name *lotor* means "one who washes." When captives are provided with water, they usually wash their food before ingesting it; when deprived of water, they usually rub their food with their dry paws. The meaning of this activity is not known.

Weasels, Marten, Skunks, and Allies (Mustelidae)

Mustelids are small to medium-sized carnivorans. Though diverse, they exhibit an underlying similarity. They have short snouts, elongate braincases, rather large auditory bullae, small eyes, and broad, rounded ears. Most have long bodies and short legs. In skunks (*Mephitis* and *Spilogale* spp.), the body and leg proportions are concealed by long, loose fur. Anal scent glands are present in most genera and well developed in skunks. Like canids, mustelids vary with the nature of their hunting. The meat eaters, such as weasels, have sharp carnassials, and their postcarnassial molars tend to be reduced or absent. In skunks, the carnassials are less developed. Dentition 3/3, 1/1, 4/4 (or 4/3 or 3/3) and 1/2 or 1/1.

Reproduction in many species of mustelids includes delayed implantation—mating and fertilization in summer or fall, with a subsequent prolonged quiescent stage during which no development occurs. Implantation follows in late December, apparently under the influence of day length, and cell division resumes. (See the section "Gestation".)

Mustelids are divided into five subfamilies, three of which occur in California. They vary from tiny weasels to the formidable Wolverine *(Gulo gulo)*. Although most species are terrestrial, the Marten *(Martes americana)* and the Fisher *(M. pennanti)* are expert tree climbers, the Badger is greatly modified for digging, the Mink *(Mustela vison)* is slightly adapted for an aquatic life, and the otters are clearly modified for swimming and foraging in the water. The skunks have many unique features but retain the basic mustelid body pattern.

The family first appeared in the Oligocene in Asia and Europe. It apparently reached North America in the Miocene, presumably across a Bering connection.

Key to Genera of Mustelidae in California

1a Premolars 4/4 2
1b Premolars fewer than 4/4 3
   2a Body with lateral stripe more lightly pigmented than rest of dorsal and lateral fur.................... *Gulo*
   2b Body without lateral stripe .... *Martes*
3a Fur conspicuously black and white dorsally 4
3b Fur dorsally an even brown or mottled gray; head may be striped.. 5
   4a With white and black stripes on back .... *Mephitis*
   4b With two indistinct rows of white patches or spots ...
       .... *Spilogale*
5a Tail at base not greatly thickened (but sometimes bushy); premolars 3/3; feet variable but not broad and webbed ....
       6
5b Tail thick and muscular at base; premolars 4/3 or 3/3; feet broad and webbed.................................... 7
   6a Dorsal fur of an even brown or black; tail slender ....
       .... *Mustela*
   6b Dorsal fur mottled gray; head with white stripe, sometimes extending onto shoulders; tail bushy ....
       .... *Taxidea*
7a Premolars 4/3 .... *Lutra*
7b Premolars 3/3.................................. *Enhydra*

SEA OTTER   *Enhydra lutris*
Pl. 5

DESCRIPTION: The most aquatic member of the weasel family and the largest of the living mustelids: the adult male is almost 2 m long, the female somewhat smaller. The Sea Otter is dark brown or black in color. The tail is thickened at the base. The hind feet are compressed into flippers, sparsely furred; the foreclaws are retractile. The female has a single pair of mammae on the abdomen. The four upper molars are broad and modified for crushing, not for shearing and cutting. Weight: 21–45 kg (males), 14–33 kg (females). Dentition: 3/2, 1/1, 3/3, 1/2.

DISTRIBUTION: Found along the coast of California (estimated population 2,000), with concentrations in Monterey County. It is established as far south as Morro Bay and rarely seen south to Baja California. The Sea Otter is associated with kelp beds in our area but may occur independent of kelp elsewhere. Strictly coastal, it rarely enters fresh water, nor does it migrate. It is found sporadically in Oregon and has been reestablished (by transplanting individuals from Alaska) in Washington and British Columbia. It also occurs on the Aleutians, the Commander Islands, and the Kuriles.

FOOD: The Sea Otter feeds on a variety of bottom-dwelling invertebrates. A large lung capacity allows it to dive to about 100 m, where it gathers abalone, sea urchins, crabs, and a few slow-moving fish. It sometimes feeds on mussels, scallops, and other bivalves. Although it takes abalone, this large snail is far less important in its diet than are sea urchins.

REPRODUCTION: Mating and birth probably occur every other year. They are not seasonal in California, but young are born in summer in the Aleutians. A single pup is born after a gestation of six to nine months, which includes a period of delay in implantation. In California birth usually takes place offshore, in kelp beds, but in Alaska it most commonly takes place on land. The young is precocial, fully furred, and active at birth. Nursing may continue until it is full grown.

STATUS: The Southern Sea Otter *(E. l. nereis)*, which is the race occurring in California, is federally listed as threatened and is fully protected in California.

COMMENTS: A complex relationship exists between densities of the Sea Otter, sea urchins, abalone, bivalves, and shallow-water fish. Whereas moderate levels of these various organisms do not conflict with one another, excessive removal of any one of them may affect densities of the others. When Sea Otters abound, they seem to curb populations of sea urchins, which, in turn, feed on kelp. Shallow-water fish find protective cover in kelp; consequently, a healthy growth of kelp allows larger populations of some fishes. Thus the Sea Otter promotes an increase of kelp— and therefore fishes—whereas a scarcity of Sea Otters allows an increase of sea urchins, which tends to depress the growth of kelp and thereby the fish populations.

The Sea Otter is one of the few mammals to use a tool. It is often seen to lie on its back on the surface, using a rock to break a shellfish on its chest. It is also reported to use rocks to dislodge abalone attached to bedrock.

The demand for Sea Otter pelts led the Russians first to Alaska and eventually to the California coast in the mid-eighteenth century. Early in the nineteenth century the Sea Otter was abundant along the coast of Oregon and Washington, where the Indians not only used the pelts for their own clothing but also traded them to merchants for beads. In 2003, the central California coast otter population, only recently recovered, experienced a sharp decline caused principally by pollution of ocean waters with the parasite-laden feces of domestic cats.

WOLVERINE
Gulo gulo

Pl. 4, Fig. 46

DESCRIPTION: Exceeds all other terrestrial members of the weasel family in body mass and ferocity. It is nearly equal to a large Fisher *(Martes pennanti)* in length, but much heavier, with a heavy body and short legs. The fur is dark brown, and the head is whitish between the eyes and ears. A wide, light band is found on each side of the body, with considerable individual variation. The feet are large and become cov-

Figure 46. Wolverine *(Gulo gulo)*.

ered with thick fur in winter. TL .9–1.1 m (males), 880–970 mm (females); T 190–260 mm (males), 170–195 mm (females); HF 180–200 mm (males), 170–185 mm (females). Weight: 13.6–16.0 kg (males), 9.0–11.5 kg (females). Dentition: 3/3, 1/1, 4/4, 1/2.

DISTRIBUTION: Mostly found in the High Sierra south of Lake Tahoe; also occurs in the northwest coast counties (Humboldt, Del Norte, Trinity). It inhabits high montane forests. Its range extends north to Oregon and Washington and across much of the coniferous forest of northern North America. It is also found in Eurasia.

FOOD: The Wolverine takes various squirrels up to the size of a marmot; it also takes Porcupines *(Erethizon dorsatum)*, although the quills may kill it. At times it eats carrion or berries.

REPRODUCTION: Mating takes place in winter, and a litter of one to four young is born in spring. Known dens of Wolverines are on the ground, in crevices under rock ledges at 3,000 m or above, well above timberline.

STATUS: The Wolverine is listed by the state as threatened and is fully protected in California.

COMMENTS: The Wolverine is nowhere common and is seldom seen. It is the most powerful predator among the Mustelidae. Hunters tell of bears and Mountain Lions *(Panthera concolor)* that retreat from their meals at the approach of a Wolverine. However, it apparently does not threaten livestock or humans. Legends of its ferocity may be partly based on the damage it can do to mountain cabins. One cabin in Plumas County was broken into and ransacked, and the distinctive hairs of a Wolverine were found where the window had been forced open. In captivity this animal may become tame and docile, even playful.

RIVER OTTER   *Lutra canadensis*
Pl. 5, Fig. 47

DESCRIPTION: One of the largest of the weasel family, with the thick, dense fur characteristic of aquatic mammals. The River Otter is dark brown, appearing black at a distance or in poor light. The toes are connected by webbing. TL .89–1.3 m, T 300–500 mm, HF 100–145 mm. Weight: 5–10 kg. Dentition: 3/3, 1/1, 4/3, 1/2.

DISTRIBUTION: Found along the margins of rivers and larger streams in the Cascade

Figure 47. River Otter
(*Lutra canadensis*).

Range and Sierra Nevada down to the Central Valley and Delta; also occurs in major drainages in the Coast Ranges north of San Francisco to Alaska. Tracks are often seen on the sandbars of larger rivers, and otter slides are frequent in the Delta. The River Otter occurs over much of North America, from Alaska to Mexico.

FOOD: This species eats crayfish, frogs, shellfish, and fish, especially rough fish (minnows and suckers). It also eats mice, birds, and birds' eggs, but it seldom forages far from water. In the San Juan Islands and other coastal areas it enters tidal regions and feeds on marine fish, mollusks, and crustaceans.

REPRODUCTION: The River Otter breeds at two years of age. Mating takes place in fall, followed by delayed implantation and birth of two to four young in a streamside burrow in April or May.

STATUS: The Southwestern River Otter (*L. c. sonorae*) is a California subspecies of special concern.

COMMENTS: The otter was one of the great incentives for the exploration of California prior to the gold rush. Commercial trapping ceased in 1961 in California; like the Marten (*Martes americana*) and Fisher (*M. pennanti*), the River Otter has greatly increased since then. In this century the River Otter has never been taken in large numbers. From 1921 to 1961, from 14 to 163 were taken annually; over this 40-year period there was no apparent trend of increase or decrease in abundance. Intensity of trapping is probably greatly influenced by abundance as well as by price, but this species holds its numbers with controlled commercial trapping.

The River Otter is sometimes a nuisance near fish hatcheries, but control is simple. This otter is usually playful and commonly constructs slides on muddy streambanks. These are worn smooth and slippery by constant use. In the mountains in winter it makes similar slides in the snow.

Some taxonomists place the River Otter in the genus *Lontra*.

Marten and Fisher *(Martes)*

In North America the genus *Martes* is represented by the Marten *(Martes americana)* and the Fisher *(M. pennanti)*, but in Eurasia it includes a number of species, more or less Marten-like in general aspect. Members of this genus are arboreal and feed mostly on tree squirrels. They live in temperate or cool areas and are active all winter. Perhaps as a result of these habits, they grow a rich fur that has always been highly valued. The fur known as sable comes from one or more Old World species of *Martes*.

Of the North American mustelids, only the Marten, the Fisher, and the Wolverine *(Gulo gulo)* have four upper and four lower premolars; the other genera have fewer. The Marten and Fisher are larger than weasels and minks *(Mustela* spp.) and are distinctive in having rather bushy tails.

Key to Species of *Martes* in California

1a Tail more than 290 mm; color grayish brown to black on rump; tail entirely black .... *Martes pennanti*
1b Tail less than 290 mm; color chocolate or yellow brown, to black on tip of tail .... *Martes americana*

MARTEN   *Martes americana*
Pl. 6

DESCRIPTION: A moderately large mustelid, chocolate brown or sometimes yellow brown, with blackish feet and tail tip, and orange yellow or buff throat and chest. The tail is long and bushy. TL .57–1.0 m (males), 540–597 mm (females); T 170–210 mm (males), 170–206 mm (females); HF 80–92 mm (males), 82–84 mm (females). Weight: 1.2–1.5 kg (males), .9–1.1 kg (females). Dentition: 3/3, 1/1, 4/4, 1/2.

DISTRIBUTION: Found in northwestern California, where it seems to favor redwood forests. It also occurs at high elevations in the northern Cascade Range and Sierra Nevada in forests of pine, fir, and hemlock above 1,200 m; on talus slopes; and in open rocky areas. Its range includes the Olympic Mountains, Cascade Range, and Blue Mountains of Washington, and high elevations in Oregon; it extends to southern Canada, the northern Rocky Mountains, and the northeastern United States.

FOOD: The Marten takes a great variety of vertebrates, especially tree squirrels and chipmunks (*Neotamias* spp.). It may prey heavily on Pikas (*Ochotona princeps*), rabbits (*Lepus* spp.), and wood rats (*Neotoma* spp.). It sometimes eats insects and many kinds of fruits and berries. Mountain ash provides a favorite food, and scats of the Marten may consist almost entirely of the seeds of this small tree.

REPRODUCTION: From two to four young are born in April or May. Mating, followed by delayed implantation, takes place shortly thereafter.

STATUS: The Humboldt Marten (*M. a. humboldtensis*) is a California subspecies of special concern.

COMMENTS: The Marten is the most frequently encountered of the larger weasels and is frequently seen by persistent enthusiasts. A skillful climber, it is as likely to be seen in the trees as on the ground. Under total protection, it has increased in numbers.

This species is sometimes considered to be the same as some Old World martens (*Martes martes, M. melampus,* and *M. zibellina*).

FISHER   *Martes pennanti*
Pl. 6

DESCRIPTION: A rather large, dark brown or blackish mustelid, somewhat grayish on the head and shoulders, with patches of white on the throat or underside. The Fisher has a heavy body and a long, thickly furred tail. As in mustelids generally, the males are much larger than the females. TL .9–1.2 m (male), 750–950 mm (females); T 381–422 mm (males), 340–380 mm (females); HF 113–128 mm (males), 89–115 mm (females). Weight: 3.5–5.5 kg (males), 2.0–2.5 kg (females). Dentition: 3/3, 1/1, 4/4, 1/2.

DISTRIBUTION: Found in northwestern California at rather low elevations and in the Cascade Range and Sierra Nevada at 1,000 m and above, north to British Columbia. It favors stands of pine, Douglas-fir (*Pseudotsuga menziesii*), and true fir, avoiding red-

wood forests. Its range extends from southern Canada through the northern Rocky Mountains and to the northeastern United States.
FOOD: The Fisher eats many sorts of small mammals, from mice to rabbits (*Lepus* spp.); tree squirrels are a favorite. In northwestern California the false truffle (a hypogeous fungus) is also an important food item. The Fisher does not capture fish but is known to eat them. A formidable predator, it has been seen to capture and kill a Gray Fox (*Urocyon cinereoargenteus*). Under duress it will attack and kill a Porcupine (*Erethizon dorsatum*).
REPRODUCTION: Mating occurs in spring or summer and is followed by an extremely long gestation of some 330 to 360 days. A single brood of one to five young is born in April or May.
STATUS: The Fisher is a California species of special concern.
COMMENTS: Although the Fisher was rare or uncommon early in the century, sightings and evidence of its presence have increased from the 1960s. It may no longer be regarded as rare in some parts of Humboldt and Trinity Counties.

Skunks

Skunks comprise a group of conspicuously black and white carnivorans infamous for their offensive odors. The color pattern is assumed to be a warning to predators. Certainly skunks have few enemies (though the Great-horned Owl [*Bubo virginianus*] does capture them); this is undoubtedly what accounts for their lack of fear. Tameness, which might indicate rabies in a fox, is typical of skunks even in good health. One can easily approach a skunk closely as it feeds in the early evening, but proximity is dangerous.

Unlike the scents used for social communication by many kinds of mammals, the secretion of a skunk is reserved to repel enemies. The fluid is ejected from anal glands that are surrounded by voluntary muscles; it can be ejected at will and fired some 3 to 4 m with accuracy. Prior to discharging its scent, a skunk turns its rear and elevates its large bushy tail. This display is sufficient to deter all but the naive or hungriest predator.

Two species of skunks are found in California.

STRIPED SKUNK   *Mephitis mephitis*
Pl. 5, Figs. 48, 49
DESCRIPTION: Instantly recognized by almost everyone, though occasionally mistaken for a black and white cat; its long fur and bushy tail

do give it a vaguely catlike aspect. Typically it has two broad stripes down the back, but these vary in width. Frequently there is a white stripe on the head. TL 575–800 mm (males), 600–725 mm (females); T 185–390 mm (males), 240–270 mm (females); HF 60–90 mm (males), 60–80 mm (females). Weight: 1.8–2.7 kg, but large males may approach 4 kg.

DISTRIBUTION: Occurs throughout most of California north to Washington, except in the extremely arid southeastern deserts. It frequently lives in well-settled areas and is often found in gardens that are not tightly fenced. Its range covers the southern half of Canada and virtually all of the United States and extends south into Mexico.

FOOD: Unlike most weasels, the skunk is a true omnivore. In well-watered lawns and gardens, it forages deliberately, searching every depression and crevice for beetle larvae, cutworms, mice, and earthworms. It also takes berries on low-growing bushes and eats underground parts of plants, such as bulbs and corms.

REPRODUCTION: The female remains in estrus for prolonged periods, as ovulation occurs only after mating, which takes place in late winter or spring. Implantation is not delayed; gestation lasts for 60 to 77 days. Four to seven (or more) young are born in May or June in a hollow log or underground chamber. They remain with the mother most of the summer.

Figure 48. Striped Skunk *(Mephitis mephitis).*

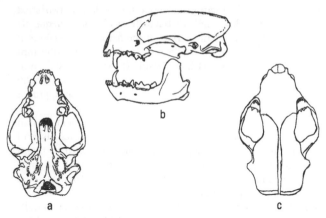

Figure 49. Skull of the Striped Skunk *(Mephitis mephitis):* (a) ventral view; (b) lateral view; (c) dorsal view.

COMMENTS: The skunk's digging leaves numerous small, cup-shaped depressions in the ground.

One occasionally sees a mother skunk in the evening with her young following in single file, a charming sight. Skunks foraging on open ground often move with a smooth, flowing motion, their long tails trailing behind. In the failing light of dusk, they seem to glide over the ground like wraiths. In spite of their friendliness and beauty, skunks are a dangerous source of rabies. In fact, the incidence of rabies in the Striped Skunk exceeds that in the domestic dog.

Although it brings a very low price, the fur has long been used extensively for trim as well as for full-length coats.

SPOTTED SKUNK   *Spilogale putorius*

Pl. 5, Figs. 50, 51

DESCRIPTION: About half the length of the Striped Skunk *(Mephitis mephitis),* and one-third or less in weight. Its fur is much softer and glossier, and its pattern is broken up into spots of white on black. TL 310–610 mm (males), 270–544 mm (females); T 80–280 mm (males), 85–210 mm (females); HF 32–59 mm (males), 30–59 mm (females). Weight: 535–800 g (males), 200–280 g (females). Dentition: 3/3, 1/1, 3/3, 1/2.

DISTRIBUTION: Found throughout most of California, north to western Washington, though it apparently avoids high mountains. It also occurs on Santa Cruz and Santa Rosa Islands, California. Its range covers most of the United States (except the northeast) and extends from Vancouver to central Mexico.

FOOD: Like the Striped Skunk, this species is omnivorous. It forages both day and night for soil-dwelling insects, worms, and mice, and it preys on small ground-nesting birds. It also eats corn, grapes, and other vegetable food.

REPRODUCTION: The Spotted Skunk dens underground, usually in the burrow of some digging mammal, such as a ground squirrel. Mating occurs in fall or winter, implantation is delayed up to 200 days, and a litter of two to six young is born in spring. Because eastern populations of the Spotted Skunk mate in spring and do not experience delayed implantation, it has been suggested that the western populations constitute a separate species. The time of implantation appears to be variable, especially in the southern part of its range, and the distinctions between the eastern and western populations may remain equivocal for some time.

STATUS: The Channel Islands Spotted Skunk *(S. p. amphiala)* is a California subspecies of special concern.

COMMENTS: The Spotted Skunk's search for food can tear up a lawn. Its fondness for insects sometimes leads it to beehives, where it can be destructive to domestic honeybees.

Figure 50. Spotted Skunk
(Spilogale putorius).

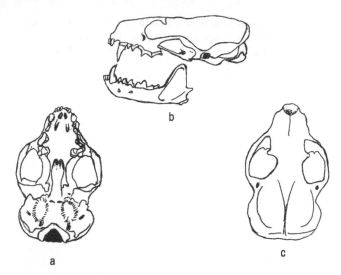

Figure 51. Skull of the Spotted Skunk *(Spilogale putorius):* (a) ventral view; (b) lateral view; (c) dorsal view.

The Spotted Skunk is known for various behaviors that may indicate nervousness or display to induce fear in the observer. Some observers have noted that it is prone to stamp its forefeet. Better known are its handstands—it sometimes stands erect on its forefeet, tail and hind legs held high over the body, perhaps to alarm its potential attacker (see pl. 5).

Weasels and Minks *(Mustela)*
These little carnivorans perhaps best illustrate the stereotype of the weasel family: the head is narrow, the neck is long and very muscular, the legs are short, and the body is elongate. The carnassials are well developed for cutting. The *Mustela* skull (fig. 52) is easily recognized by the large auditory bullae. The dentition is 3/3, 1/1, 2–3/3–2, 1/2.

Key to Species of *Mustela* in California

1a Color brown dorsally, much lighter (even white) ventrally; may be white in winter . . . . , ... 2

1b Color dark brown or black, scarcely any lighter ventrally; no
    seasonal color change . . . . *Mustela vison*
  2a Tail usually not more than 40 percent as long as head
      and body; hind foot less than 35 mm (males) or 28
      mm (females) . . . . *Mustela erminea*
  2b. Tail usually more than 40 percent as long as head and
      body; hind foot more than 40 mm (males) or 30 mm
      (females) . . . . *Mustela frenata*

SHORT-TAILED WEASEL or ERMINE   *Mustela erminea*
Pl. 6

DESCRIPTION: The smallest weasel in the Pacific Coast states, generally much smaller than the Long-tailed Weasel. It is usually
brown or yellow brown, though it has a white winter coat in the
mountains. Its tail is rather short, and black on the distal third.
TL 225–275 mm (males), 190–230 mm (females); T 55–75 mm
(males), 50–63 mm (females); HF 27–34 mm (males), 23–27 mm
(females). Weight: 50–60 g (males), 28–35 g (females); specimens
in the mountains average smaller.

DISTRIBUTION: Found throughout the foothills and mountains of
both the Cascade Range and Sierra Nevada in the northern half
of the state; also in the Coast Ranges from Marin County north
through Oregon and Washington to Alaska. It prefers coniferous
forests. Its range extends throughout North America from
Greenland to the southwestern United States. It is also found in
Eurasia; in England it is called the Stoat.

FOOD: This weasel mostly eats mice and small, ground-nesting
birds. Its prey ranges from deer mice and voles up to rodents the
size of a chipmunk or wood rat. It also eats some reptiles and amphibians and rarely fruits and berries. It climbs well and takes the
eggs and young of small birds.

REPRODUCTION: Mating takes place in late summer and is followed
by delayed implantation. Four to eight young are born in spring,
weighing about 1.7 g. Females may mate in their first summer;
males become sexually mature the following summer.

COMMENTS: The Short-tailed Weasel is active day and night, foraging over 50 to 100 acres.

Although this species is, in its white winter coat, the Ermine of
commerce, the pelts bring only a dollar or two. An Ermine cape
becomes expensive because of the large number of skins used in a

single garment. The skin is thin and the fur quality is vastly inferior to that of the Mink *(Mustela vison)*. The development of the white winter pelage may be influenced by genetic factors, but experimentally it can be induced by low temperature. Molt itself is induced by day length.

LONG-TAILED WEASEL   *Mustela frenata*
Pl. 6, Fig. 52

DESCRIPTION: A chocolate brown weasel with a black-tipped tail, a pale yellow or white underside, and frequently, a distinctive white or light-colored facial mask, sometimes in the form of a whitish patch between the eyes. It is both longer and heavier than the Short-tailed Weasel *(M. erminea)* and has a relatively long tail. Its winter pelage is white in montane populations but brown at lower elevations where there is not normally snow on the ground. TL 350–450 mm (males), 335–395 mm (females); T 125–180 mm (males), 120–145 mm (females); HF 42–50 mm (males), 32–41 mm (females). Weight: 225–345 g (males), 115–220 g (females). (See fig. 52.)

DISTRIBUTION: Found statewide except in extremely arid country. It favors fairly open areas, rock piles, and stacks of firewood, though it can be found almost anywhere except the streamside habitat apparently preempted by the Mink *(Mustela vison)*. Its range extends from southern Canada to southern Bolivia, up to 3,300 m in the United States.

FOOD: This weasel forages at any hour and is opportunistic in its choice of prey, though it clearly prefers warm-blooded vertebrates. It takes a variety of small mice, pocket gophers *(Thomomys* spp.), ground squirrels *(Ammospermophilus* and *Spermophilus* spp.), chipmunks *(Neotamias* spp.), and small birds. It prefers mice and other mammals of approximately that size; its small head and long neck permit the exploration of crevices and hiding places of mice. However, it is capable of killing rabbits *(Lepus* spp.) several times its own weight. It also eats bees.

REPRODUCTION: Mating normally takes place in July, after the young are weaned. The prolonged copulation resembles that of the Mink: a pair of Long-tailed Weasels may remain clasped together for two hours or more and may repeat the performance

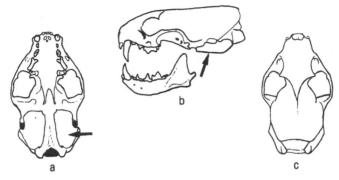

Figure 52. Skull of the Long-tailed Weasel *(Mustela frenata):* (a) ventral view; (b) lateral view; (c) dorsal view (arrows to auditory bullae).

the same day. After a prolonged delay, implantation occurs the following winter; four to eight young are born in June, some 225 to 300 days after mating. Embryonic growth takes four weeks or slightly longer. The young are nearly naked and weigh approximately 3 g at birth. They take milk for five weeks or so. The nest can be almost any concealed shelter, hollow log, or squirrel nest and is frequently lined with the fur of prey. Females become sexually mature in their first summer, in advance of their male siblings, and while older males are still sexually active.

COMMENTS: The Long-tailed Weasel's predatory prowess is the result of technique, not size: typically it clutches the back or neck of large prey and then bites the soft parts of the skull, especially about the ears, taking advantage of the sharp posterior edges of its canines.

Although this weasel sometimes kills grown chickens, it more commonly feeds upon rats that infest chicken coops. In view of the well-established rat-killing habits of weasels generally, it is foolish to destroy them on sight on the assumption that they are about to prey on domestic fowl.

Like other members of this family, the Long-tailed Weasel is playful. A captive enjoyed chasing dead Starlings *(Sturnus vulgaris)* swung before it on a string; when finally allowed to catch its "prey," it performed a little dance, jumping high into the air while doing twists and aerial somersaults, the meaning of which is not apparent.

MINK
Pl. 6

Mustela vison

DESCRIPTION: A dark brown or black weasel with small white spots about the chin or throat and dense, glossy fur. Its general form is weasel-like, with a rather bushy tail. TL 491–720 mm (males), 473–560 mm (females); T 160–211 mm (males), 157–203 mm (females); HF 60–75 mm (males), 58–64 mm (females). Weight: .88–1.3 kg (males), 540–750 g (females).

DISTRIBUTION: Found in watercourses and marshes from the San Joaquin Valley north to the Delta and throughout most of the northern half of California, Oregon, and Washington. It frequents tidal margins in the Delta, mudflats around San Francisco Bay, and streamsides to elevations of 2,000 m or higher. Its range extends throughout North America to Alaska, and it has been introduced into Eurasia.

FOOD: The Mink subsists on streamside invertebrates, such as crayfish; vertebrates, such as frogs and Muskrats *(Ondatra zibethicus)*; and carrion. It sometimes captures ducks and coots.

REPRODUCTION: Mating takes place in late winter; three to 10 young are born in June or July. The den is usually near water, under a log or beneath the ground.

COMMENTS: As a predator the Mink is of neutral value. It preys on mice and carrion, but also takes some game birds and rabbits *(Lepus* spp.) and may occasionally molest domestic fowl and even domestic trout.

This species used to be important as a furbearer, and it survived well under sustained trapping. In 1978–1979, Mink pelts brought the trapper $10 to $20 apiece, for a total of more than $7,000 for the state. In 1996–1997 the price averaged only $7.63, for a total of less than $1,000.

BADGER
Pl. 5, Figs. 53, 54

Taxidea taxus

DESCRIPTION: A large mustelid rather obviously modified for a semifossorial life, with powerful forefeet for digging and a stout, flattened body. The body is silver gray, the head patterned with gray and white. The tail is short and moderately furred. TL

Figure 53. Badger *(Taxidea taxus)*.

600–730 mm, T 100–135 mm, HF 92–126 mm. Weight: up to 11.4 kg (large males), 4.5 kg (females). Dentition: 3/3, 1/1, 3/3, 1/2. (See fig. 54.)

DISTRIBUTION: Most commonly found in the Great Basin region of California, Oregon, and Washington, and sporadically common in the Sacramento Valley, fluctuating with populations of squirrels (Sciuridae) or pocket gophers (*Thomomys* spp.). A creature of open areas, including deserts, it ranges from south-central

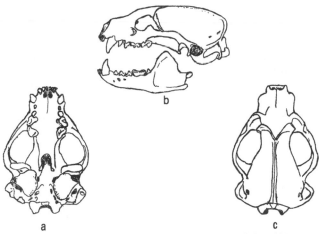

Figure 54. Skull of the Badger *(Taxidea taxus):* (a) ventral view; (b) lateral view; (c) dorsal view.

Canada over the western and central United States to central Mexico.

FOOD: The Badger mostly eats ground squirrels (*Ammospermophilus* and *Spermophilus* spp.) and pocket gophers.

REPRODUCTION: Mating takes place in late summer, but implantation is delayed until December or January. One to four young are born in March or April in an extensive burrow system.

COMMENTS: In its almost constant pursuit of its quarry, the Badger tears up a great deal of ground and may damage rodent-infested levees. On open range, ranchers object to its burrows because they may endanger horses. In the 1950s, Badgers were so common in alfalfa fields heavily infested with pocket gophers that their diggings sometimes damaged the blades of cutting machines.

Red-tailed Hawks (*Buteo jamaicensis*) may hover over and watch foraging Badgers and have been known to stoop down and snatch ground squirrels fleeing the mammalian predator. Coyotes (*Canis latrans*) sometimes attend a digging Badger for the same reason.

The Badger is sometimes an important furbearer. The pelage is soft, durable, and very beautiful. Before the advent of the electric razor and brushless shaving cream, the fur of the Badger provided half the adult population with shaving brushes.

Cats (Felidae)

Throughout the world there is little basic diversity among cats. Our two native species, the Mountain Lion (*Panthera concolor*) and the Bobcat (*Felis rufus*), have typical feline structure.

The cat's skull reveals its diet. The dentition is reduced: 3/3, 1/1, 2–4/2–3, 1/1. The molars are reduced. The bladelike upper and lower carnassials meet so that their adjacent lateral surfaces form a shearing tool, in contrast to the cheek teeth in bears, which form a crushing or grinding surface. The canine teeth are long and pointed.

Cats have a short snout, in contrast to the pointed snout of dogs, and large, elliptical eyes, reflecting their nocturnal habit (figs. 55, 57). They are digitigrade, and all California species have retractable claws: they can be withdrawn to prevent their becoming dulled when walking or extended to grapple with prey.

The cat family is first recorded from the Miocene in Europe.

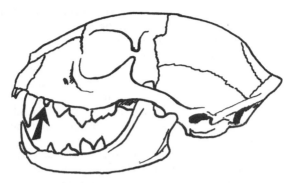

Figure 55. Skull of the Feral House Cat *(Felis sylvestris)*. Note small upper premolar (arrow) behind the canine tooth, characteristic for this species.

FERAL HOUSE CAT   *Felis sylvestris*
Fig. 55

Most agricultural areas have populations of feral house cats. These are continuously augmented when people abandon unwanted pets, which are frequently able to survive in the wild. They seem to subsist mostly on wild rodents taken at some distance from human dwellings, and they probably take some carrion after the shooting of upland game birds. They are mentioned here to remind readers that the "wild mammal" they glimpse may be only a feral house cat.

Linnaeus applied the specific name *catus* to the domestic cat. This cat presumably derives from *Felis sylvestris,* a wild species in Eurasia and North Africa. (See fig. 55.)

MOUNTAIN LION   *Panthera concolor*
Pl. 4 Fig. 56

DESCRIPTION: The Mountain Lion (also called Puma or Cougar) is the largest cat in California. Its sandy color and long (1 m), usually black-tipped tail are distinctive. The kits have sometimes indistinct brownish spots, which are lost in the first six months of life. TL 1.5–2.5 m, T 550–775 mm, HF 220–275 mm. Weight:

40–100 kg; males usually larger than females. Dentition: 3/3, 1/1, 3/2, 1/1. (See. fig. 56.)

DISTRIBUTION: Found throughout most of the forested and brushy regions of western North America. It survives within the city limits of Berkeley, Hayward, Richmond, and Sacramento. It is the most widely ranging mammal in the New World, found from the Canadian coniferous forest south to Patagonia.

FOOD: The Mountain Lion preys heavily on deer (Cervidae) and other mammals. It takes squirrels (Sciuridae), rabbits (*Lepus* spp.), and even mice and is one of the few predators to take skunks (*Mephitis* and *Spilogale* spp.) and Porcupines (*Erethizon dorsatum*). It has been known to molest domestic stock but is a lesser problem than the Coyote (*Canis latrans*) in this regard. Most urban lions seem to live on house cats (*Felis sylvestris*).

REPRODUCTION: Gestation takes about 90 days. A litter of one to six kits is born blind and helpless, usually in spring. The Mountain Lion is sexually mature at 25–35 kg.

STATUS: The Yuma Mountain Lion (*P. c. browni*) is a California subspecies of special concern.

COMMENTS: The Mountain Lion is not uncommon today and is

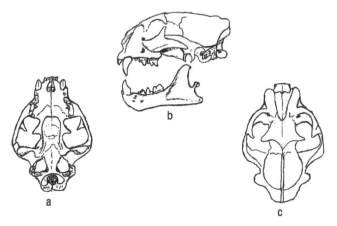

Figure 56. Skull of the Mountain Lion *(Panthera concolor):* (a) ventral view; (b) lateral view; (c) dorsal view.

among the many wild mammals that are found as road kills on mountain highways. This magnificent cat survives close to civilization for two reasons: it is very shy and seldom seen, and it seldom conflicts with human activities. It tends to avoid open areas; its presence is known more from tracks than from sightings. It is an extremely valuable predator, for it is a check on numbers of deer. However, it has been implicated in the reduction in numbers of the few remaining Bighorn Sheep *(Ovis canadensis)* in California. For many years it was assumed to have disappeared from eastern North America except for a population in Florida. In recent years, however, the unmistakable tracks have been reported from Georgia, Pennsylvania, New York, Maine, and New Brunswick. It has survived in these areas, undetected, for as long as 75 years. In the Pacific states it has become common and has extended its range. Although it is rarely dangerous, attacks on humans—including hikers and joggers—have increased in recent years. The New Mexico Department of Fish and Game pays a bounty of $350 on Mountain Lions.

BOBCAT   *Felis rufus*
Pl. 4, Fig. 57

DESCRIPTION: A spotted cat with a short, white-tipped tail, tufted ears, and broad whiskers. It is rather long-legged for a cat. TL .7–1.0 m, T 95–150 mm, E (from crown) 60–75 mm. Weight: 5–15 kg; males average larger than females; in eastern North America, Bobcats may weigh more than 30 kg. (See fig. 57.) Dentition: 3/3, 1/1, 2/2, 1/1.

DISTRIBUTION: Occurs statewide in California, from Death Valley to the high mountains, equally at home in brushland, foothill chaparral, sagebrush, and forests. It is also widespread in Oregon, Washington, and British Columbia; its range extends from southern Canada to central Mexico.

FOOD: The Bobcat is an opportunist whose diet varies with availability more than with any apparent preference. It takes many rabbits *(Lepus* spp.), squirrels (Sciuridae), mice, and pocket gophers *(Thomomys* spp.), as well as small reptiles and birds. It occasionally molests sheep, but it is not nearly as troublesome as the domestic dogs or the Coyote *(Canis latrans)*. In eastern North America it is a major predator of the White-tailed Deer *(Odocoileus virginianus)*, but eastern Bobcats are larger than California specimens.

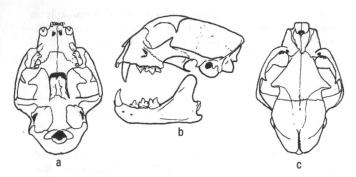

Figure 57. Skull of the Bobcat *(Felis rufus):* (a) ventral view; (b) lateral view; (c) dorsal view.

REPRODUCTION: A litter of one to six kits is born in spring or summer after a gestation of some 50 days. If a fertile mating does not occur at the first estrous period, estrus recurs one or more times. This accounts for the extended period over which young are born.

COMMENTS: The Bobcat remains relatively common despite heavy trapping and controls on sheep pastures. In contrast to the secretive Mountain Lion *(Panthera concolor)*, it is sometimes quite bold and may not run away at the first sight of human observers. It may occur close to buildings and has been known to crouch next to water troughs and strike down bats as they fly low to drink. Its spotted fur and erect pointed ears make it one of the most beautiful mammals in our fauna.

Some taxonomists place Bobcats in the genus *Lynx,* which is sometimes considered a subgenus of *Felis.* As with the Mountain Lion, morphological features, distinctive in the flesh, are not present in fossil material.

Pinnipeds (Pinnipedia)

Paleontologists generally consider Pinnipedia—walruses (Odobenidae), seals (Phocidae), and sea lions (Otariidae)—to be a superfamily (or suborder) of Carnivora. Biochemical information and external parasites (sucking lice) indicate that the pinnipeds are monophyletic, derived from a single ancestor. Details of the skull suggest that they are derived from very early relatives of bears, the Arctoids. Presumably this descent occurred some 22 million years ago, as Pinnipedia dates from the early and middle Miocene.

Because walruses do not occur in California, we consider here only the seals (Phocidae) and the sea lions (Otariidae). Phocids have a long history in both the Atlantic and Pacific Oceans, whereas otariids are primarily Pacific in distribution. The common ancestry and the long history of aquatic existence could certainly account for many physical and physiological resemblances between these groups. Although it is convenient and reasonable to refer to them collectively as pinnipeds, seals and sea lions have a number of differences, some of them obvious even when these animals are viewed from a distance.

Pinnipeds are properly considered to be semiaquatic or amphibious, for they bear their young on land. Infants, moreover, may not enter water immediately. Pinnipeds, however, find locomotion on land difficult and awkward, and they do not obtain food on land. In the water they are swift and graceful, and they do all of their hunting beneath the surface.

Seals mate after the young are weaned. Female sea lions come into estrus immediately following birth of the young and usually mate within hours after parturition. Such postpartum estrus occurs among many groups of mammals, including some artiodactyls and many rodents. Like many species of mustelids, sea lions and seals experience delayed implantation. This delay represents an adaptation to the temporary presence of both sexes together on land and provides for birth of the young nearly a year after mating, when the females return to their breeding grounds (rookeries). Actual embryonic development takes much less than one year.

In many species, the mothers remain with the young continuously until weaning, though such intensive parental care may prevent the adults from feeding. Most seals have a rather brief nursing period, during which the mother usually does not feed. Lactation in seals is shortened by the provision of extremely fat-rich milk that enables the pup to grow rapidly. Most sea lions have less rich milk and a longer lactation period, during which the mother may feed.

Pinnipeds' senses also reflect their long history in a marine environment. Although the external ears are reduced in sea lions and absent in seals, pinnipeds hear quite well underwater. An ability to echolocate is reported but disputed; in any case pinnipeds very likely recognize sounds produced by potential food items. They have much larger eyes than land mammals of comparable body size, allowing more light to enter the eye when they forage in murky depths. Most pinnipeds, moreover, seem to hunt at night. When light is bright, the pupil can be greatly reduced for aerial vision. Both the lens and the cornea are thickened—an adjustment to the increased refractive index of water—that probably results in some degree of myopia when on land. Apparently the senses of smell and taste are not nearly as acute as in terrestrial carnivorans.

The diving ability of pinnipeds results not only from a streamlined body form but also from less obvious physiological adaptations. Just before diving, a pinniped exhales and the heartbeat slows. Circulating blood is concentrated to the brain and heart, which have a great need for oxygen. The Harbor Seal can remain submerged for nearly a half hour and forage to depths of about 100 m. After surfacing, heavy breathing and rapid circulation restore oxygen to the tissues.

Sea Lions or Eared Seals (Otariidae)

Sea lions have rather long limbs (flippers), nude at the tips and with fewer than five toenails on each foot. Males are much larger than females. As in most groups of land mammals, the testes lie within a scrotum. The ears are small but clearly visible, pointed and protruding from the outline of the head. The outer incisor is like the canine of a carnivoran; the molars and premolars are somewhat triangular and expanded in the middle (fig. 60b).

Sea lions, including fur seals, are better able than seals to progress on land. They can rotate their hind limbs anteriorly, which provides some thrust in movement on land and also assists, in a small way, in raising the rear of the animal off the ground. They use their forelimbs for propulsion both on land and in the water.

Sea lions have many natural enemies. In addition to being a favorite food of the Killer Whale *(Orcinus orca)*, they are susceptible to neurotoxins produced by dinoflagellates. When these protozoans become extremely abundant, fish and fish-eating vertebrates may pick up their toxins. Researchers at Moss Landing linked the death of more than 400 sea lions on the California coast to one toxic bloom of dinoflagellates.

The four California species are rather similar in many respects. They do differ in shape, color, size, and distribution, but these differences are not always conspicuous, and field identification may not always be certain. Polygyny is characteristic of the family.

GUADALUPE FUR SEAL   *Arctocephalus townsendi*
Pl. 7, Fig. 58

DESCRIPTION: The smallest of the sea lions known to occur in California waters; males reach 2 m in length, females typically reach 1.5 m. This sea lion is dark brown when wet, somewhat grayish when dry; old males tend to become a light yellow or gold on the

Figure 58. Guadalupe Fur Seal *(Arctocephalus townsendi)*.

chest. The underfur is dense. The species is additionally distinguished by very long front flippers and a rather long, slender snout. Weight: 150 kg (males), 45 kg (females). Dentition: 3/2, 1/1, 4/4, 2/1.

DISTRIBUTION: Today known to breed only on Guadalupe Island, off Baja California, but in early times it bred on islands at least north to the Channel Islands or perhaps the Farallon Islands. It is sometimes seen off San Miguel and San Nicholas Islands off southern California; it may be expected to occur more frequently as populations increase.

FOOD: Its diet is unknown, but it probably consumes medium-sized fish and shellfish.

REPRODUCTION: This fur seal repairs to caves for birth and breeding. One bull and a small number of cows form the breeding unit. Pups are born in spring, and mating follows within a week or so. Presumably the pup nurses for a prolonged period at sea, for the population tends to disperse at the end of summer.

STATUS: This species is federally listed and listed by the state as threatened. It is fully protected in California.

COMMENTS: This species belongs to a group which is otherwise confined to the Southern Hemisphere. It has sometimes been considered the same species as a fur seal *(Arctocephalus philippi)* known from Juan Fernández Islands, off the coast of Chile, that may now be extinct. Until the 1950s the Guadalupe Island population was considered to be extinct; since its rediscovery, it has been increasing steadily, and an estimated 1,000 or more exist today.

NORTHERN FUR SEAL   *Callorhinus ursinus*
Pl. 7, Fig. 59

DESCRIPTION: A dark brown sea lion of medium size. Males are more than 2 m long, females are 1.5 m. The flippers, especially the hind flippers, are quite long, with nails only on the dorsal surface. Weight: up to 225 kg (males), 50 kg (females). Dentition: 3/2, 1/1, 4/4, 2/1.

DISTRIBUTION: Breeds on the Pribilof Islands west to the Commander Islands, and on islands off the coast of Asia. A recently established colony on San Miguel Island greatly expands its breeding range. During the nonbreeding season, populations are dispersed. Although bulls remain mostly in the Gulf of Alaska,

PLATES

PLATE 1   **Shrews and Moles**

Vagrant Shrew *Sorex vagrans*   PAGE 121
Length: 90–120 mm. Weight: 5–7 g. Variable with location; usually brown in summer, gray in winter. Tail bicolored.

Fog Shrew *Sorex sonomae*   PAGE 119
Length: 120–158mm. Weight: 14–18 g. Reddish, light brown, tail uniformly colored.

Trowbridge's Shrew *Sorex trowbridgii*   PAGE 120
Length: 95–132 mm. Weight: 4.8 g. Brown in summer, gray in winter. Tail bicolored.

Water Shrew *Sorex palustris*   PAGE 118
Length: 144–158 mm. Weight: 8–14 g. Black or dark brown. Fur dense, velvety, water-repellent; hind feet fringed with stiff white hairs.

Desert Shrew *Notiosorex crawfordi*   PAGE 113
Length: 81–90 mm. Weight: 3–5 g. Gray or brown. Tail short; ears large; teeth faintly reddish.

Shrew-mole *Neurotrichus gibbsii*   PAGE 125
Length: 110–125 mm. Weight: 10–13 g. Dark or black. Forefeet moderately expanded; tail 30 percent of total length, constricted at base. As in shrews but not other moles, fur directed toward rear.

Townsend's Mole *Scapanus townsendii*   PAGE 127
Length: 195–240 mm. Weight: 115–150 g. Black or dark gray. In California, found only in far northwestern corner.

Broad-footed Mole *Scapanus latimanus*   PAGE 125
Length: 135–190 mm. Weight: 66–78 g. Light gray or black. Forehand as wide as long, or wider; tail less than 25 percent total length.

Overleaf:
Mountain Lion *(Panthera concolor)* chasing **Black-tailed Deer** *(Odocoileus hemionus).*

PLATE 1 Shrews and Moles

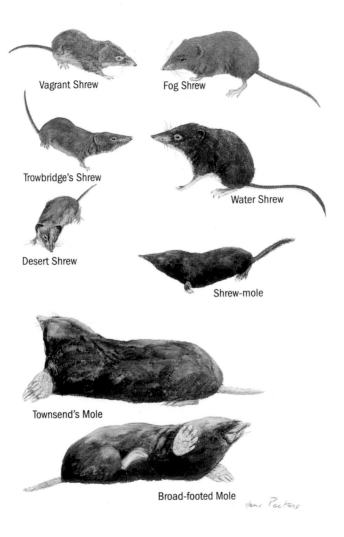

Vagrant Shrew

Fog Shrew

Trowbridge's Shrew

Water Shrew

Desert Shrew

Shrew-mole

Townsend's Mole

Broad-footed Mole

Hans Peeters

PLATE 2   Bats

Long-eared Bat *Myotis evotis*   PAGE 148
When laid forward, ears project far beyond nose.

California Bat *Myotis californicus*   PAGE 146
Keeled calcar; tail tip does not protrude.

Pallid Bat *Antrozous pallidus*   PAGE 139
Pale, beige fur; pink-brown face; large, separated ears.

Silver-haired Bat *Lasionycteris noctivagans* PAGE 142
Blackish to deep brown fur with white tips, "frosted."

Western Pipistrelle *Pipistrellus hesperus*   PAGE 153
Tiny. Tragus bent forward.

Red Bat *Lasiurus blossevillii*   PAGE 144
Rufous-red fur.

Hoary Bat *Lasiurus cinereus*   PAGE 144
White-tipped fur. Broad yellow collar, round ears.

Big Brown Bat *Eptesicus fuscus*   PAGE 139
Long brown fur, black membranes.

California Leaf-nosed Bat
Macrotus californicus   PAGE 135
Flattened, triangular, fleshy "horn" above nostrils.

Townsend's Long-eared Bat
Plecotus townsendii   PAGE 154
Huge ears joined at base. Flaplike lumps on muzzle.

Spotted Bat *Euderma maculatum*   PAGE 140
Black with three white patches on back.

Hog-nosed Bat *Choeronycteris mexicana*   PAGE 134
Long tubular muzzle with fleshy "horn" above nostrils.

Western Mastiff Bat *Eumops perotis*   PAGE 155
Huge. Large ears, joined at base, project forward. Free tail.

Guano Bat *Tadarida brasiliensis*   PAGE 158
Small. Ears separate. Free tail.

PLATE 2   Bats

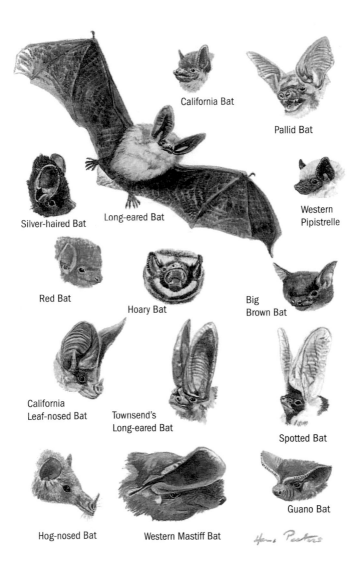

California Bat

Pallid Bat

Silver-haired Bat

Long-eared Bat

Western
Pipistrelle

Red Bat

Hoary Bat

Big
Brown Bat

California
Leaf-nosed Bat

Townsend's
Long-eared Bat

Spotted Bat

Hog-nosed Bat

Western Mastiff Bat

Guano Bat

PLATE 3 Foxes

Gray Fox *Urocyon cinereoargenteus* PAGE 164
Length: .8–1.1 m. Weight: 3–5 kg, rarely to 7 kg. Silvery gray.
Patches of yellow, russet, brown, or white on throat and belly; legs
short. Black-tipped tail has middorsal erectile crest of stiff black
hairs, a feature shared only by the Island Fox *(U. littoralis)*.

Kit Fox *Vulpes macrotis* PAGE 165
Length: 730–840 mm. Weight: 1.7–2.5 kg. Uniformly gray. Tail
black at tip; ears very large. Smaller than the Red Fox *(V. vulpes)*,
with much larger ears; legs longer than in the Gray Fox *(Urocyon
cinereoargenteus)*.

Red Fox *Vulpes vulpes* PAGE 166
Length: .88–1.0 m. Weight: 5–8 kg. Bright reddish or yellowish.
Much black on legs; tail white at tip. Variations include all-black
"silver fox" and brown-and-gray "cross fox."

PLATE 3   Foxes

Gray Fox

Kit Fox

Red Fox

H. Peters

PLATE 4   Large Carnivorans

Bobcat *Felis rufus*   PAGE 197
Length: .7–1.0 m. Weight: 5–15 kg. Spotted. Tail short and white
tipped; ears tufted; whiskers broad; legs long. Males larger than females; in eastern North America, may weigh more than 30 kg.

Mountain Lion *Panthera concolor*   PAGE 195
Length: 1.5–2.5 m. Weight: 40–100 kg. Sandy. Long, usually black-
tipped tail. Kits under six months have brown spots. Males usually
larger than females; largest cat in California.

Coyote *Canis latrans*   PAGE 162
Length: 1.0–1.3 m. Weight: 8–20 kg. Gray, sandy, or brown. Shape
roughly that of German shepherd; some, reaching 25 to 33 kg,
could be mistaken for the Gray Wolf *(C. lupus)*.

Wolverine *Gulo gulo*   PAGE 179
Length: .9–1.1 m (males), 880–970 mm (females). Weight:
13.6–16.0 kg (males), 9.0–11.5 kg (females). Dark brown. Head
whitish between eyes and ears; variable wide light band on each
side. Body heavy, legs short; large feet thick furred in winter.
Largest and most ferocious terrestrial member of weasel family.

Black Bear *Ursus americanus*   PAGE 170
Length: 1.4–1.9 m. Weight: 45–140 kg. Black, brown, or cinnamon, with white or pale patch on throat or belly. Fur coarse; tail
minute. Claws on forefeet same length as those on hind feet, not
longer as in the Grizzly Bear, or Brown Bear *(U. arctos)*.

PLATE 4 Large Carnivorans

Bobcat

Mountain Lion

adult

juvenile

Coyote

Wolverine

black phase

brown phase

Black Bear

© Hans Peeters 2000

PLATE 5   Opossum, Small Carnivorans (Part I)

Raccoon *Procyon lotor*   PAGE 175
Length: 780–930 mm. Weight: 4–8 kg. Face masked with black; tail
ringed. Body heavier, tail much shorter, than in the Ringtail *(Bas-
saricus astutus)*.

Ringtail *Bassariscus astutus*   PAGE 174
Length: 620–800 mm. Weight: .9–1.2 kg. Eyes dark and huge; tail
long and ringed. Soles mostly or partly furred; claws partly retractable.

Badger *Taxidea taxus*   PAGE 192
Length: 600–730 mm. Weight: Up to 11.4 kg (males), 4.5 kg (females). Silver gray. Head marked with black and white; tail short
and moderately furred. Body stout and flattened.

Virginia Opossum *Didelphis virginiana*   PAGE 109
Length: 700–900 mm. Weight: 1.5–3.1 kg. Light gray. Fur long and
loose; body ratlike but larger; teeth numerous, mostly small. Prehensile tail; large canines, opposable thumbs on hind feet; marsupial pouch in female.

River Otter *Lutra canadensis*   PAGE 180
Length: .89–1.3 m. Weight: 5–10 kg. Dark brown. Fur dense,
water-repellent fur; toes webbed.

Striped Skunk *Mephitis mephitis*   PAGE 184
Length: 575–800 mm (males), 600–725 mm (females). Weight:
1.8–2.7 kg. Black, with two broad white stripes down back that
vary in width. Head often has white stripe; tail bushy.

Spotted Skunk *Spilogale putorius*   PAGE 186
Length: 310–610 mm (males), 270–544 mm (females). Weight:
535–800 g (males), 200–280 g (females). Black with white spots.
Length half that of the Striped Skunk *(Mephitis mephitis)*, weight
one third, fur softer and glossier.

Sea Otter *Enhydra lutris*   PAGE 177
Length: Up to 2 m. Weight: 21–45 kg (males), 14–33 kg (females).
Dark brown or black. Tail thick at base; hind feet compressed into
flippers, sparsely furred; foreclaws retractable.

PLATE 5 Opossum, Small Carnivorans (Part I)

Ringtail

Raccoon

Badger

Virginia Opossum

River Otter

Striped
Skunk

Spotted Skunk

Sea Otter

©Hans Peeters

PLATE 6   Small Carnivorans (Part II)

Fisher *Martes pennanti*   PAGE 183
Length: .9–1.2 m (males), 750–950 mm (females). Weight: 3.5–5.5
kg (males), 2.0–2.5 kg (females). Dark brown or blackish. Head
and shoulders grayish; white patches on throat or underside. Body
heavy; tail long and thickly furred.

Marten *Martes americana*   PAGE 182
Length: .57–1 m (males), 540–597 mm (females). Weight: 1.2–1.5
kg (males), 0.9–1.1 kg (females). Chocolate brown or yellow
brown. Feet and tail tip blackish; throat and chest orange yellow;
tail long and bushy.

Long-tailed Weasel *Mustela frenata*   PAGE 190
Length: 350–450 (males), 335–395 (females). Weight: 225–345 g
(males), 115–220 g (females). Chocolate brown. Tail black tipped;
underside pale yellow or white; whitish mask or patch between
eyes. Body longer and heavier than in the Short-tailed Weasel, or
Ermine, *(M. erminea)*. In mountains, face lacks mask in summer
and whole body turns white in winter.

Mink *Mustela vison*   PAGE 192
Length: 491–720 mm (males), 473–560 mm (females). Weight:
.88–1.3 kg (males), 540–750 g (females). Dark brown or black. Fur
dense and glossy; small white spots on chin or throat. Body weasel-
like, tail bushy. Seen here preying on Muskrat *(Ondatra zibethicus)*.

Short-tailed Weasel
Mustela erminea   PAGE 189
Length: 225–275 mm (males), 190–230 mm (females). Weight:
50–60 g (males), 28–35 g (females). Brown or yellow brown, but
white in winter in mountains, as seen here. Tail short and black on
distal third. Smallest weasel in Pacific Coast states.

PLATE 6   Small Carnivorans (Part II)

Fisher

Marten

Long-tailed
Weasel

Mink

Short-tailed
Weasel

PLATE 7   Seals and Sea Lions

Guadalupe Fur Seal *Arctocephalus townsendi*   PAGE 201
Length: Up to 2 m (males), 1.5 m (females). Weight: 150 kg
(males), 45 kg (females). Dark brown when wet, grayish when dry.
Old males have yellow or gold on chest. Underfur dense; front
flippers long; snout long and slender. Smallest sea lion in California waters.

Harbor Seal *Phoca vitulina*   PAGE 208
Length: Up to 1.7 m. Weight: Up to 130 kg. Variable, from nearly
white to almost black; usually brown or gray and spotted. Body
heavy, as though neckless. Females slightly smaller.

Northern Fur Seal *Callorhinus ursinus*   PAGE 202
Length: 2 m or more (males), 1.5 m (females). Weight: Up to 225
kg (males), 50 kg (females). Dark brown. Front and hind flippers
quite long, with nails only on dorsal surface.

California Sea Lion *Zalophus californianus*   PAGE 204
Length: 2.5 m (males), 2 m (females). Weight: 250 kg (males), 100
kg (females). Dark brown or blackish, darker when wet. Ears short
and pointed. Males darker, with prominent foreheads.

Steller Sea Lion *Eumetopias jubatus*   PAGE 204
Length: More than 3 m (males), about 2 m (females). Weight: Up
to 1,000 kg (males), 250 kg (females). Straw or yellow brown;
whitish when submerged. Ears small but distinct. Pups are black.

Northern Elephant Seal
Mirounga angustirostris   PAGE 206
Length: 5 m or more (males), 3 m or more (females). Weight: Up
to 3,500 kg (males), 900 kg (females). Gray or brown. Body seal-
like, but much larger; male has pendulous, inflatable snout. Young
are black.

PLATE 7 Seals and Sea Lions

Guadalupe
Fur Seal

male

female

Harbor Seal

female

male

Steller Sea Lion

California Sea Lion

female

male

female

male

Northern Fur Seal

young
males

Northern Elephant Seal

female

PLATE 8   Sheep, Pronghorn, and Deer

Bighorn Sheep *Ovis canadensis*   PAGE 250
Length: 1.4–1.6 m. Weight: 70–190 kg. Dorsum brown; rump buff
or white; belly white. Lambs born nearly white. Horns large, coiled
in rams and curved in ewes; fur and tail short.

Pronghorn *Antilocapra americana*   PAGE 249
Length: 1.2–1.4 m. Weight: 20–70 kg. Dorsum buff to russet; venter white. Face and throat have blackish marks. Prominent, shiny
black horns, often two pronged in males but unbranched in females. White hairs on rump can be raised to "flash."

Axis Deer *Axis axis*   PAGE 241
Length 1–1.7 m. Weight: Up to 85 kg (males). Reddish brown with
middorsal dark stripe, permanent white spots. Tail dorsally dark,
ventrally white.

Fallow Deer *Dama dama*   PAGE 242
Length: 1.4–1.8 m. Weight: 68–92 kg (males), 43–72 kg (females).
Variable, from nearly white to almost black. Back and sides spotted
year-round, more distinctly in summer. Rump white with black
border; legs and belly white; diagonal white line on flank. Antlers
palmate.

Black-tailed Deer *Odocoileus hemionus*   PAGE 246
Length: 1.1–2.0 m. Weight: 50–220 kg (males). Dorsum reddish in
summer, gray brown in winter. Ears dark; face marked with gray;
tail black tipped or black dorsally. Antlers dichotomous, often
with small medial spike near base. Mule Deer is larger interior
form with black-tipped tail.

Roosevelt Elk *Cervus elaphus roosevelti*   PAGE 243
Length: Nearly 3 m. Weight: Up to 450–550 kg (males). Dorsum
dark brown. Head and neck darker and heavily maned; rump
much lighter. Antlers on male with five to seven tines, including
well-developed basal tines. Largest of California's three subspecies.

Tule Elk *Cervus elaphus nannodes*   PAGE 243
Length: About 2.5 m. Weight: 195–250 kg (males). Coloring and
antlers as in the Roosevelt Elk *(C. e. roosevelti)*.

PLATE 8 Sheep, Pronghorn, and Deer

Pronghorn

Bighorn Sheep

Black-tailed Deer

Mule Deer form

Axis Deer

Fallow Deer

Roosevelt Elk

Tule Elk

PLATE 9   Ground Squirrels and Marmot

Antelope Ground Squirrel
Ammospermophilus leucurus   PAGE 259
Length: 211–233 mm. Weight: 74–103 g. Dorsum ash gray, sometimes cinnamon in summer. Lateral stripe white; tail light with
dark margin.

Nelson's Antelope Ground Squirrel
Ammospermophilus nelsoni   PAGE 261
Length: 218–240 mm. Weight: 75–110 g. Fawn brown. Dorsum
buff, sometimes yellowish; lateral stripe white.

Golden-mantled Ground Squirrel
Spermophilus lateralis   PAGE 264
Length: 235–295 mm. Weight: 136–245 g. Mantle yellow or orange;
lateral stripe white, bounded by black stripes. No stripe on head, as
in chipmunks.

Beechey Ground Squirrel *Spermophilus beecheyi* PAGE 262
Length: 375–500 mm. Weight: 300–650 g. Gray brown, mottled by
light flecks. Fur coarser and browner winter. Tail long and bushy.
Mantle large, from faint shadow to black; pale lateral stripes rare.

Round-tailed Ground Squirrel
Spermophilus tereticaudus   PAGE 266
Length: 204–266 mm. Weight: 116–133 g. Uniform pale cinnamon
brown; venter white. Tail hair sparse and evenly dispersed, as in rats.

Mojave Ground Squirrel
Spermophilus mohavensis   PAGE 265
Length: 210–230 mm. Weight: 85–130 g. Uniform pink brown,
lacking spots or stripes. Tail moderately furred and flattened.

Belding's Ground Squirrel *Spermophilus beldingi* PAGE 263
Length: 268–296 mm. Weight: 126–305 g. Gray brown, with
darker brown dorsal saddle; venter and legs pinkish. Body stout;
tail short, ventrally reddish, scantily furred.

Rock Squirrel *Spermophilus variegatus*   PAGE 266
Length: 434–510 mm. Weight: 580–795 g. Variegated gray brown,
with no stripes, patches, or mantle. Tail long and well furred.

Yellow-bellied Marmot *Marmota flaviventris*   PAGE 269
Length: 470–700 mm. Weight: 1.5–4.0 kg. Dorsum russet and
grizzled; venter yellow; white patch on nose. Body heavy; tail short
and moderately bushy; eyes small.

PLATE 9 Ground Squirrels and Marmot

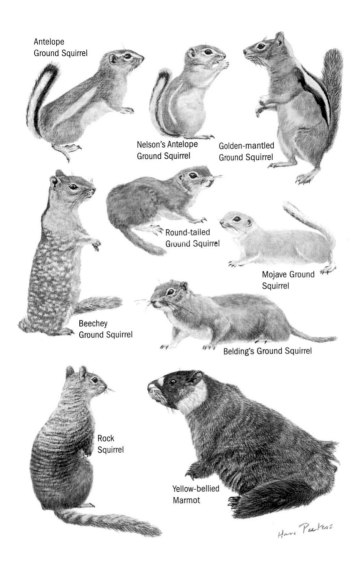

Antelope Ground Squirrel

Nelson's Antelope Ground Squirrel

Golden-mantled Ground Squirrel

Round-tailed Ground Squirrel

Mojave Ground Squirrel

Beechey Ground Squirrel

Belding's Ground Squirrel

Rock Squirrel

Yellow-bellied Marmot

Hans Peeters

PLATE 10   Tree Squirrels

Northern Flying Squirrel *Glaucomys sabrinus*   PAGE 267
Length: 250–310 mm. Weight: 100–200 g. Brownish gray. Forelimbs and hind limbs joined by flaps of skin; tail flattened, with
very long hairs on sides and very short hairs on top and bottom.

Western Gray Squirrel *Sciurus griseus*   PAGE 270
Length: 500–575 mm. Weight: 735–900 g. Silver gray. Dorsum
grizzled salt-and-pepper, sometimes brown; venter white. Tail
large and bushy, rarely yellowish. Grayer than the Fox Squirrel *(S.
niger)*; larger and grayer than the Eastern Gray Squirrel *(S. carolinensis)*, with larger tail.

Douglas's Squirrel *Tamiasciurus douglasii*   PAGE 273
Length: 330–370 mm. Weight: 200–300 g. Dorsum chestnut
brown or olive; venter tawny. Black stripe on sides in summer; ears
tufted in winter.

Eastern Gray Squirrel *Sciurus carolinensis*   PAGE 270
Length: 445–500 mm. Weight: 500–625 g. Grayish. Dorsum yellow
or reddish brown; venter tawny or whitish. Tail long and bushy.

Fox Squirrel *Sciurus niger*   PAGE 272
Length: 475–580 mm. Weight: 590–700 g. Usually russet, though
variable. Tail long and bushy; belly bright yellowish cream; head
often irregularly colored in California.

PLATE 10 Tree Squirrels

Northern Flying Squirrel

Western Gray Squirrel

Douglas's Squirrel

Eastern Gray Squirrel

Fox Squirrel

PLATE 11   Chipmunks

Yellow-pine Chipmunk *Neotamias amoenus*   PAGE 278
Length: 188–202 mm. Weight: 36–50 g. Brown sides and underside of the tail redder than in the Least Chipmunk *(N. minimus)*
or Alpine Chipmunk *(N. alpinus);* tail shorter than in the Least
Chipmunk. Lateral stripes white.

Merriam's Chipmunk *Neotamias merriami*   PAGE 278
Length: 233–277 mm. Weight: 60–82 g. Drab. Stripes light gray;
head dark gray; tail long.

Lodgepole Chipmunk *Neotamias speciosus*   PAGE 284
Length: 197–218 mm. Weight: 30–64 g. Russet brown sides with
white outer dorsal stripes; light patch behind ears.

Long-eared Chipmunk *Neotamias quadrimaculatus* PAGE 281
Length: 200–250 mm. Weight: 52–100 g. Lateral stripes white or
nearly so; underside of tail russet. Ears longer, and with brighter
patches behind, than in the Lodgepole Chipmunk *(N. speciosus)*.

Sonoma Chipmunk *Neotamias sonomae*   PAGE 283
Length: 220–227 mm. Weight: 42–65 g. Lateral stripes grayish;
head stripes brownish or reddish; underside of tail reddish.
Darker than the Redwood Chipmunk *(N. ochrogenys)*.

Least Chipmunk *Neotamias minimus*   PAGE 279
Length: 184–203 mm. Weight: 27–38 g. Dorsal stripes indistinct;
dark stripes wider than light stripes; shoulders and upper back
gray. Underside of tail yellowish, not reddish as in the Yellow-pine
Chipmunk *(N. amoenus)*.

Shadow Chipmunk *Neotamias senex*   PAGE 282
Length: 229–258 mm. Weight: 70–98 g. Dull. Stripes grayish; ears
short and lacking bright patches behind.

Panamint Chipmunk *Neotamias panamintinus*   PAGE 280
Length: 190–214 mm. Weight: 55–70 g. Brightly colored. Rump
gray; tail ventrally reddish.

Uinta Chipmunk *Neotamias umbrinus*   PAGE 284
Length: 210–225 mm. Weight: 52–71 g. Dark, mostly grayish, with
gray on top of head. Lateral stripes white; darker stripes obscure.

Alpine Chipmunk *Neotamias alpinus*   PAGE 277
Length: 166–195 mm. Weight: 30–42 g. Yellowish drab. Grayish
lateral stripes obscure; stripes on head brown; ears short.

PLATE 11   Chipmunks

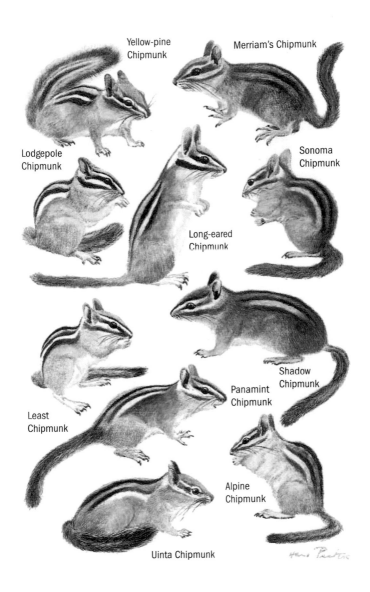

Yellow-pine Chipmunk

Merriam's Chipmunk

Lodgepole Chipmunk

Sonoma Chipmunk

Long-eared Chipmunk

Least Chipmunk

Shadow Chipmunk

Panamint Chipmunk

Alpine Chipmunk

Uinta Chipmunk

PLATE 12   Voles and Pocket Gophers

Red Tree Vole *Arborimus longicaudus*   PAGE 342
Length: 158–186 mm. Weight: 24–27 g. Chestnut red or brick red.
Tail long and well furred; claws curved; ears partly concealed in fur.

California Red-backed Vole
Clethrionomys californicus   PAGE 343
Length: 155–165 mm. Weight: 17–33 g. Chestnut red, but with
blackish dorsal hairs making it darker than the Red Tree Vole *(Arborimus longicaudus);* venter buff or cream. Tail faintly bicolored.

Sagebrush Vole *Lemmiscus curtatus*   PAGE 344
Length: 108–140 mm. Weight: 20–30 g. Light gray. Tail short; soles
of feet furred.

Creeping Vole *Microtus oregoni*   PAGE 349
Length: 129–154 mm. Weight: 18–22 g. Tail less than twice length
of hind foot. Five plantar tubercles.

California Meadow Vole *Microtus californicus*   PAGE 346
Length: 157–211 mm. Weight: 35–72 g. Tail faintly bicolored,
more than twice length of hind foot. Six plantar tubercles.

Long-tailed Vole *Microtus longicaudus*   PAGE 347
Length: 155–221 mm. Weight: 21–56 g. Tail faintly bicolored,
more than twice length of hind foot and more than 33 percent
total length. Six plantar tubercles.

Botta's Pocket Gopher *Thomomys bottae*   PAGE 291
Length: 190–300 mm. Weight: 102–209 g. Usually dull brown
but sometimes yellow, buff, or black for camouflage with surrounding soil.

Mazama Pocket Gopher *Thomomys mazama*   PAGE 292
Length: 175–262 mm. Weight: 90–120 g. Reddish brown. Dorsum especially reddish; black marks on head including patches
around eyes; feet gray. Ears shorter than in the Mountain Pocket
Gopher *(T. monticola).*

Mountain Pocket Gopher *Thomomys monticola*   PAGE 293
Length: 190–220 mm. Weight: 50–105 g. Brown, with white on
back of forefeet and black around ears. Tail has more hairs than in
other species. Ears pointed.

PLATE 12   Voles and Pocket Gophers

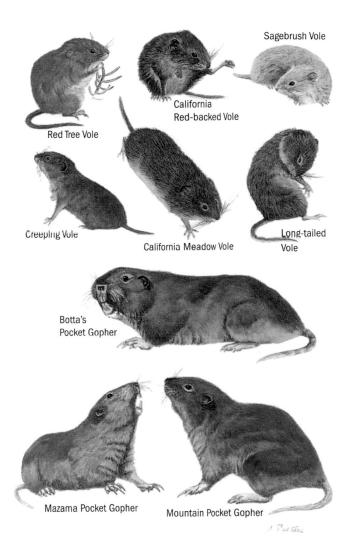

Sagebrush Vole

California
Red-backed Vole

Red Tree Vole

Creeping Vole

California Meadow Vole

Long-tailed
Vole

Botta's
Pocket Gopher

Mazama Pocket Gopher

Mountain Pocket Gopher

PLATE 13   Kangaroo Rats and Kangaroo Mice

Narrow-faced Kangaroo Rat *Dipodomys venustus* PAGE 309
Length: 293–332 mm. Weight: 66–74 g. Very dark overall. Dark
ventral stripe wider than lateral white stripe in terminal half of
tail; five toes on hind foot. Ears larger than in other species, except
Big-eared Kangaroo Rat *(D. elephantinus)*.

Heermann's Kangaroo Rat *Dipodomys heermanni* PAGE 302
Length: 250–340 mm. Weight: 56–74 g. Dark overall. Five toes on
hind foot.

Panamint Kangaroo Rat
Dipodomys panamintinus   PAGE 308
Length: 285–334 mm. Weight: 64–81 g. Ash gray to dark or cinnamon brown. Five toes on hind foot. Darker than the Desert Kangaroo Rat *(D. deserti)*.

San Joaquin Kangaroo Rat *Dipodomys nitratoides* PAGE 306
Length: 211–253 mm. Weight: 39–47 g. Dark overall. Tail short
and buff tipped; four toes on hind foot.

Desert Kangaroo Rat *Dipodomys deserti*   PAGE 301
Length: 305–377 mm. Weight: 83–138 g. Very pale buff. Tail lacks
black tip and usually dark ventral stripe; four toes on hind foot.

Merriam's Kangaroo Rat *Dipodomys merriami*   PAGE 304
Length: 220–260 mm. Weight: 39–52 g. Buff. Four toes on hind
foot; lacks minute fifth toe found in *D. ordii*.

Dark Kangaroo Mouse
Microdipodops megacephalus   PAGE 311
Length: 140–177 mm. Weight: 10–17 g. Dorsum brownish, black-
ish, or gray; venter paler, but with hairs dark or smoky at base.
Upper tail darker than body; hind foot smaller than in *M. pallidus*.

Pale Kangaroo Mouse *Microdipodops pallidus*   PAGE 312
Length: 85–90 mm. Weight: 10–17 g. Dorsum pale pink; venter
white; both dorsal and ventral hairs white at base. Tail no darker
than body, with no black tip.

PLATE 13   Kangaroo Rats and Kangaroo Mice

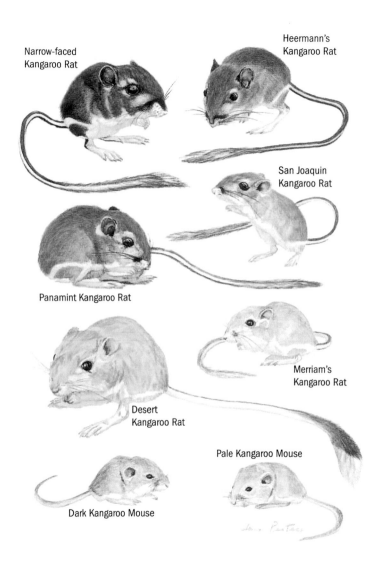

Narrow-faced
Kangaroo Rat

Heermann's
Kangaroo Rat

San Joaquin
Kangaroo Rat

Panamint Kangaroo Rat

Merriam's
Kangaroo Rat

Desert
Kangaroo Rat

Pale Kangaroo Mouse

Dark Kangaroo Mouse

PLATE 14   Pocket Mice

San Joaquin Pocket Mouse

Perognathus inornatus   PAGE 319

Length: 128–163 mm. Weight: 15–18 g. Dorsum brown or buff orange, with dark guard hairs. Lateral line indistinct; hind heel with hair.

Desert Pocket Mouse *Chaetodipus penicillatus*   PAGE 318

Length: 153–221 mm. Weight: 14–20 g. Yellow gray. Fur coarse but not spiny; tail faintly bicolored, often with tuft and crest and ringed; hind heel naked. Ears pointed.

Little Pocket Mouse *Perognathus longimembris*   PAGE 320

Length: 112–138 mm. Weight: 7–10 g. Pink buff. Tail, bicolored or pale, heavily furred on distal third, tufted on distal 3–7 mm; hind heel with hair covering third of sole.

Great Basin Pocket Mouse *Perognathus parvus*   PAGE 322

Length: 160–195 mm. Weight: 16–28 g. Buff or ashy, with silky fur. Tail long, bicolored, with slight apical crest; hind heel furred; ears small with inner lobe at base.

California Pocket Mouse *Chaetodipus californicus* PAGE 316

Length: 190–235 mm. Weight: 16–21 g. Dorsum of mixed yellow and black, with spiny hairs on sides and rump. Hind heel naked. Ears long.

San Diego Pocket Mouse *Chaetodipus fallax*   PAGE 317

Length: 176–200 mm. Weight: 15–18 g. Brownish; spiny hairs white on dorsum and sides, black on rump. Lateral line buff. Tail bicolored and crested; hind heel naked.

Long-tailed Pocket Mouse *Chaetodipus formosus* PAGE 317

Length: 172–211 mm. Weight: 19–25 g. Dark. Tail bicolored and tufted at tip.

Bailey's Pocket Mouse *Chaetodipus baileyi*   PAGE 315

Length: 201–230 mm. Weight: 15–20 g. Gray buff. Tail bicolored and tufted; heel naked.

Spiny Pocket Mouse *Chaetodipus spinatus*   PAGE 319

Length: 164–225 mm. Weight: 19–29 g. Yellow brown; whitish spiny hairs on rump. Tail bicolored and crested; heel naked. Ears small.

PLATE 14 Pocket Mice

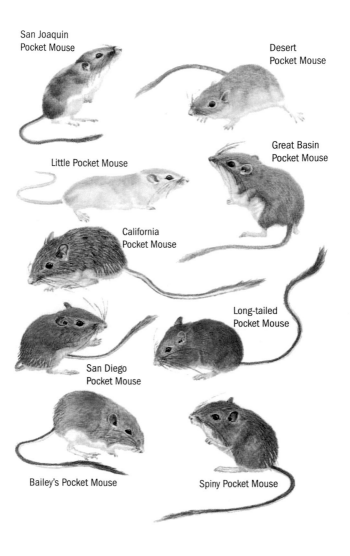

San Joaquin
Pocket Mouse

Desert
Pocket Mouse

Great Basin
Pocket Mouse

Little Pocket Mouse

California
Pocket Mouse

Long-tailed
Pocket Mouse

San Diego
Pocket Mouse

Bailey's Pocket Mouse

Spiny Pocket Mouse

PLATE 15   Rats

Desert Wood Rat *Neotoma lepida*   PAGE 327
Length: 282–305 mm. Weight: 100–190 g. Dorsum pale gray, with
small white patch on throat; venter white, but with hairs gray at
base. Tail bicolored.

White-throated Wood Rat *Neotoma albigula*   PAGE 324
Length: 282–400 mm. Weight: 145–200 g. Grayish. Throat pure
white; tail bicolored.

Bushy-tailed Wood Rat *Neotoma cinerea*   PAGE 325
Length: 335–425 mm. Weight: 220–435 g. Dorsum pale gray; tail
long and bushy.

Brown Rat *Rattus norvegicus*   PAGE 355
Length: 300–475 mm. Weight: 200–525 g. Dorsum brownish; venter gray. Large ears naked; tail bare and scaly; nose blunt. Tail
shorter than in the Roof Rat *(R. rattus)*.

Roof Rat *Rattus rattus*   PAGE 356
Length: 325–435 mm. Weight: 160–350 g. Variable, from light to
dark brown; venter pale bluff. Body slender. When black or nearly
so, called Black Rat.

Hispid Cotton Rat *Sigmodon hispidus*   PAGE 338
Length: 224–365 mm. Weight: 100–225 g. Blackish or dark brown,
with grizzled fur. Browner than the Brown Rat *(R. norvegicus)*,
with tail shorter than in the Roof Rat *(R. rattus)*.

Muskrat *Ondatra zibethicus*   PAGE 351
Length: 456–553 mm. Weight: .7–1.8 kg. Chocolate brown, with
long, glossy guard hairs. Dense, water-repellent underfur fluffy
when dry; hind feet partly webbed with fringe of hairs; tail laterally compressed; eyes small. A rat-sized vole.

PLATE 15   Rats

Desert
Wood Rat

White-throated
Wood Rat

Bushy-tailed Wood Rat

Roof Rat

Brown Rat

Hispid Cotton Rat

Muskrat

Hans Peeters

PLATE 16   Mice

House Mouse *Mus musculus*   PAGE 354
Small eyes. Gray brown to dark brown with paler venter.

Pacific Jumping Mouse *Zapus trinotatus*   PAGE 288
Dark band down dorsum; yellowish sides; tail long.

Harvest Mouse *Reithrodontomys megalotis*   PAGE 337
Small, delicate; upper incisors deeply grooved.

Deer Mouse *Peromyscus maniculatus*   PAGE 335
Gray-brown or reddish. Tail mid-length, sharply bicolored.

Cactus Mouse *Peromyscus eremicus*   PAGE 335
Dorsum pale buff; venter white. Tail long. Sonoran Desert.

Brush Mouse *Peromyscus boylii*   PAGE 332
Tail bicolored, usually longer than head and body.

Piñon Mouse *Peromyscus truei*   PAGE 336
Tail bicolored, usually shorter than head and body.

Cañon Mouse *Peromyscus crinitus*   PAGE 334
Wide pale brown stripe down dorsum.

Parasitic Mouse *Peromyscus californicus*   PAGE 333
Very large mouse; unicolored tail longer than head and body.

Northern Grasshopper Mouse
Onychomys leucogaster   PAGE 329
Stocky. Dorsum sandy; venter white. Tail very short, bicolored.

Southern Grasshopper Mouse
Onychomys torridus   PAGE 330
Stocky. Dorsum buff to reddish; venter white. Tail very short, bi-
colored.

PLATE 16 **Mice**

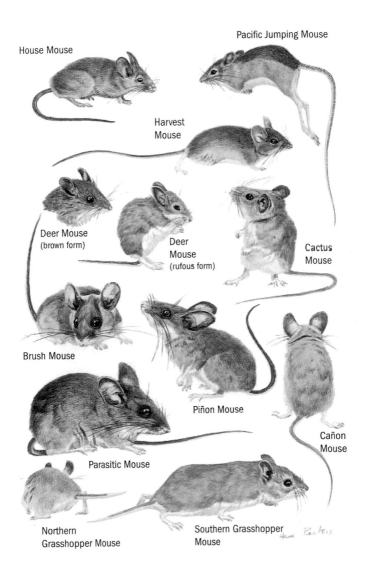

House Mouse

Pacific Jumping Mouse

Harvest
Mouse

Deer Mouse
(brown form)

Deer
Mouse
(rufous form)

Cactus
Mouse

Brush Mouse

Piñon Mouse

Cañon
Mouse

Parasitic Mouse

Northern
Grasshopper Mouse

Southern Grasshopper
Mouse

PLATE 17   Pika and Rabbits

Pika *Ochotona princeps*   PAGE 359
Length: 150–210 mm. Weight: 120–130 g. Dorsum gray buff; tail minute; ears short and broad, projecting above fur.

Snowshoe Hare *Lepus americanus*   PAGE 361
Length: 365–390 mm. Weight: .9–1.1 kg. White in winter, with short, dark-tipped ears and large feet.

Brush Rabbit *Sylvilagus bachmani*   PAGE 365
Length: 300–360 mm. Weight: 560–840 g. Legs short; ears moderately pointed; tail gray; vibrissae black. Females usually larger than males.

Nuttall's Cottontail *Sylvilagus nuttallii*   PAGE 366
Length: 335–392 mm. Weight: 650–980 g. Large cottontail; brown, not gray, dorsum; inside of ears furred.

Audubon's Cottontail *Sylvilagus audubonii*   PAGE 364
Length: 370–400 mm. Weight: 750–900 g (males), .88–1.25 kg (females). Dorsum gray; legs long; fur short. Cotton tailed. Ears larger, and hind feet more slender, than in Nuttall's Cottontail *(S. nuttallii);* ears and feet also more sparsely furred.

Black-tailed Jackrabbit *Lepus californicus*   PAGE 362
Length: 495–550 mm. Weight: 1.5–3.6 kg. Legs long; ears very long; tail black or partly black; rump partly black dorsally and grayish ventrally.

Pigmy Rabbit *Brachylagus idahoensis*   PAGE 368
Length: 230–295 mm. Weight: 350–460 g. Legs and ears very short; buff under tail. Very short legs and ears, with buff under the tail. Females usually larger than males; smallest rabbit in United States.

White-tailed Jackrabbit *Lepus townsendii*   PAGE 363
Length: 545–650 mm. Weight: 2.15–4.3 kg. Gray in summer, white in winter. Tail white year round; hind foot more than 140 mm long.

PLATE 17   Pika and Rabbits

Pika

Snowshoe Hare

Brush Rabbit

Nuttall's
Cottontail

Black-tailed
Jackrabbit

Audubon's Cottontail

Pigmy Rabbit

White-tailed
Jackrabbit

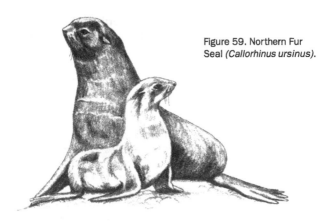

Figure 59. Northern Fur Seal *(Callorhinus ursinus).*

cows and pups migrate; they commonly get as far south as our coast, although they are seldom seen near shore.

FOOD: Foraging at night, the Northern Fur Seal may take advantage of the upward nocturnal movement of many fish and squid, its staples. In addition to several kinds of squid and cuttlefish, this pinniped takes many species of rockfish, herring, and mackerel.

REPRODUCTION: A breeding unit comprises one bull and a group of cows. Like other sea lion bulls, the Northern Fur Seal bulls arrive at the rookeries first and set up territories. They announce their presence to the cows, which arrive just before the pups are born. Mating occurs shortly after the birth of the pups. By fall the cows (already long pregnant) and pups have begun their migration.

COMMENTS: Because of the great commercial importance of this sea lion, its habits have been rather intensively studied. The Northern Fur Seal, together with the Sea Otter *(Enhydra lutris),* was in large measure responsible for early exploration of the North Pacific coast. Its habit of polygyny results in a large surplus of nonbreeding, bachelor bulls, which originally formed the basis of the sealing industry. Because these bulls do not reproduce, many can be removed without affecting the annual productivity of the population. In the eighteenth and nineteenth centuries, however, these furbearers were taken in a totally indiscriminate manner, with both sexes being killed throughout the year. Consequently there were drastic declines in both species. A treaty among Canada, Japan, the Soviet Union, and the United

States not only limited the numbers of Northern Fur Seals that could be killed but confined the take to the bachelor bulls. Commercial sealing no longer occurs.

NORTHERN SEA LION or STELLER SEA LION

Eumetopias jubatus

Pl. 7

DESCRIPTION: A large sea lion: adult males reach more than 3 m in length, females reach about 2 m. The pups are black, but the adults are straw, yellow brown, or even whitish when submerged. The ears are small but distinct. Weight: up to 1,000 kg (males), 250 kg (females). Dentition: 3/2, 1/1, 4/4, 1/1.

DISTRIBUTION: Breeds from the northern Channel Islands north to the Aleutians and Pribilofs and west to Kamchatka; the islands in the Sea of Okhotsk, north of Hokkaido; and intervening islands. Año Nuevo Island has a breeding colony. This species is sometimes seen on the California coast but is less common inshore than the California Sea Lion *(Zalophus californianus)*. After breeding, southern populations move to the north, whereas northerly populations tend to move south.

FOOD: We have little specific knowledge on the food of this offshore species; it eats squids, crabs, bivalves, and a variety of medium-sized fish.

REPRODUCTION: Females collect into groups of 20 to 30 within the territory of a large bull. Bulls defend their territories, and battles may be vigorous and bloody. Cows apparently have no loyalty toward either a particular bull or his plot of rock but move about from one harem to another. When breeding, bulls do not eat for some six weeks, but cows apparently feed at night. A single pup is born in late spring or early summer; mating follows within about a week. Nursing may last for about a year. Females may breed in their third year, and bulls do not breed until several years later.

STATUS: This species is federally listed as threatened.

CALIFORNIA SEA LION

Zalophus californianus

Pl. 7, Fig. 60

DESCRIPTION: A rather dark brown or blackish sea lion, darker when wet. Males are darker than females. Males reach 2.5 m in length, females reach some 2 m. The ears are short and pointed;

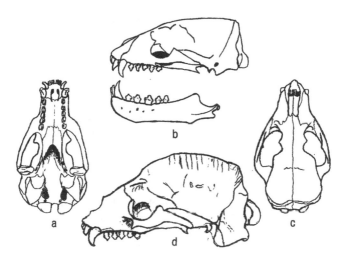

Figure 60. Skull of the California Sea Lion *(Zalophus californianus)*:
(a) ventral view, female; (b) lateral view, female; (c) dorsal view, female;
(d) adult male, showing sagittal crest.

males have prominent foreheads. Weight: 250 kg (males), 100 kg (females). Dentition: 3/2, 1/1, 4/4, 2–1/1; the molars and premolars are nearly triangular (see fig. 60b).

DISTRIBUTION: Breeds from the Channel Islands south to Baja California and occurs in the Gulf of California. It is most commonly observed during the nonbreeding period, when individuals may move north as far as British Columbia and are found in many areas along our coast. This sea lion habitually seeks certain beaches where it is not disturbed. It is known also to ascend coastal rivers. Separate populations (subspecies) occur on the Galápagos Islands and in the Sea of Japan.

FOOD: This species takes squid and a variety of fish.

REPRODUCTION: This sea lion is known for its large and conspicuous aggregations during the breeding season. Females assemble on offshore islands, and bulls move in. Bulls are territorial, but females move from one harem to another. A single pup is born in early summer; mating follows within 10 days or two weeks. Thus the female may be pregnant for much of the lactation period, which may last for the better part of a year. Gradually, during summer, the pups venture more and more into the sea. From the end of summer until the following spring, mothers and young may move in small or large groups.

COMMENTS: Although this species' predation on commercial fisheries may be negligible, it is known to damage salmon in gill nets and to damage the nets themselves by becoming entangled in them.

This is the animal famous as the "trained seal." Usually only the female is used for this purpose. In the wild, this sea lion frequently can be heard before it is seen. Its loud barking is a familiar sound that aids in distinguishing it from the Steller Sea Lion (*Eumetopias jubatus*).

Seals (Phocidae)

Seals have smaller, less flexible limbs than sea lions. The forelimbs are furred. In *Phoca*, the limbs have some toenails or claws; in *Mirounga*, the claws are rudimentary or absent. Ear openings are present, but external pinnae are lacking. The sexes are similar in size in most species.

Seals are perhaps more adapted to an aquatic life than sea lions and fur seals. Their body form is more compact, with the neck not apparent, and their limbs are smaller and weaker. Consequently, seals can do little more than wriggle along the rocks. Their forelimbs lie adpressed to the sides of the body when swimming, and propulsion results from movement of the hind flippers and tail region. The testes are housed permanently within the body cavity, as is also the case with cetaceans.

NORTHERN ELEPHANT SEAL   *Mirounga angustirostris*
Pl. 7, Fig. 61

DESCRIPTION: A very large seal: males reach more than 5 m in length, females 3 m. Adults are gray or brown; the young are black. The immense size and the seal body form identify this species. The male is also distinguished by its pendulous but inflatable snout. Weight: up to 3,500 kg (males), 900 kg (females). Dentition: 2/1, 1/1, 4/4, 1/1.

DISTRIBUTION: Breeds on coastal islands from Baja California north to the Farallon Islands and Año Nuevo. It ranges north to Alaska during the nonbreeding season.

FOOD: This species apparently eats bottom fish, such as ratfish and cusk eels, as well as squid.

Figure 61. A month-old Northern Elephant Seal *(Mirounga angustirostris)* weaner (foreground), having been left by its mother, and a doomed so-called super weaner grossly overfattened by nursing from more than one female.

REPRODUCTION: This seal differs from most in that a male dominates a group of breeding females. Adults gather on coastal islands in December, and the single pup is born within days, usually in early January. Nursing is completed in about a month, during which period the pup more than triples its natal weight as a result of the fat-rich milk characteristic of many pinnipeds. Until the pup is weaned, the mother does not feed; by the time lactation ceases, therefore, she has lost a great deal of weight. After her pup is weaned, she mates; gestation lasts 11 months (including the delay in implantation), until she returns once again to the rookery. Mating takes place on land, in contrast to the aquatic sexual pursuit characteristic of most seals. This may be because constant territorial defense by the bulls prevents their entering the water.

STATUS: This species is fully protected in California.

COMMENTS: This seal is known to dive at least 200 m.

Like the Guadalupe Fur Seal *(Arctocephalus townsendi),* the Northern Elephant Seal was once very scarce. Its large size made it a source of a great amount of oil, and commercial exploitation almost caused its extinction. A small population remained on Guadalupe Island and prospered, and breeding populations have now spread northward. With continued protection, we can expect this northward movement to continue.

One other species of elephant seal, the Southern Elephant Seal *(Mirounga leonina),* is found in the Southern Hemisphere.

HARBOR SEAL   *Phoca vitulina*
Pl. 7

DESCRIPTION: This seal is extremely variable in color, from nearly white to almost black, though usually some shade of brown or gray and spotted. It is chunky in shape and appears neckless. Males reach up to 1.7 m in length and more than 130 kg in weight; females are slightly smaller.

DISTRIBUTION: Occurs commonly along the coasts of California, Oregon, and Washington, frequently close to shore, and sometimes on isolated beaches; it is found in San Francisco Bay. It may forage close to shore and commonly hauls out on rocky islets close to beaches. It is mostly coastal and is known to enter rivers for 160 km or more. This seal is well distributed about the northern seas of both hemispheres, although there are local variations. Probably the Atlantic and Pacific populations were in contact until the emergence of the Isthmus of Panama some 3 million years ago.

FOOD: The Harbor Seal takes bivalves, crabs, squid, octopus, and a wide variety of medium-sized fish, including herring, flounders, and cod.

REPRODUCTION: A single pup is born in spring. As in most seals, the period of nursing is rather brief; mating occurs in the water after the pup is weaned. A delay in implantation regulates development so that the next pup is born just about one year later. Breeding assemblages may contain a smaller number of both males and females than nonbreeding groups.

COMMENTS: Salmon make up a very small amount of its natural diet, but the Harbor Seal may damage gill nets trying to remove captured fish.

This little seal seems to have remained common throughout the encroachment of civilization. It tends to be timid and enter the water as people approach. Its small size is not a great inducement to commercial exploitation.

CETACEA

These large mammals are almost totally adapted for aquatic life, except that they retain lungs and breathe air. Because of their great weight, they generally cannot breathe without the support of water; their mass collapses their lungs, and they quickly die. They have streamlined bodies, with the tail broadened horizontally into two lobes, or flukes. They always have forearms in the form of paired fins, or flippers, but never have external hind limbs. Some species have a dorsal fin, which may be small or large.

Aspects of streamlining include the absence of hair and external ears. The nostrils have moved, over evolutionary time, to the front or top of the head. The skin is thin but covers a thick insulating layer of fat, the so-called blubber, which was widely processed for use in oil lamps before petroleum was discovered. The nipples are normally retracted into long grooves and protrude only during periods of suckling. Scarcely any external mammary swelling occurs during lactation. The expulsion of milk is under the conscious control of striated muscles. The testes lie within the body, and the penis can be entirely retracted within the body.

The cetaceans are divided into two distinct groups, sometimes called suborders. Because they seem so different, some researchers have felt that they had separate origins, but DNA analysis indicates a common origin.

The Odontoceti, or toothed whales, have at least two, and sometimes many, well-developed teeth in one or both jaws. Usually the mandibles are fused for part of their anterior length and bear short, peglike teeth. The nostrils are joined to form a single blowhole, which is placed dorsally but sometimes near the front of the head. The nasal passages have diverticula (blind side pockets), which may produce sound.

The Mysticeti are the whalebone or baleen whales. They lack teeth (except as embryos), but many long, thin plates of a stiff substance called baleen hang down from their upper jaws. The mandibles are loosely connected anteriorly, which allows them some independent movement. Paired nostrils or blowholes are on top of the head.

Most of the mysticetes, except for the Balaenidae, and some of the toothed whales have grooves along the outer surface of the throat, parallel to the floor of the mouth, allowing the mouth cavity to expand. This expansion is under conscious control and makes the oral cavity, together with the tongue, which functions as a piston, into a gigantic suction device. Suction, together with filter feeding, is the means by which baleen whales obtain food. Suction feeding is also known in some odontocetes. Odontocetes usually feed on squid and schooling fish; they frequently feed at night, apparently capturing their food by echolocation, which allows them to locate prey at depths to which light does not penetrate.

Although vocalization is known in many cetaceans, prolonged songs are characteristic only of the Mysticeti. Such songs are heard only in winter, during the breeding period, and presumably have some sexual significance.

Details of reproduction are not thoroughly known in all whales, though early whalers did keep some records of the breeding patterns of the beasts they killed. At least some whales are photosensitive and experience testicular growth seasonally as the nights shorten and the days lengthen. Love play, in which females indicate receptivity, precedes mating, perhaps because in a watery environment odors are not associated with estrus and are probably not detected. Mating has been observed rarely. Whales and dolphins are among the few mammals in which coition is venter to venter. It takes place in either a vertical or a horizontal position, with the heads projecting above the surface, and may or may not include pelvic thrusts. It lasts from 10 to 30 seconds. Curiously, in odontocetes embryos implant in the left uterine horn, and in mysticetes in the right. Embryonic growth is extremely rapid, considering the size of the newborn. It is most rapid in the Mysticeti, which mate in lower latitudes, migrate to polar regions, and give birth approximately a year later. The infant also grows very rapidly, partly because of the very rich milk of the mother.

The earliest fossils of whales, dating from the Eocene, are known from India and Pakistan. These initial whales apparently evolved in freshwater. Their teeth were heterodont, differentiated into incisors, canines, premolars, and molars, whereas those of modern whales are homodont. It is generally accepted, on the basis of fossil evidence, that whales arose from artiodactyls, early carnivorous ungulates that arose in the Paleocene. Some molecular evidence, however, which is not compatible with fossil evidence, suggests that whales have their closest living relatives in the hippopotamuses, and that hippos and whales are closer to each other than either is to the Artiodactyla. This is an area of active research among molecular biologists and paleontologists.

Sperm Whales, Dolphins, and Porpoises (Odontoceti)

The Odontoceti, or toothed whales, are mostly small species. The huge Sperm Whale *(Physeter catodon)* is a conspicuous exception and has long been of commercial significance. The dolphins and porpoises are numerous in both species and number; they are extremely varied in distribution and, presumably, in biology. Seven to nine families are in Odontoceti, depending upon which classification you consult. Four of these families have been recorded from our coast.

Dolphins and Allies (Delphinidae)

This is a large, loosely knit family of variable form and size. The first two vertebrae are fused, and most species have a dorsal fin. Most also have a distinct beak. The mandibles are frequently fused anteriorly for up to one-quarter of the length of the jaw. Usually there are many similar conical teeth. The snout may be bluntly rounded or gradually pointed, or a distinct, slender beak may be clearly set apart from the brow area. In some countries, dolphins are commonly captured for meat. Thirty-two species exist, of which 11 have been observed in our waters. Most are pelagic, living in the open sea.

COMMON DOLPHIN   *Delphinus delphis*

Fig. 62

DESCRIPTION: A small (2.5 m) dolphin with a distinct beak. It is black or brownish dorsally with a narrow dark line from the corner of the mouth to the base of the flipper. The dorsal fin is falcate (sickle shaped), pigmented, and broader and shorter than the flippers. Both jaws have many similar conical teeth.

Figure 62. Common Dolphin *(Delphinus delphis).*

DISTRIBUTION: Found throughout the seas of the world, except in polar regions. It is more common off the southern coast of California than farther north, but its range extends north to Alaska.
FOOD: This dolphin eats a variety of fish and squid.
REPRODUCTION: Gestation takes 10 to 13 months; young are born in summer, and there is a long period of lactation. Pregnancies recur every two to three years.
COMMENTS: This species pursues its prey at some depth (perhaps 10 to 25 m) at night.

This is one of the most familiar small dolphins and one very likely to be seen swimming at the prow of ships. It is usually not found in shallow waters. It may occur, sometimes together with other kinds of dolphins, in herds in the hundreds or even thousands.

PILOT WHALE   *Globicephala macrorhynchus*
Fig. 63
DESCRIPTION: A medium-sized dolphin reaching a maximum length of about 6 m. A lightly pigmented saddle sits just behind the dorsal fin; that fin is falcate and placed in the anterior third of the body. The flippers are slender, pointed, and slightly curved; each is about three times the length of the mouth. The head is rather large, and the snout is bluntly rounded. Each jaw contains seven to 12 pairs of teeth. The Pilot Whale is similar in some aspects to the False Killer Whale *(Pseudorca crassidens),* but the False Killer Whale has a dorsal fin just slightly anterior to the midpoint of the body and relatively short flippers. Risso's Dol-

Figure 63. Pilot Whale *(Globicephala macrorhynchus).*

phin *(Grampus griseus)* differs in having a dorsal fin higher than its basal length.

DISTRIBUTION: Found most commonly in the southern waters of the coastal states and commonly in Puget Sound. Its range extends worldwide; the same or similar species occur in the western Pacific and south of the equator. *Globicephala melas* (also known as *Globicephala melaena),* a widespread species, has been recorded once from California.

FOOD: It eats mostly squid and cuttlefish, as well as some fish.

REPRODUCTION: Mating seems to be concentrated in summer; gestation occupies 15 to 16 months. The single young is not weaned until nearly two years after birth but begins to capture its own food before that time.

COMMENTS: The Pilot Whale presumably gets its name from the frequency with which entire schools, or pods, swim ashore as a unit. Some cultures have exploited this tendency by herding the whale ashore. It provides a substantial amount of red meat. It is also known as the Blackfish or Pilot Blackfish.

RISSO'S DOLPHIN *Grampus griseus*
Fig. 64

DESCRIPTION: A gray or dark (but not black), medium-sized (4 m) dolphin with a blunt head. The skin is commonly scarred, presumably from fighting. Young are usually darker than adults, apparently because the accumulation of scars results in a loss of pigmentation. This dolphin has a falcate dorsal fin higher than its basal length and located just anterior to the middle of its body. The flippers are moderate, falcate, and longer than the mouth. The head is rather large and the snout bluntly rounded.

Figure 64. Risso's Dolphin (Grampus griseus).

The lower jaw bears two to seven pairs of teeth, the upper jaw none.

DISTRIBUTION: Occurs in warm and temperate regions of the world but is usually pelagic, not coastal. It is found along the west coast of the United States but is not common north of southern California.

FOOD: This dolphin feeds largely on squid.

REPRODUCTION: Little is known of its reproductive pattern.

WHITE-SIDED DOLPHIN   *Lagenorhynchus obliquidens*
Fig. 65

DESCRIPTION: A small (2 m) dolphin with graceful lines and a falcate fin placed midway on the back. It has a conspicuous yellow brown or green black dorsum, a mostly white venter, and a narrow gray stripe on either side. The dorsal fin is usually largely white except on the anterior margin. The beak is short. The upper and lower jaw each have 18 to 20 conical teeth.

DISTRIBUTION: Commonly found along the California coast and may occur in large numbers north to the Straits of Juan de Fuca. This dolphin migrates south in winter and north in spring. It occurs throughout the northern Pacific Ocean.

FOOD: It subsists on squid and small fish.

REPRODUCTION: The calf is born in summer, but little is known of the details of the breeding cycle.

Figure 65. White-sided Dolphin (Lagenorhynchus obliquidens).

NORTHERN RIGHT WHALE DOLPHIN *Lissodelphis borealis*
Fig. 66

DESCRIPTION: A medium-sized (3 m) dolphin, black dorsally and white ventrally. The young is duller than the adult and lacks the characteristic contrasting black-and-white pattern. This dolphin is distinctive in lacking a dorsal fin and in having a lower jaw that protrudes beyond the snout. Its body is slender, and its flukes are small. The upper and lower jaws bear many small conical teeth.

DISTRIBUTION: Occurs across the Pacific, south at least to Baja California, but is not common close to shore. It moves to offshore waters in the North Pacific in summer.

FOOD: This dolphin feeds on squid and small fish.

REPRODUCTION: Little is known of its breeding pattern.

COMMENTS: It may swim in schools of hundreds, sometimes together with other species.

Figure 66. Northern Right Whale Dolphin (Lissodelphis borealis).

KILLER WHALE   *Orcinus orca*
Fig. 67

DESCRIPTION: Perhaps the most familiar of whales, easily recognized by its conspicuous black-and-white pattern and large white blotch over the eye. Behind the dorsal fin is a lightly pigmented saddle. In males, the dorsal fin is triangular, stands erect, and is about 2 m high; in females it is lower and falcate. Males may reach 9 m in length and females reach 7 m. The head is rather large, and the snout is bluntly rounded. Both jaws have a series of large, similar teeth, 10 to 12 to a side.

DISTRIBUTION: Found in all seas. It is not uncommon along the coastal states, is common in Puget Sound, and ranges far into polar waters.

FOOD: This whale is unusual in that it captures birds and mammals. It is often observed killing and eating seals and small dolphins, but squids and many fish are also important food items. Occasionally it has been known to kill large baleen whales as well as Sperm Whales *(Physeter catodon)*.

REPRODUCTION: Breeding is apparently nonseasonal, but little is known of this whale's reproductive biology.

Figure 67. Killer Whale *(Orcinus orca)*.

FALSE KILLER WHALE   *Pseudorca crassidens*
Fig. 68

DESCRIPTION: An almost entirely black whale some 6 m long. Its moderate-sized dorsal fin is placed slightly in front of the mid-

Figure 68. False Killer Whale *(Pseudorca crassidens)*.

point of the body. Its flippers are short, not longer than the mouth, and have a swelling on the anterior margin. The head is rather large, and the snout is bluntly rounded. Each jaw has eight to 11 conical teeth.

DISTRIBUTION: Seen mostly in the tropics, but reaches the northern shores of California and perhaps farther north. Whether migratory or nomadic, it sporadically appears in some numbers. It occurs worldwide in warm oceans.

FOOD: This species feeds mostly on cuttlefish and squid but also on bonito, snappers, and jacks.

REPRODUCTION: Apparently mating is nonseasonal. Very little is known of this whale's breeding pattern.

PACIFIC SPOTTED DOLPHIN   *Stenella attenuata*
or SPINNER DOLPHIN

Fig. 69

DESCRIPTION: A smallish dolphin of more or less uniform gray or dull color with poorly defined clusters of spots about the eye and on the sides behind the dorsal fin. The fin and flippers are of moderate size. This species has up to 160 small conical teeth. The mandibles are fused for one-quarter of their length.

DISTRIBUTION: Occurs commonly as far north as Baja California;

Figure 69. Pacific Spotted Dolphin *(Stenella attenuata)*.

known from southern California. This is a subtropical and tropical dolphin and is both coastal and pelagic.

FOOD: It subsists mostly on small fish of many species and also takes squid and cuttlefish.

REPRODUCTION: Calving is nonseasonal; gestation is estimated to take approximately one year.

BLUE AND WHITE DOLPHIN *Stenella coeruleoalba*
Fig. 70

DESCRIPTION: A strikingly marked black (or dark) and white dolphin; distinctive stripes run from the eye to the base of the flipper and from the eye to the belly. This dolphin is about 3 m long. Its flippers are short and slightly curved; its falcate fin is placed midway on the body. Its beak is distinct but short.

DISTRIBUTION: Characteristic of warm waters; apparently prefers pelagic waters. It is known from strandings along the coasts of Washington, Oregon, and California but is more common from Baja California southward. The same or a similar species occurs in the western Pacific.

FOOD: It eats small fish, squid, and pelagic crustaceans.

REPRODUCTION: Mating is nonseasonal and followed by a gestation of about one year. The young takes milk for more than one year, and reproduction may occur about every three years.

Figure 70. Blue and White Dolphin *(Stenella coeruleoalba)*.

COMMENTS: The Blue and White Dolphin, also called the Striped Dolphin, is sometimes seen in large numbers, from hundreds to one or two thousand.

ROUGH-TOOTHED DOLPHIN   *Steno bredanensis*
Fig. 71

DESCRIPTION: A small (2.5 m) dolphin, dark gray or black dorsally and pink and spotted ventrally. The dorsal fin is conspicuous and placed midway along the back. This dolphin has a long, slender snout and 24 to 32 teeth per side in each upper and lower jaw. The teeth have rough crowns and fine vertical ridges.

DISTRIBUTION: Occurs widely in the warm seas of the world. It is apparently rare in California but has been found on the coast of Marin County.

FOOD: It subsists on small fish, squid, and pelagic octopuses.

REPRODUCTION: Its reproductive patterns are unknown.

COMMENTS: This cetacean is poorly known and appears nowhere

Figure 71. Rough-toothed Dolphin *(Steno bredanensis)*.

to be common. It has been observed in small groups, sometimes with other species. It has been seen to swim slowly near the surface; this may be a feeding maneuver.

BOTTLENOSED DOLPHIN   *Tursiops truncatus*
Fig. 72

DESCRIPTION: A dull-colored dolphin without a sharp distinction between dorsal and ventral pigmentation. It is about 4 m long. The beak is distinct and rather stout. The fin is falcate and placed at or slightly anterior to the midpoint of the back. This species has about 20 conical teeth in each of the four jaws.

DISTRIBUTION: Occurs in small groups along our southern coast and in most of the coastal waters of the Pacific Ocean. This species is found virtually worldwide and known under several names. It is characteristic of warm and temperate coastal waters.

FOOD: It takes a broad spectrum of marine fish and invertebrates, especially squid and shellfish.

REPRODUCTION: One young is born after a gestation lasting approximately a year. Mating may occur every second or third year. Weaning is a gradual process, usually complete in about a year.

Figure 72. Bottlenosed Dolphin *(Tursiops truncatus)*.

Porpoises (Phocoenidae)

These small cetaceans have blunt, rounded heads. Their teeth are laterally compressed and spatulate (spadelike). Presumably this shape has a dietary significance, but its special function remains a mystery. Some species, including the two known from California, have a well-developed dorsal fin.

HARBOR PORPOISE   *Phocoena phocoena*
Fig. 73

DESCRIPTION: A small (1.5–1.8 m) cetacean with a rather small dorsal fin placed in the middle of the back. The dorsum is black, and the venter is white or whitish. Weight: 45–55 kg.

DISTRIBUTION: Found worldwide in temperate or cold coastal waters. It migrates to the southern part of its range in fall. It is common along the California coast south to Monterey and sometimes enters San Francisco Bay.

FOOD: This porpoise takes a broad spectrum of small fish such as herring, sardines, small cod, and whiting and also captures squid and crustaceans.

REPRODUCTION: A single young may be born every other year; mating sometimes follows immediately after birth of the young, so that lactation and embryonic development are concurrent.

COMMENTS: Though it is frequently abundant and sometimes assembles in schools of 50 to 100 or more, this little cetacean is shy and not easily observed. Throughout the world it seems to be declining, for unknown reasons. It is seldom pursued by humans but may sometimes become entrapped in gill nets. It is known to

Figure 73. Harbor Porpoise *(Phocoena phocoena)*.

be preyed upon by the Great White Shark *(Carcharodon carcarias)* and the Killer Whale *(Orcinus orca)*.

DALL'S PORPOISE *Phocoenoides dalli*
Fig. 74

DESCRIPTION: A conspicuous black-and-white porpoise about 2 m in length. The triangular dorsal fin is sometimes whitish at the tip. A large, clear-cut white patch covers the middle third of the sides. Weight: 200 kg or more.

DISTRIBUTION: Found from Baja California to Alaska. It is frequently encountered in interisland waters between Washington and Alaska but occurs also in deeper seas westward to Japan and the coast of Eurasia.

FOOD: It eats mostly cuttlefish and squid, as well as small fish.

REPRODUCTION: Mating is apparently not seasonal. This porpoise bears a single young.

COMMENTS: This is one of the small cetaceans that swim next to moving ships. It is also one of the fastest species, said to attain speeds of up to 30 knots. In Asian waters it is captured for food.

Figure 74. Dall's Porpoise *(Phocoenoides dalli)*.

Sperm Whales (Physeteridae)

This family includes the Sperm Whale *(Physeter catodon)* and the much smaller Pigmy Sperm Whale *(Kogia breviceps)*. The head is blunt, and the jaw is ventral. The mandibles are fused for about half their anterior length. All members of this family have distendable throat grooves. Only the left blowhole functions in breathing; the right breathing passage produces sound. The lower jaw has eight to 25 teeth per side, depending on the species; the upper jaw lacks teeth.

PIGMY SPERM WHALE   *Kogia breviceps*

DESCRIPTION: In many respects a miniature edition (about 3 m long) of the Sperm Whale *(Physeter catodon)*. It is dark above, lighter ventrally. The dorsal fin is small and located posterior to the midpoint of the back. The head is smaller, relatively, than that of the Sperm Whale *(Physeter catodon)*. The lower jaw has about 15 curved teeth.

DISTRIBUTION: Found worldwide except in polar waters.

FOOD: It feeds on marine invertebrates, especially crabs, squid, and octopuses.

REPRODUCTION: Little is known of the breeding of this uncommon species. Females have been known to be pregnant and nursing at the same time.

COMMENTS: Little is known of this species' occurrence except when it is found dead on beaches. It is seldom seen but apparently not scarce.

The Dwarf Sperm Whale *(Kogia simus)* was long confused with the Pigmy Sperm Whale *(K. breviceps)*. The Dwarf Sperm Whale's dorsal fin is higher than the Pigmy Sperm Whale's and located midway along the back. Other distinctive aspects of its biology are not known. The Dwarf Sperm Whale has been recorded from the coastal waters of California.

SPERM WHALE   *Physeter catodon*
Fig. 75

DESCRIPTION: The Sperm Whale may reach up to 18 m in length, with males much larger than females. It is distinctive for its large, bulbous head, which has a vertical front margin and an oil-filled spermaceti organ. The function of this cavity is not clearly established, though it may be involved in hearing or echolocation. Its posterior boundary is a vertical bony crest, which is curved in a way that directs reflected sound to a single point. The mandibles are fused for about half their length and bear a row of about 20 teeth.

Figure 75. Sperm Whale *(Physeter catodon)*.

DISTRIBUTION: Found worldwide but favors warmer seas. Females seem to avoid polar regions; males are prone to wander widely.

FOOD: The Sperm Whale is one of the most specialized cetaceans in its selection of food. It has a penchant for squid of various sizes, including the Giant Squid (*Architeuthis* spp.), but also takes octopuses and a variety of medium-sized fishes.

REPRODUCTION: Mating usually occurs in summer. Gestation lasts some 15 to 16 months. Because the Sperm Whale does not migrate annually between feeding and calving grounds, there is no need to restrict gestation to a period of slightly less than a year. The calf is some 4 m long at birth and may nurse for up to two years. Births are four to five years apart for a given female.

STATUS: This species is federally listed as endangered.

COMMENTS: The Sperm Whale sometimes assembles in large numbers, especially females with young. It frequently hunts in water 500 m and deeper; in the region of perpetual darkness, it locates its prey by the sounds they produce and by echolocation. Food in its stomach generally lacks tooth marks and apparently is taken in by suction. The primary function of the teeth may be defense.

The horny beaks of squid accumulate in the Sperm Whale's stomach and are eventually regurgitated as a floating mass called ambergris. This material is used as a base to carry fragrances and has long been of great value in the perfume industry. Despite its replacement by more readily available materials, ambergris still has value. Sperm Whales were widely hunted.

Beaked Whales (Ziphiidae)

These are medium-sized whales, 4 to 12 m in length. The dorsal fin is small and placed well behind the midpoint of the body. The snout is long and slender; in some species, the lower jaw projects beyond its tip. The throat has two longitudinal grooves that diverge posteriorly. The number of teeth varies, but usually there are only one or two on each side of the lower jaw and none in the upper jaw. The teeth are not used for holding food but apparently function in male-male combat. In females, they frequently do not erupt through the gum.

This distinctive family is poorly known. Its members seem to be more pelagic than coastal, and most species are known from stranded carcasses. Some are rather large and have long been taken commercially. Beaked whales feed largely on squid, which are generally found intact, and without tooth marks, in their stomachs. Dr. John Heyning of the Los Angeles County Natural History Museum and Dr. James Mead of the U.S. Museum of Natural History determined that beaked whales ingest squid and bottom-dwelling invertebrates by suction. The whale can retract its tongue, which acts as a piston. This action, together with expansion of the mouth cavity (aided by the throat grooves), reduces pressure within the mouth, permitting rapid ingestion of food by suction. These workers suggested that some other toothed whales, perhaps the Sperm Whale *(Physeter catodon)* and Harbor Porpoise *(Phocoena phocoena)*, are also suction feeders.

A beaked whale *(Mesoplodon hectori)* recently stranded on a southern California beach was previously known only from the Southern Hemisphere and probably does not regularly occur in our waters.

BAIRD'S BEAKED WHALE *Berardius bairdii*

DESCRIPTION: This whale is dark or blackish dorsally and somewhat lighter ventrally, frequently with long scars. It reaches up to 12 m in length, with the females somewhat larger than the males. The dorsal fin is low, much longer at the base than it is high, and placed well behind the midpoint of the body. The snout is narrow and round in cross section, hence the alternate name "Bottlenosed Whale." Males have a pair of large teeth in each lower jaw, with the front teeth placed in front of the snout.

DISTRIBUTION: Generally occurs in the North Pacific, from Alaska to California, in both temperate and cold waters. It is apparently mostly pelagic but is occasionally stranded on the coast.

FOOD: This whale largely eats squid but also takes other deep-water invertebrates and fish.

REPRODUCTION: Its 17-month gestation is perhaps the longest of any whale. Mating occurs in fall and parturition two springs later.

COMMENTS: Because of its great size, Baird's Beaked Whale has occasionally been the object of commercial pursuit. In Japan it is one of several whales sought for their flesh.

STEJNEGER'S BEAKED WHALE   *Mesoplodon stejnegeri*

DESCRIPTION: A blackish whale; the beak may be white. This species reaches up to 5 m in length. The low dorsal fin is placed slightly behind the midpoint of the body. In males, the single pair of stout teeth protrude outside the snout, not in front of it.

DISTRIBUTION: Known from several strandings on our coast but apparently more frequent in colder waters. It occurs across the North Pacific west to the Sea of Japan. *Mesoplodon carlhubbsi,* which has been found on the California coast, may be the same species.

FOOD: Unknown.

REPRODUCTION: Unknown.

CUVIER'S BEAKED WHALE   *Ziphius cavirostris*
Fig. 76

DESCRIPTION: A darkish or black whale, frequently with numerous scars; adults have white faces. This species reaches up to 7 m in length. The dorsal fin is triangular and placed in the rear third of the back. The lower jaw protrudes beyond the tip of the snout,

Figure 76. Cuvier's Beaked Whale *(Ziphius cavirostris).*

and the sole pair of teeth emerges in front of the snout. The mouth is small, reaching about halfway to the eye.

DISTRIBUTION: Found worldwide in warm and temperate waters.

FOOD: It takes squid and small fish.

REPRODUCTION: Virtually unknown.

COMMENTS: This is the most common beaked whale along the Pacific coast.

Whalebone Whales or Baleen Whales (Mysticeti)

This group includes the largest creatures the world has ever seen. The baleen whales have supported whale fisheries in many nations, but dwindling stocks of most species have prompted protective measures. Today most countries support the recommendation of the International Whaling Commission for a total ban on the commercial pursuit of the large baleen whales, but some whaling does persist under the guise of scientific studies. There are three families of Mysticeti, and all have representatives in California.

Gray Whale (Eschrichtidae)

This family has only one species. It is sometimes considered a subfamily of the Balaenopteridae.

GRAY WHALE   *Eschrichtius robustus*

Fig. 77

DESCRIPTION: This is a large (up to 15 m), rather slender whale with a rough, knobby dorsal ridge. The lack of a dorsal fin distinguishes it from the rorquals. It is grayish in color, both dorsally and ventrally. The mouth is short and curved or arched. The flippers and flukes are broad. The throat has two to four grooves allowing for expansion of the skin in that area. The baleen is much shorter than in other mysticetes.

DISTRIBUTION: Found in the North Pacific. It occurred in the North Atlantic at least until 1,500 years ago and perhaps as recently as the early 1700s. It summers in the far north, from the Bering Sea north to the Arctic Ocean, and migrates approxi-

Figure 77. Gray Whale *(Eschrichtius robustus)*.

mately 11,000 km south to winter calving areas off Baja California and Korea, about the longest mammalian migration known.

FOOD: It sucks in bottom-dwelling amphipods and strains them from the water with the upper side of the baleen plates.

REPRODUCTION: Pregnancy lasts about 13.5 months. In the Gulf of California, this whale enters shallow waters, where a single calf is born, some 5 m long and 450 kg in weight. The infant nurses for six months and is weaned the following summer, while in the productive arctic seas. Pregnancies probably occur every other year.

STATUS: This species was removed from the federal endangered list in 1994.

COMMENTS: Unlike other whalebone whales, the Gray Whale forages at or near the bottom of shallow waters. It takes in food through the right side of the mouth; thus that side tends to become worn, and the baleen plates become shorter than those on the left. As a consequence of this style of feeding, it gouges oblong depressions in the sea floor.

This species, once drastically reduced in numbers by commercial whaling, has made a spectacular recovery. At the close of the last millennium its population was an estimated 26,000.

Rorquals (Balaenopteridae)

These are medium-sized to large, rather slender whales with a small curved dorsal fin and distinctive longitudinal grooves on the outer surface of the throat and other parts of the ventral surface; rorqual means "furrow whale" in Norwegian. These ventral grooves allow for expansion of the mouth cavity, which allows ingestion of large amounts of food. Rorquals pursue concentrations of small euphasiid shrimp (krill) and schools of herring and mackerel. They sometimes approach their prey while swimming on their sides and drive krill into their mouths by beating their flukes. Modern whaling methods have greatly reduced the populations of these whales. Possibly as a result of this decline, the large species of finback whales (*Balaenoptera* spp.) now grow faster and reach sexual maturity earlier. Prior to 1930 these large whales first bred at about 10 years of age; by 1960, however, the age of sexual maturity had dropped to five or six years. Females are larger than males.

Five of the six known species of rorquals occur in the coastal waters of California.

MINKE WHALE   *Balaenoptera acutorostrata*
Fig. 78

DESCRIPTION: A rather small rorqual, about 10 m long. It is blue or blue gray dorsally, whitish ventrally to near the base of the tail. The flippers are white in the middle and dark at the base and tip (or may have a white band). The dorsal fin is placed relatively far forward. The snout is sharp and pointed, rather like a broad knife blade. The baleen is yellowish; it is black or dark in other rorquals.
DISTRIBUTION: Occurs worldwide; in the Pacific Ocean, it ranges north into the Bering Sea in summer and south of California in winter.
FOOD: This whale primarily eats krill but is more prone than other rorquals to take anchovies and other small fish.
REPRODUCTION: Its reproductive pattern resembles that of the Blue Whale *(B. musculus)*, except that the Minke Whale may produce young every year, mating shortly after parturition.

Figure 78. Minke Whale *(Balaenoptera acutorostrata)*.

Figure 79. Sei Whale *(Balaeonoptera borealis)*.

SEI WHALE   *Balaenoptera borealis*
Fig. 79

DESCRIPTION: This whale may reach 20 m in length but is usually much smaller. It is blue black dorsally and lighter ventrally, except under the flippers. The dorsal fin is about one-third the distance from the tail; because of its position, this fin is more visible than in the Blue Whale *(B. musculus)*. The head has an arched rostrum, or snout.

DISTRIBUTION: Found worldwide but favors warm waters.

FOOD: This whale feeds on various planktonic crustaceans, such as copepods, amphipods, and krill. Its baleen is finer than that of other rorquals, and food items are smaller.

REPRODUCTION: Its reproductive pattern resembles that of the Blue Whale. About 1 percent of pregnancies result in twins. The young are about 5 m long at birth.

STATUS: This species is federally listed as endangered.

COMMENTS: This rorqual dives less deeply than some others and feeds nearer the surface.

It is more common on the Pacific coast than the Blue Whale but less common than the Fin Whale *(B. physalus)*.

BLUE WHALE   *Balaenoptera musculus*
Fig. 80

DESCRIPTION: The largest animal known ever to have existed, reaching a maximum length of around 31 m. It is bluish dorsally and sometimes, due to the growth of microorganisms, pale yellow ventrally. The dorsal fin is small and set far back on the body. When the whale breaks the water, this fin remains beneath the surface until just before the flukes are raised for the next dive.

Figure 80. Blue Whale *(Balaenoptera musculus)*.

DISTRIBUTION: Roams worldwide and is highly migratory. It is not generally common in coastal waters in our latitude. It summers in the North Pacific.

FOOD: It primarily eats krill.

REPRODUCTION: Pregnancies occur every other year. Gestation is adjusted to this whale's migratory pattern. Mating occurs in low latitudes in winter, after which the whales move to the plankton-rich arctic waters; birth occurs some 10 or 11 months later in warm waters at low latitudes. The young is 4 m long at birth. Weaning occurs in the high latitudes some six or seven months later. This timing results in rapid growth of the embryo and nursing young, both in the highly productive waters of the north. A newborn Blue Whale doubles its natal weight in about one week.

STATUS: This species is federally listed as endangered.

FIN WHALE   *Balaenoptera physalus*
Fig. 81

DESCRIPTION: A large rorqual, up to 27 m in length but usually less. The dorsum is black or blackish with a large, dark chevron mark just behind the head. The left side of the lower jaw is dark, the right side light colored. The venter is light, sometimes yellowish (due to the growth of yellowish algae called diatoms). The dorsal fin is placed more anteriorly than in the Blue Whale. The rostrum is flat.

DISTRIBUTION: Occurs worldwide. It migrates into the Bering Sea in summer and south to the Gulf of California in winter.

FOOD: This whale apparently feeds beneath the surface in the areas where krill are most concentrated. It also takes small fish (herring) and copepods.

REPRODUCTION: Its reproductive pattern is probably not greatly

Figure 81. Fin Whale *(Balaenoptera physalus)*.

unlike that of the Blue Whale *(B. musculus)*. It usually has one young: twinning occurs in about 1 percent of pregnancies. The newborns are 6 to 7 m long and weigh 1,500 kg. Postpartum estrus and mating have been noted in some Fin Whales.

STATUS: This species is federally listed as endangered.

COMMENTS: This is the most common rorqual off our coast.

HUMPBACK WHALE   *Megaptera novaeangliae*
Fig. 82

DESCRIPTION: Immediately recognizable by its extremely long and rather slender flippers, which it often raises well out of the water. It is dark dorsally and light ventrally. The head and flippers have a rough surface. It reaches up to 16 m in length, with a stout body, broad flukes, and a middorsal fin. It is also recognized by its habit of breaching, or jumping out of the water.

DISTRIBUTION: Found worldwide, with clear-cut migration routes to low latitudes in winter and polar seas in summer. It roams north to the Bering Sea and the Gulf of Alaska in summer and south to California and Hawaii in winter. Recently it has occurred regularly near the Farallons.

FOOD: It eats largely krill but not infrequently fish ranging from anchovies to cod and salmon.

REPRODUCTION: Mating occurs in winter with rather conspicuous

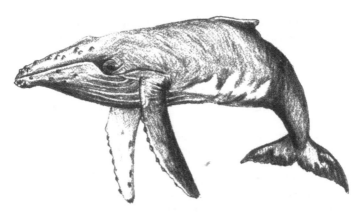

Figure 82. Humpback Whale *(Megaptera novaeangliae).*

displays of courtship. Conception is biennial, as in other large rorquals. The single calf weighs some 900 kg and is 5 m long. It nurses for about 11 months.

STATUS: This species is federally listed as endangered.

COMMENTS: This whale is known to employ a so-called bubble net, a unique method of driving prey into dense schools: one or two individuals circle below a school of fish and release a constant stream of bubbles that rise as a coil, apparently frightening the prey into a dense aggregation. The whale then rises quickly to the surface with its mouth open, engulfing some of the prey.

This species has received much publicity because of recordings of its songs. These extended vocalizations may last 20 or 30 minutes and are individually distinctive. Apparently only males sing, and all sing the same song, which changes from year to year. The significance of their squeaky music is not clearly established, but it is heard mostly in winter, when the humpbacks have assembled for breeding.

Right Whales (Balaenidae)

These are large whales, with the head making up about one-third of their total length. They have rather short, rounded flippers and no dorsal fin. The mouth is strongly arched. The throat has no longitudinal grooves. The baleen plates are very long, but the rows do not join anteriorly. These whales typically feed by swimming with their mouths open, thereby engulfing small crustaceans and some small fish (up to the size of a herring). The closely spaced baleen plates retain the food while water passes out. One species is known from our waters.

NORTHERN RIGHT WHALE — *Eubalaena glacialis*
Fig. 83

DESCRIPTION: A large, black whale, some 15 to 18 m long, sometimes with white patches on the throat. Raised rough patches on the head are more common in males than in females and presumably play a role in male-male aggression.

DISTRIBUTION: Scarce in modern times; consequently, its distribution is poorly known. It summers in the Gulf of Alaska and the Aleutians to the Bering Strait. Its winter range is known only from fragmentary observations, but it has been seen as far south as Baja California and Hawaii. In recent years it has been observed in the Santa Barbara Channel.

FOOD: It eats small surface-dwelling copepods, krill, and some small fish.

REPRODUCTION: Breeding takes place perhaps every third or fourth year. After a gestation of about one year, a calf some 6 m in length is born.

STATUS: This species is federally listed as endangered. It is fully protected in California.

COMMENTS: This whale is very rare today but was once very common and the most eagerly sought-after species. Because it floats after being killed (in contrast to the fin whales, which sink at death), it was called the right whale and was the whale most easily dealt with in the days of open-boat whaling and handheld harpoons.

Figure 83. Northern Right Whale *(Eubalaena glacialis)*.

PERISSODACTYLA

Perissodactyls, commonly known as odd-toed ungulates, are those living forms with either one or three toes. They have well-developed limbs and lack bony horns or antlers. Their teeth are modified for crushing plant material.

Only one family, Equidae (horses), occurs in California. Modern horses and other perissodactyls represent a small remnant of a once major order.

Horses (Equidae)

Horses have a greatly developed middle toe (the hoof). They have short body hair, but the neck has a ridge of stiff, heavy hairs (the mane), and the tail has long hairs, at least at the end. The legs and neck are long. Dentition: 3/3, 0–1/0–1, 3–4/3, 3/3.

Horses have a long history in North America. They flourished well before the development of the various families of Artiodactyla, but as the artiodactyls diversified through evolutionary time, the horses declined, becoming extinct in North America around the close of the Ice Age. Feral horses in the western states descend from those introduced by early Europeans. Ironically, native horses are no longer found in the Western Hemisphere, where the family flourished for many millions of years.

Today, two feral species are found within our borders: the "wild" Feral Horse *(Equus caballus)* and the Burro *(E. asinus)*. In limited numbers, each has desirable features. Each can also harm the environment and, indirectly at least, depress populations of native birds and mammals. In the Wild Horse Annie Act of 1959, Congress halted the pursuit of wild horses and Burros by motorized vehicles on public lands, a small protective step. In 1971 the Wild Free-roaming Horse and Burro Act extended this protection, authorizing the Department of the Interior and the Department of Agriculture to control the method and extent of capture of these feral mammals. They are now protected on public lands and cannot be taken without permission.

BURRO
Equus asinus

DESCRIPTION: A small equid with large ears and a short, erect mane. A slender, dark brown dorsal strip has lateral branches at the shoulder.

DISTRIBUTION: Inhabits arid lands from Inyo to Imperial Counties up to 3,500 m or more. It is native to northeast Africa.

FOOD: Basically a plant feeder but very adaptable, the Burro begs food from motorists and eats almost anything offered.

REPRODUCTION: Breeding is not known to be seasonal. The female bears one young at a time.

COMMENTS: The Burro is a quaint addition to our fauna. Present populations descend from stock discarded by miners. Although not common, the species seems very well established and has few enemies.

The Burro eats some of the native plants in Death Valley. It competes with the less aggressive native Bighorn Sheep (*Ovis canadensis*), feeding heavily on some of the most desirable perennial grasses and forbs and driving the sheep from waterholes. The Burro is undoubtedly at least partly responsible for the low densities of this sheep and its actual disappearance from some of its original range.

FERAL HORSE
Equus caballus

DESCRIPTION: Familiar to everyone. Wild horses vary widely in color and size. Many are spotted. They tend to be rather small, weighing less than 375 kg, though some are fine specimens of 500 kg or more.

DISTRIBUTION: Complete distributional pattern unknown, but small herds are established in many scattered localities. Substantial populations have persisted in Modoc and Lassen Counties. A large population is also found at the Naval Weapons Center in Inyo and Kern Counties.

FOOD: Horses eat both grasses and forbs and browse the leaves of many species of shrubs. They depend upon the presence of standing water.

REPRODUCTION: Gestation lasts 330 to 345 days. Most births are in April and May. Approximately 3 percent of births are twins.

COMMENTS: The equine social organization is fairly rigid, comprising a dominant male, up to 10 females, and occasionally immature males. In winter, small social groups assemble into larger groups (herds). Herds are few, widely scattered, and usually rather small. Large herds exist in some northeastern parts of the state and may compete with domestic stock and damage wildlife habitat. Similarly, horses at the Naval Weapons Center are undesirably numerous; periodically the Bureau of Land Management removes excess horses and makes them available for adoption.

ARTIODACTYLA

Artiodactyls are herbivores with well-developed third and fourth toes. The side toes (second and fifth) are rather small and do not bear weight. Artiodactyls are blessed with very effective digestive systems; many have a stomach of two, three, or four chambers with populations of bacteria and protozoa to digest cellulose. Males frequently have horns or antlers. The upper incisors are frequently lacking (see fig. 91); the cheek teeth are modified for the species' different diets. Their flesh is often palatable, and many are important game mammals.

The artiodactyls today are far more diverse than the perissodactyls. The order includes deer, pigs, cattle, and sheep in the New World and antelopes and their allies elsewhere. Artiodactyls can be separated into two major groups: the Bunodontia, comprising rather omnivorous forms such as pigs and hippopotamuses, and the Selenodontia, embracing the various antelopes, deer, cattle, sheep, camels, and giraffes. Another taxonomic arrangement divides them into three suborders: Suiformes, the pigs and hippos; Tylopoda, the camels; and Ruminantia, the deer, giraffes, cattle, sheep, and antelopes.

Artiodactyls exist in a variety of habitats, including deserts and high mountains, and their feeding habits vary accordingly. Deer occur in forests and are mostly browsers or mixed feeders, whereas such forms as the Pronghorn (*Antilocapra americana*) occupy open areas and are both browsers and grazers. A few forms are also frugivorous (fruit eaters).

Most of the California artiodactyls are so-called short-day breeders: gonadal growth occurs in fall, when days are becoming shorter, and mating follows in fall or winter, so that young are born in spring, at the time of rapid plant growth. The young are generally precocial, walking and taking solid food within hours after birth. Most artiodactyls bear a single young, with occasional twinning.

Early ungulates were rather small generalists that subsisted on most anything edible. Their cheek teeth were bunodont (with tubercles for crushing hard objects such as nuts), and their legs were not greatly modified for running. Ungulate-like forms are known from fossil beds of the Late Cretaceous of China and North America but were gone by the middle Eocene. These early forms, which were derived from Late Cretaceous carnivorous or omnivorous ungulates, were the size of rabbits or dogs. In the early Paleocene these archaic ungulates constituted a large part of the mammalian fauna. Fossil artiodactyls date from the early Eocene. In North America camels had a long history, but most of them were forest-dwelling browsers. With the increase of grasslands and decline of woodlands in the Miocene, large grazing forms became a major part of the artiodactyl fauna. Most of the history of deer occurred in the Old World, and the fossil record dates from the early Pliocene in North America. Pronghorns have been in North America since the early Miocene, at which time there were many species.

Pigs (Suidae)

Pigs have four toes, the median pair functional. They have a truncate, oval snout. The stomach has partial divisions but is non-ruminant. The incisors and canines are well developed. Dentition: 3/3, 1/1, 4/4, 3/3.

This family is not native to the New World but has been widely introduced and is adaptable, so that feral populations now exist in many regions.

FERAL PIG or WILD BOAR   *Sus scrofa*
Fig. 84

DESCRIPTION: A pig with a barrel-shaped (laterally compressed) body and short legs. (Feral animals tend to have longer legs than domestic strains.) It reaches up to 1 m in height and 265 kg in weight. Its color varies. Both sexes have large canine teeth.

Figure 84. Wild Boar, or Feral Pig *(Sus scrofa).*

DISTRIBUTION: Found in open woodlands and grasslands and chaparral throughout much of the state, except at higher elevations in the mountains. Widely scattered populations are found in the North and South Coast Ranges; in foothills on the western slopes of the Sierra Nevada, south to Riverside County; and on the Channel Islands.

FOOD: These pigs largely eat plant materials. Near irrigated land they favor green perennial grasses and forbs, especially clover; in foothills they graze on grasses in spring, eat such native foods as acorns and manzanita berries, and browse on native bushes. When acorns are scarce, they may root in soil for bulbs. They take little animal food but may eat small reptiles, ground-nesting birds, mice, and the like.

REPRODUCTION: Breeding is not clearly seasonal, but most mating takes place in fall. Usually five or six young are in a litter, many fewer than in domestic hogs, perhaps reflecting a lower nutritional level in the feral animals. The young are partly altricial and at first confined to a nest. A sow experiences estrus shortly after giving birth; if she does not conceive at that time, mating occurs when the young are weaned. Adult boars engage in vigorous competition for a sow in estrus. This rivalry does not necessarily result in hostility: when the dominant boar has finished mating, the sow mates with the next subordinate boar, and so on until the dominant boar recovers his strength and repeats his performance. Because estrus recurs until the sow conceives, most adult sows are either pregnant or lactating. Feral hogs in California average two litters a year.

COMMENTS: California hosts both introduced Wild Boars and domestic pigs that have become wild; the two are the same species and interbreed readily. Feral Pigs have existed in California since the early 1800s. The Wild Boar, which already had some domestic blood, was first brought to California in 1924 by George Gordon Moore, who obtained a dozen animals from North Carolina and released them on the San Francisquito Ranch near Carmel. Some eight years later two dozen more were released in the nearby Los Padres National Forest. Consequently the Wild Boars in our state all contain some genetic contribution from domestic stock, but not all feral hogs contain infusions of Wild Boar stock. The total population may exceed 70,000.

Today the Wild Boar and Feral Pig are important big game mammals. The annual take is an estimated 28,000 to 36,000, second only to that of the Black-tailed Deer and Mule Deer (*Odo-*

coileus hemionus). Feral Pigs do extensive damage to rangelands, plowing up meadows and woodlands in their search for food. They also foul waterholes and streams by wallowing.

Deer (Cervidae)

Deer are small to large ungulates with two functional toes and outer and inner dewclaws that do not touch the ground. The legs are long, and the tail is relatively short. The males usually have bony antlers, which are shed annually, and rarely the females have these as well. The ears are long and rather mobile. Several dermal glands (suborbital, tarsal, metatarsal, and interdigital) are present, and some of them produce scent. In many species, the hairs are hollow. The North American species normally lack upper canines. Dentition (for species in California): 0/3, 0–1/1, 3/3, 3/3 (fig. 91).

This family includes such dissimilar species as Moose *(Alces alces),* Wapiti, or Elk, *(Cervus elaphus),* Caribou *(Rangifer tarandus),* and the familiar Black-tailed Deer and Mule Deer *(Odocoileus hemionus).* Most regions of California, except for most parts of the Central Valley, have at least one species; the Black-tailed Deer and Mule Deer are the major large game species of our state.

AXIS DEER or CHITAL   *Axis axis*

Pl. 8, Fig. 85

DESCRIPTION: A medium-sized, reddish brown deer with permanent white spots, a middorsal dark stripe, and a white tail and venter. The antlers are up to 1 m long and directed posteriorly and upward, with a small brow tine and a terminal fork. Grown bucks reach about 1 m in height at the shoulders. Weight: 45–85 kg.

DISTRIBUTION: Found mostly in pastures and open grassy areas on the Point Reyes Peninsula.

FOOD: During the rainy season, this deer largely eats grasses; it takes some forbs and browse at other times.

REPRODUCTION: Breeding apparently occurs throughout the year in California, but most fawns are born in spring. Does may conceive at less than one year of age. A single fawn is the rule.

COMMENTS: Native to India, Nepal, and Sri Lanka, this deer was introduced to our fauna from San Francisco Zoo stock in the 1940s. The population has increased continuously since then and is expanding its range.

Figure 85. Axis Deer *(Axis axis)*.

FALLOW DEER   *Dama dama*
Pl. 8, Fig. 86

DESCRIPTION: A medium-sized deer of highly variable color, from nearly white to almost black. Its back and sides are spotted at all seasons, though in the winter its coat may darken and its spots tend to be less clear. There is a distinct diagonal white line on the flank; the rump is white with a black border; and the legs and belly are white. The antlers are distinctly palmate or expanded over the distal half. Scent glands are found on the inside of the hock (the first leg joint above the hoof), behind the hock, and beneath the eye. Weight: 68–92 kg (bucks), 43–72 kg (does).

DISTRIBUTION: Generally found on open lands on the Point Reyes Peninsula and in central Mendocino, Tehama, and San Mateo Counties. Small populations may occur elsewhere.

FOOD: It consumes grasses, forbs, and browse.

Figure 86. Fallow Deer *(Dama dama)*.

REPRODUCTION: Rut occurs in late summer and fall, when bucks are attracted to the scent of urine from does in estrus. In California, most mating occurs in October. Usually a single young is born in June.

COMMENTS: The Fallow Deer is native to southern Turkey. It has been widely introduced throughout the world. In the Old World it is a popular game mammal. The beauty of its large, palmate antlers and spotted coat have made it a favorite in many public and private parks.

WAPITI or ELK
Pl. 8, Figs. 87, 88

Cervus elaphus

■ Roosevelt Elk
■ Tule Elk

DESCRIPTION: A large, brown or pale yellow tan deer with a conspicuous mane. The head and neck are darker brown than the back; the rump is light, almost white from a distance. Bulls have large, posteriorly projecting antlers with five to seven (usually six) tines, including well-developed basal tines. The species is unusual among North American deer in having an upper canine tooth, which has traditionally provided the Elks Club with a pendant. It is the largest California deer and occurs in three subspecies in our state. The Tule Elk *(C. e. nannodes)* (fig. 88) is the smallest of these:

Figure 87. Wapiti, or Elk *(Cervus elaphus)*. Bull with growing antlers "in velvet." Antlers are annually shed and regrown.

bulls range in weight from 195 to 250 kg. Bulls of the Roosevelt Elk *(C. e. roosevelti)* usually weigh from 450 to 550 kg but are sometimes smaller. The Rocky Mountain Elk *(C. e. nelsoni)* is intermediate between these two in size and is paler than the Roosevelt Elk. Weights vary with age of the individual and condition of the range.

DISTRIBUTION: Originally widespread in the Pacific states. The Tule Elk ranged from north of Red Bluff, in Shasta County, south throughout much of the Central Valley and west to the coast. The Roosevelt Elk was found from the northwest corner of the state south to about Fort Ross, as well as in much of western Oregon and western Washington. The Rocky Mountain Elk is native to northeastern California and regions to the north and east. Today there are about 470 Tule Elk in the Owens Valley from Bishop southward; other herds have been established from Mendocino and Lake Counties south to Marin County and on Grizzly Island in the Sacramento–San Joaquin River Delta. South of San Francisco the Tule Elk has been released in scattered localities from Alameda and Contra Costa Counties south to San Luis Obispo and Kern Counties. Some of these herds are small, but they seem

Figure 88. Tule Elk *(Cervus elaphus nannodes):* note "bugling" posture of bull during rut.

to be increasing annually. The Rocky Mountain Elk occurs at Shasta Lake, part of its original range. The Roosevelt Elk occurs in some separated locations in northwestern California.

FOOD: All three subspecies eat grasses and forbs, moving to new growth with the change in seasons. They also browse on the terminal growth of many broad-leaved and some coniferous trees; browse becomes more important as low annual and perennial herbaceous plants dry out in summer. In addition, they eat mast, such as acorns.

REPRODUCTION: Wapiti become sexually mature in their second year, or sometimes as yearlings. Bulls mature at approximately the same age as cows but are much less likely to mate until they are older. A large bull dominates a small group of cows and expels other bulls. Rut and mating occur in late summer; a single calf is born the following spring after a gestation of about 250 days. Twinning is rare.

COMMENTS: The Wapiti is the North American representative of the Red Deer of Europe. Populations of this stately deer are strictly managed by the DFG. Supervised hunts of the Tule Elk have occurred in the Owens Valley, but not elsewhere. Both the Rocky Mountain and the Roosevelt Elk have been subjected to small quota-type hunts in the past.

BLACK-TAILED DEER and MULE DEER

Odocoileus hemionus

Pl. 8, Figs. 89, 91

DESCRIPTION: A medium-sized deer, though its size varies with sex, age, locality, and the morphological differences among the six subspecies known from California. Different geographic populations of this species are generally distinguished in the field. The subspecies in the Coast Ranges is smaller and is known as the Black-tailed Deer; the larger Sierran subspecies is called the Mule Deer. Both are dorsally reddish in summer and gray brown in winter. The ears are dark; there is much gray about the face. The tail is black-tipped and sometimes has a black dorsal surface; it is relatively short (about 130 mm long). Whether it is held down or erect, a considerable amount of the black is clearly visible. The antlers are dichotomously branched, often with a small spike medially, near the base.

Figure 89. Mule Deer *(Odocoileus hemionus)*.

DISTRIBUTION: Found in forests, brushfields, and meadows throughout most of the state, except in most of the San Joaquin Valley and some southeastern desert areas. Its range covers much of western North America, from northern Mexico (including Baja California) north throughout the Rocky Mountains and the coniferous forests of western Canada.

FOOD: Both a grazer and a browser, this species consumes a broad variety of grasses, forbs, and leaves of shrubs and small trees, as well as acorns.

REPRODUCTION: In late summer or early fall a swelling in the neck of the mature buck marks the start of the rut, which extends into January in some herds but ends in November in others. Does breed first when about a year and a half old; bucks normally become sexually mature a year later. One or two fawns are born in May or June and continue to nurse until late summer.

COMMENTS: This species is California's most important big game mammal and a familiar sight to most hikers. It is frequently so abundant as to overbrowse its habitat and reduce the availability of its food. Deterioration of the range lowers survival of the young as well as the strength and size of those that do survive. In California the Mule Deer is migratory, moving to lower elevations in fall. Its winter range in the Sierra foothills, however, is becoming more urbanized. The increase of permanent homes, with dogs, is an encroachment on its breeding range.

WHITE-TAILED DEER   *Odocoileus virginianus*
Figs. 90, 91

DESCRIPTION: A medium-sized deer, rich chestnut red in summer and gray in winter. The tail is snow white ventrally, about 250 mm long, and commonly held erect when in retreat. The antlers have unbranched tines emerging dorsally from the beam.

DISTRIBUTION: Occasionally reported in a narrow area where northeastern California joins Nevada. Although these reports and some antlers indicate that the White-tailed Deer may have existed in our state in the past, there is no evidence that it still does. Its nearest known populations are near Roseburg, in southwestern Oregon, and near the Snake River in the northeastern region of that state. It is also found in a coastal region of Oregon, an open-country area in eastern Oregon, and lowlands in eastern Washington. Its range extends from northern South America

Figure 90. White-tailed Deer *(Odocoileus virginianus)*.

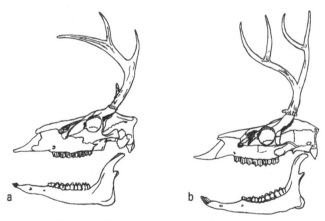

Figure 91. Skulls of (a) White-tailed deer *(Odocoileus virginianus)*;
(b) Mule Deer *(O. hemionus)*.

north through most of the United States (except the Great Basin)
and much of the Canadian coniferous forest.

FOOD: This species probably subsists on grasses, forbs, and leaves
of shrubs.

REPRODUCTION: Little is known about this deer's reproduction in
our area.

Sheep, Goats, and Cattle (Bovidae)

Members of the Bovidae usually have a pair of permanent, hollow, unbranched horns, which may be long and coiled or twisted. In most species both sexes are horned. The upper canine is small or absent, but a lower canine may be present.

In North America the Bovidae include such large ungulates as the Bison, Pronghorn *(Antilocapra americana)*, Bighorn Sheep *(Ovis canadensis)*, and Mountain Goat *(Oreamnos americanus)*. Many species also exist in the Old World, especially Africa. The diverse antelopes and gazelles of the Old World all belong to the family.

PRONGHORN   *Antilocapra americana*
Pl. 8, Fig. 92

DESCRIPTION: A medium-sized, antelope-like creature with prominent, often shiny black horns, usually two pronged in the male but unbranched in the female. Both sexes shed their horn sheaths, the males after breeding, the females aseasonally. The general body color is buff to russet dorsally and whitish ventrally, with blackish facial and throat markings. The rump has conspicuous patches of white, which can be flashed by the erection of hairs. Weight: 20–70 kg.

DISTRIBUTION: Found in the high-elevation sagebrush of northeastern California and also in the sagebrush plains about Big Pine. It is also seen in the open plains east of the Cascade Range in Oregon. Its range extends from central Mexico and Baja California north through the Great Basin to southern Canada.

FOOD: The Pronghorn consumes an extensive range of leaves of shrubs, including sagebrush and bitterbrush. Forbs predominate in the summer diet.

REPRODUCTION: Mating takes place from late summer to early fall. Dominant bucks are territorial, and many immature males do not mate. A strong buck may defend a territory containing some eight to 15 does during the breeding season. One or two young are born in May or June after a gestation of about 250 days. The young are extremely precocial and stand to suckle even before the embryonic fluids are dry. Sexual maturity normally occurs at about 16 months.

Figure 92. Pronghorn *(Antilocapra americana)*.

COMMENTS: Early in the twentieth century the Pronghorn was extirpated in the Big Pine region; the population there today is the result of an introduction in 1949 by the DFG. Originally the Pronghorn was common in the Central Valley from at least the Sutter Buttes south to the desert. It survived in the Antelope Valley of Los Angeles County until late 1933 and has been reintroduced on the Carrizo Plain.

The Pronghorn is a very popular game animal, and some 40,000 are taken annually in the western United States. Despite this hunting pressure, the population continues to increase.

BIGHORN SHEEP
Ovis canadensis

Pl. 8, Fig. 93

DESCRIPTION: A large sheep with relatively short pelage. Its back is brown, its rump is buff or white, and its belly is white. Lambs are born nearly white and darken with age. Both sexes have large horns, curved in ewes and coiled in rams. The tail is short. Both tarsal and facial glands are present. TL 1.4–1.6, T 70–115 mm, HF 390–420 mm, E 105–120 mm. Weight: 70–190 kg.

Figure 93. Bighorn Sheep *(Ovis canadensis).*

DISTRIBUTION: Found from the high elevations (3,800 to 4,500 m) of the southern Sierra Nevada to the lower elevations of the Owens Valley (2,200 m). It also occurs in the desert mountains of southeastern California, the San Gorgonio Mountains (near Los Angeles), and the Warner Mountains. It was once killed off in Oregon but was recently introduced in Sherman and Wallowa Counties and is now found east of the Cascade Range in Oregon and Washington. It ranges from northern Mexico (including Baja California) north through the Rocky Mountains to southern British Columbia.

FOOD: Strictly a grazer, this species subsists on grasses, sedges, and forbs.

REPRODUCTION: Mating takes place mostly from mid-November to mid-December but may vary locally. If the Bighorn Sheep demonstrates the same environmental responses as domestic sheep, gonadal growth and breeding are stimulated by short days. Following a gestation of approximately 180 days, a single lamb is born. Older rams (three years or more) dominate the mating, but in their absence young rams are sexually active. Ewes mate in their second fall and drop their first lamb when just two years of age.

STATUS: The California Bighorn Sheep *(Ovis canadensis californiana)*, also known as the Sierra Nevada Bighorn Sheep, is federally listed as endangered, listed by the state as endangered, and fully protected in California. The Peninsular Bighorn Sheep *(O. c. cremnobates)* is federally listed as endangered, listed by the state as threatened, and fully protected in California.

COMMENTS: Bighorns are among the most prized of all North American game mammals. The horns of old rams are most impressive, and their homes are frequently difficult to reach. Today no hunting of Bighorn Sheep is allowed in California, and there is very little loss to poachers in most areas, but populations in general have not responded to protection. A recent decline in populations in the southern Sierra Nevada has been more plausibly attributed to an increase in densities of the Mountain Lion *(Panthera concolor)* than to poaching.

Previously the Bighorn was more common and widespread in our state. Aborigines apparently extirpated it in California deserts before European settlement. John Muir found a large number of heads and horns of Bighorn Sheep in a cave near Sheep Rock, on the northern slope of Mount Shasta. This was an area frequented by the Bighorn in winter, and it is likely that the remains represented the accumulation of sheep killed by Indians.

RODENTIA

Mice, squirrels, and their allies are diverse but share certain readily apparent features. All have a single pair of both upper and lower incisors that grow continuously and are rooted deep in the skull and the dentary bones. The incisors frequently have orange or yellow enamel on the anterior surface and no enamel on the inner surface. Canine teeth are absent, and a large toothless space (the diastema) separates the incisors from the cheek teeth (premolars and molars).

The cheek teeth of some rodents are either bunodont or hypsodont (high crowned). The dentary bone is attached to the skull at the glenoid fossa, an anterio-posterior depression (running front to back). This connection allows the lower jaw to move sideways as well as forward and backward, permitting gnawing. In most groups of rodents (mice and squirrels), the upper and lower cheek teeth meet on only one side of the jaw at a time. In the Porcupine (*Erethizon dorsatum*) they nearly meet on both sides simultaneously.

There are many families of rodents, and taxonomists change them occasionally, sometimes reducing families to subfamilies. The difficulty arises from the great adaptability of rodents; it is very difficult to know whether similarities in teeth reflect phylogenetic relationships or dietary specializations. The problem is greater in fossil groups, because many are known only from teeth, without postcranial bones.

The first true rodents are known from the late Paleocene in North America, Europe, and Asia; the earliest may be *Heomys,* known from the Paleocene of China. It has been suggested that rodents branched off from a primitive primate, perhaps in the early Paleocene. The fossil evidence suggests that rodents arose in eastern Asia and dispersed from Asia to North America across a Bering connection to western Europe (Europe and Asia were separated at that time) by the late Eocene.

Rodents and lagomorphs (rabbits, hares, and pikas) have some obvious features in common. Both groups have a diastema rather than a canine tooth. This is also true of the multituberculates, a primitive group that survived up to the early Eocene in North America. The anatomical parallels suggest that rodents may have outcompeted the earlier multituberculates. Arguments

for affinity or nonaffinity of the rodents and lagomorphs remain unsettled, but the early fossil evidence from China suggests a common origin. An early Paleocene genus called *Euromylus,* known from Asia, has features of both rodents and lagomorphs.

Porcupine (Erethizontidae)

Porcupines are short, stout rodents with many stiffened and barbed hairs. They are modified for tree climbing; the tail is prehensile in some species. The skull is heavy and deep, with a large infraorbital foramen (fig. 95b), but no postorbital process. The incisors are deep orange.

Figure 94. Porcupine
(Erethizon dorsatum).

Another group of porcupines occurs in the Old World, but this family is confined to the New. It is essentially a South American group that entered North America when the two continents were joined, some 3 million years ago.

PORCUPINE   *Erethizon dorsatum*
Figs. 94, 95

DESCRIPTION: A large, heavy-set, stout-limbed, spiny rodent. Its color is blackish, suffused with yellow buff. The dorsal pelage includes many easily detached spinous hairs, which provide an offensive covering, and the thick tail is armed with dozens of stout quills. TL .79–1.3 m, T 145–300 mm, HF 75–101 mm, E 25–42 mm. Weight: 10–18 kg. (See fig. 94).

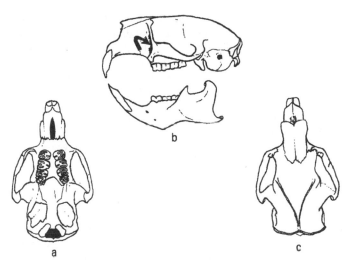

Figure 95. Skull of the Porcupine *(Erethizon dorsatum):* (a) ventral view; (b) lateral view (arrow to infraobital foramen); (c) dorsal view.

DISTRIBUTION: Found in virtually all coniferous forests of the western United States, especially in cutover or burned areas and meadows. It occurs in the Sutter Buttes, California, and occasionally moves into lowlands. Its range extends from northern Mexico to Alaska.

FOOD: Porcupines eat the bark and cambium of conifers, especially pines, and rarely the bark and cambium of hardwoods. Departing from a woody diet in spring and summer, they may be seen eating grass in mountain meadows. They also eat the roots, stems, berries, leaves, and nuts of many kinds of plants. A rich microbial population in the gut breaks down cellulose to sugars, starches, and vitamins.

REPRODUCTION: Mating takes place in late fall or early winter. A single precocial young with soft dorsal spines is born some 210 days later. The infant takes solid food within hours and is weaned at an early age.

COMMENTS: In North America the Porcupine is common and very tame. Like skunks, it has few enemies. Its barbed quills can penetrate not only skin but also heavy shoe leather. Although few predators attack Porcupines, some large carnivores kill and eat them under duress. The prey may have its revenge after death, however, for occasionally such carnivores as Bobcats have been found dead with their entire bodies riddled with the quills of their last meal.

Although Porcupines are slow moving and inoffensive, they can be dangerous, and hikers should not molest them. Old tales tell of Porcupines throwing their quills. This does not actually happen, but sometimes it might seem to happen. The naturalist Vernon Bailey recounted an attempt to overthrow a Porcupine. After he had repeatedly flipped it over with a stick, many quills became loosened. Finally the animal shook itself vigorously, sending quills out to a distance of almost 2 m.

Mountain Beaver (Aplodontidae)

The Aplodontidae are medium-sized, stocky rodents with a minute tail, small eyes, and reduced external ears. The skull is wide and shallow at the rear, which makes it triangular when viewed dorsally (fig. 97c); it has no postorbital process. The cheek teeth grow continuously. Dentition: 1/1, 0/0, 2/1, 3/3.

The only living species, the Mountain Beaver, is the most ancient living rodent and the sole survivor of a long line of very primitive rodents. The family dates from the late Oligocene of both Asia and North America, but its relatives date from the Eocene. It is known from fossils from the past 20 million years. Its ancestors were once much more widely distributed in Eurasia and North America. It is not related to the true beavers (Castoridae).

MOUNTAIN BEAVER   *Aplodontia rufa*
Figs. 96, 97

DESCRIPTION: A rabbit-sized rodent with the family characters given above. Its entire body is dark, grizzled brown, except for a white spot by each ear. The soles of the feet are naked. TL 300–465 mm, T 20–35 mm, HF 32–60 mm, E 15–20 mm. Weight: .8–1.0 kg. (See fig. 97.)

DISTRIBUTION: Found in creekside thickets along the north coast, and near mountain creeks up to 2,300 m in the Sierra Nevada south to the Mono Lake region. It is sporadically common on brush-covered hillsides, usually close to water. It ranges north

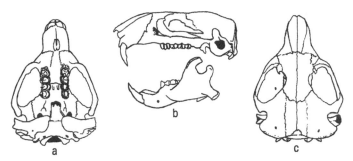

Figure 96. Skull of the Mountain Beaver *(Aplodontia rufa):* (a) ventral view; (b) lateral view; (c) dorsal view.

through western Oregon and western Washington to southern British Columbia.

FOOD: This species eats the leaves of a broad spectrum of forbs and the terminal twigs of such shrubs and small trees as deerberry, maple, dogwood, and alder. It also eats wild berries.

REPRODUCTION: One litter, usually of two to three young, is born annually. The young are nearly naked and blind at birth.

STATUS: The Point Arena Mountain Beaver *(A. r. nigra)* is federally listed as endangered and is a California subspecies of special concern. The Sierra Nevada Mountain Beaver *(A. r. californica)* and the Point Reyes Mountain Beaver *(A. r. phaea)* are California subspecies of special concern.

COMMENTS: The Mountain Beaver is active throughout the year; its burrows in moist earth often disclose its presence.

Figure 97. Mountain Beaver *(Aplodontia rufa).*

Squirrels (Sciuridae)

Squirrels are small to medium-sized rodents that usually have rather long, bushy tails. The ears are erect and sometimes tufted. The postorbital processes are well developed.

Squirrels are among the most familiar of native mammals, for most of them, including tree squirrels (*Sciurus* and *Tamiasciurus* spp.), chipmunks (*Neotamias* spp.), and ground squirrels, are diurnal. Only flying squirrels (*Glaucomys* spp.) are nocturnal.

Squirrels are known from fossil beds of the late Eocene of North America, the early Oligocene of Europe, and the early Miocene of Africa.

Key to Genera of Squirrels (Sciuridae) in California

1a Forelimbs and hind limbs not joined to each other and to sides of body by flap or fold of skin 2

1b Forelimbs and hind limbs joined to each other and to sides of body by fold of skin; pelage very soft, light brownish gray; tail flattened, with long hairs on side and short hairs on top and bottom; with peglike upper premolar (fig. 100b) . . . . . .
. . . . *Glaucomys*

    2a Head without distinctly contrasting light and dark stripes 3

    2b Head with contrasting light and dark stripes; skull with laterally directed postorbital process (fig. 103c) .
. . . . *Neotamias*

3a Tail bushy, broader than thigh; postorbital process directed distinctly toward rear rather than to side (figs. 101c, 102c) .
. . . . 4

3b Tail pelage not bushy but narrower than thigh; postorbital process usually directed laterally, rear margin approximately at right angles to longitudinal axis of skull (if directed toward rear, pale lateral stripe is present) (figs. 98c, 99c) 5

    4a Body without discrete dark lateral stripe; with or without small peglike first premolar (see fig. 101a) . . . .
. . . . *Sciurus*

    4b Body with distinct black lateral stripe; without small peglike premolar (see fig. 101a) . . . . *Tamiasciurus*

5a Skull dome-shaped, markedly convex dorsally; postorbital

process below dorsal margin of skull when viewed from side
....6

5b Skull flat; postorbital process above dorsal margin of skull
when viewed from side .... *Marmota*

    6a Interorbital constriction narrower than postorbital
constriction (see fig. 98c)........ *Ammospermophilus*

    6b Interorbital and postorbital constrictions about equal
(see fig. 99c) .... *Spermophilus*

Antelope Ground Squirrels and Ground Squirrels (*Ammospermophilus* and *Spermophilus*)

These are small or large squirrels of various color patterns: even, flecked, spotted, or striped (though there are no stripes on the head). The tail is moderately or scantily furred, seldom bushy. The pelage is coarse. A ground squirrel's weight fluctuates markedly from season to season. Therefore, weights given in the following species accounts usually indicate a broad range and are not very critical in identification. Dentition: 1/1, 0/0, 1–2/1, 3/3 (see figs. 98a, b, 99a, b).

Most of these squirrels live in open spaces and do not live in trees, but they sometimes climb them. Often they live near large rocks, stumps, or hillocks from which they can view the surrounding countryside. Many species are abundant and may be pests in forests or on ranches. In forests they may eat conifer seeds set out by foresters to reseed logged or burned areas. Some kinds of ground squirrels carry plague. Thus, it is prudent to avoid close contact with wild ground squirrels. If ill or dead squirrels are seen, county or state health authorities should be notified. The many species of ground squirrels have long been known under the generic name *Citellus,* and mammalogists in Europe and Asia still use this name. Much of the important literature on these rodents lists them as species of *Citellus.*

ANTELOPE GROUND SQUIRREL
Ammospermophilus leucurus

Pl. 9, Fig. 98

DESCRIPTION: A small, ash-colored ground squirrel with a conspicuous white lateral stripe. The dorsum may be cinnamon brown, especially in summer. The tail is light-colored with a dark margin. The interorbital breadth is less than the postorbital breadth. (See fig. 98c.) TL 211–233 mm, T 63–71 mm, HF 35–40 mm, E 8–11 mm. Weight: 74–103 g.

DISTRIBUTION: Found in arid regions of sagebrush, greasewood, and other Great Basin shrubs, far from trees. It occurs east of the Sierra Nevada in the Great Basin and the Mojave and Colorado Deserts. Its range extends from southeastern Oregon and Idaho east to Colorado and south to Arizona, New Mexico, and Baja California.

FOOD: This squirrel eats much green plant material when available and the fruits and seeds of many desert plants, including the prickly pear, buffalo berry, mesquite, and saltbrush. It also takes insects and small vertebrates, such as mice and lizards.

REPRODUCTION: One, or sometimes two, litters of five to eight young are born in spring but may remain in the nest until August.

COMMENTS: The Antelope Ground Squirrel is a conspicuous member of the rodent fauna of the western edge of the Great Basin. Although it does not hibernate, inclement weather does seem to reduce its activity. This squirrel appears to be somewhat intolerant of others of its own kind; it occurs widely scattered, not clustered in colonies. Nevertheless, individuals sometimes

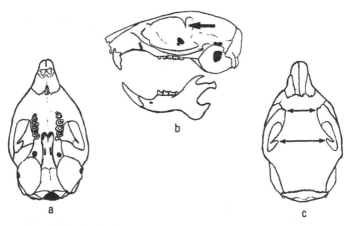

Figure 98. Skull of the Antelope Ground Squirrel (*Ammospermophilus leucurus*): (a) ventral view; (b) lateral view (arrow to postorbital process); (c) dorsal view (arrows to interorbital and postorbital constrictions).

huddle together at night, thus reducing their energy expenditure by an estimated 40 percent.

NELSON'S ANTELOPE GROUND SQUIRREL
Ammospermophilus nelsoni

Pl. 9

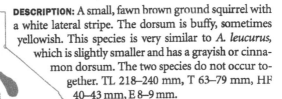

DESCRIPTION: A small, fawn brown ground squirrel with a white lateral stripe. The dorsum is buffy, sometimes yellowish. This species is very similar to *A. leucurus,* which is slightly smaller and has a grayish or cinnamon dorsum. The two species do not occur together. TL 218–240 mm, T 63–79 mm, HF 40–43 mm, E 8–9 mm.

DISTRIBUTION: Found in open grassy areas, sometimes with scattered brush. It occurs in the San Joaquin Valley and Coast Range foothills from southwest Merced County through western Kern County and San Luis Obispo County, California, as well as in the Cuyama Valley and the Carrizo Plain–Elkhorn Plain in San Luis Obispo County, north to San Benito County.

FOOD: This ground squirrel eats green leaves in winter and spring and insects and seeds throughout the year.

REPRODUCTION: Mating activity begins in winter, and mating itself in mid-February. Young are born in mid-March after a gestation of almost four weeks. Litters range from six to 12.

STATUS: This species is listed by the state as threatened.

Key to Species of *Spermophilus* (Ground Squirrel) in California

1a Dorsum striped or variegated 2
1b Dorsum rather evenly pigmented 3
   2a Dorsum with conspicuous lateral stripes . . . . . *lateralis*
   2b Dorsum variegated, not striped 4
3a Tail short, moderately bushy 5
3b Tail very short haired, cylindrical; Sonoran and Mojave Deserts . . . . *tereticaudus*
   4a With black or brown patch framed in gray on nape and sometimes shoulders . . . . *beecheyi*
   4b Without black or brown patch on nape; Mojave Desert . . . . *variegatus*

5a Dorsum brown, set off from gray sides; length 253–300 mm; Great Basin .... *beldingi*
5b Dorsum and sides blending into pale venter; size less than 250 mm. 6
   6a Body grayish; Great Basin .... *mollis*
   6b Body pink brown; Mojave Desert .... *mohavensis*

BEECHEY GROUND SQUIRREL   *Spermophilus beecheyi*
Pl. 9, Fig. 99

DESCRIPTION: A large ground squirrel, gray brown mottled by light flecks. It commonly has a large dark mantle, ranging from a slightly darkened area to a black patch, and rarely adjoined by faint light lateral stripes. Its tail is long and bushy. Its size, dark mantle, and usual absence of stripes distinguish this species from all other ground squirrels in California. The interorbital and postorbital widths are about equal. (See fig. 99c.) TL 375–500 mm, T 135–195 mm, HF 50–64 mm, E 17–22 mm. Weight: 300–650 g.

DISTRIBUTION: This squirrel is common in fields of stubble, along roadsides, on well-grazed pastures, and in open oak woodland. It is found in most of California, from sea level to 2,280 m or perhaps above, except in the Great Basin and the southeastern desert regions. Its range extends from western Washington south through Baja California.

FOOD: This species consumes a broad spectrum of seeds, berries, and leaves of grasses, forbs, and woody plants, as well as corms, tubers, and road-killed carrion.

REPRODUCTION: A single litter of three to 10 young is born annually. The season of birth varies with the locality.

COMMENTS: This is almost strictly a ground-dwelling species that apparently needs an open area to provide a clear view of the surrounding cover. It avoids ungrazed grasslands where the cover is sufficiently tall to obstruct the view. Along roadsides where the grass has become tall, it is frequently seen on fenceposts or awkwardly grappling wire fencing.

Adult males of this species enter hibernation in late summer, adult females enter later.

Young of the year remain active until fall. Hibernation lasts

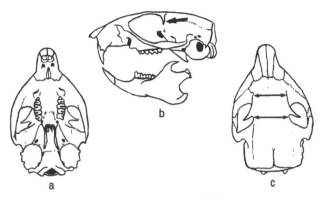

Figure 99. Skull of the Beechey Ground Squirrel *(Spermophilus beecheyi):* (a) ventral view; (b) lateral view (arrow to postorbital process); (c) dorsal view (arrows to interorbital space and postorbital process).

until early spring. During its periodic arousals, this squirrel appears briefly aboveground.

This species was formerly called the California Ground Squirrel.

BELDING'S GROUND SQUIRREL   *Spermophilus beldingi*
Pl. 9

DESCRIPTION: A stout, stocky ground squirrel, gray brown with a darker brown dorsal saddle. The venter and legs are pinkish. The rather short tail is reddish ventrally and scantily furred. TL 268–296 mm, T 60–75 mm, HF 42–47 mm, E 9–10 mm. Weight: 126–305 g.

DISTRIBUTION: Commonly resides in high mountain meadows in the Sierra Nevada up to 3,000 m and in the sagebrush flats of the Great Basin in northeastern California. It ranges from western Oregon and southern Idaho to north-central Nevada.

FOOD: This squirrel subsists on the leafy parts of growing plants, including grasses and forbs.

REPRODUCTION: Like many mammals, this species is quite promiscuous: in the very brief period of sexual receptivity (less than five hours), females may mate with up to five males, and in most cases, the annual litter of four to 12 young is sired by more than one male.

COMMENTS: Belding's Ground Squirrel enters its annual dormant cycle in summer and emerges in late winter or spring when green grasses and forbs flourish.

Unlike most ground squirrels, it typically sits quite erect, either on its haunches or stretching up to the length of its hind legs. This is its method of keeping a constant lookout, for it rarely climbs trees.

GOLDEN-MANTLED GROUND SQUIRREL
Pl. 9

Spermophilus lateralis

DESCRIPTION: A medium-sized, brightly colored ground squirrel distinguished by a bright white lateral stripe bounded by black stripes and by a yellow or orange mantle over the head and shoulders. Unlike chipmunks, which it otherwise vaguely resembles, the Golden-mantled Ground Squirrel has an unstriped head. TL 235–295 mm, T 83–102 mm, HF 37–41 mm, E 13–16 mm. Weight: 136–245 g.

DISTRIBUTION: Occurs in coniferous forests from the Klamath and Warner Mountains south through the Sierra Nevada from 1,800 to 3,000 m. It is also found in the San Bernardino Mountains and into the sagebrush flats of the Great Basin east of the Sierra Nevada. Its range extends from central British Columbia south through extreme northeastern Washington, eastern Oregon, and the Rocky Mountain region into Arizona and New Mexico.

FOOD: This squirrel eats many leaves and seeds of forbs and grasses. It occasionally makes heavy use of nuts. It also consumes roots, bulbs, and other underground plant parts, including underground fungi, as well as some insects.

REPRODUCTION: A single litter of four to eight young is born about 30 days after the end of hibernation. Young emerge from the nest when they are about six weeks old.

COMMENTS: The Golden-mantled Ground Squirrel is a very common and easily observed forest dweller. It is active through the summer holiday season, and its beautiful and unique colors make it one of the best known California squirrels. It is a profound hibernator, and much research on hibernation has used it as a model.

MOJAVE GROUND SQUIRREL   *Spermophilus mohavensis*
Pl. 9

DESCRIPTION: A rather small, pale, evenly pink brown ground squirrel without distinctive spots or stripes. Its tail is moderately furred but flattened. It may occur together with the Antelope Ground Squirrel (*Ammospermophilus leucurus*), which has distinctive white stripes on the sides. TL 210–230 mm, T 57–72 mm, HF 32–38 mm. Weight: 85–130 g.

DISTRIBUTION: Found in open areas in the eastern and northern Mojave Desert. It reaches to the foothills of the southern Sierra Nevada as far as Harper Dry Lake and Searles Dry Lake, California.

FOOD: This species eats many seeds and vegetative parts of desert plants. Fruits of the Joshua Tree are a favorite.

REPRODUCTION: Breeding takes place from early March; litters contain four to six young.

STATUS: This species is listed by the state as threatened.

COMMENTS: The Mojave Ground Squirrel is a relatively shy, secretive animal and seems not to be abundant anywhere. It is a hibernator and thus is dormant when most people visit its desert habitat. It emerges in March in the southern part of the desert but may remain dormant until May farther north.

PIUTE GROUND SQUIRREL   *Spermophilus mollis*

DESCRIPTION: A small, gray ground squirrel without spots. TL 167–270 mm, T 32–72 mm, HF 29–38 mm, E 8–9 mm.

DISTRIBUTION: Found in the open sagebrush flats of the Great Basin in east-central California, the southeast corner of Oregon, and north of the Yakima River, Washington.

FOOD: It consumes fresh, leafy vegetation as well as sprouting seeds, bulbs, and insects.

REPRODUCTION: A single litter of five to 12 young is born in April.

COMMENTS: This little squirrel may be seen as early as February, when precipitation has stimulated plant growth. By early summer, it has begun to enter estivation, and adult males have gone underground by June, when herbaceous plant growth has dried up. This pattern is biologically reasonable for a species that seems to depend so heavily on fresh herbaceous plants for sustenance, but it illustrates the arbitrary distinction between estivation and hibernation.

ROUND-TAILED GROUND SQUIRREL
Spermophilus tereticaudus
Pl. 9

DESCRIPTION: A small, evenly colored, pale cinnamon brown ground squirrel with a white venter. The tail hair is evenly distributed and very sparse, unlike that of other ground squirrels in the state; the tail appears more ratlike than squirrel-like. TL 204–266 mm, T 60–107 mm, HF 32–40 mm, E 5–6 mm. Weight: 116–133 g.

DISTRIBUTION: Occurs on sandy and coarse soils in the Mojave and Colorado Deserts. Its range extends from southern Nevada through southern Arizona to northern Mexico.

FOOD: This squirrel eats green leaves, fruits, and seeds of desert plants.

REPRODUCTION: A litter of six to 12 is born in June.

STATUS: The Palm Springs Round-tailed Ground Squirrel *(S. t. chlorus)* is a candidate for federal listing as endangered and is a California subspecies of special concern.

COMMENTS: This species prefers open areas but is known to climb low shrubs.

ROCK SQUIRREL
Spermophilus variegatus
Pl. 9

DESCRIPTION: A large, mottled or variegated gray brown ground squirrel lacking stripes, patches, and the black mantle normally present in Beechey Ground Squirrel *(S. beecheyi)*. The tail is long and well furred. This species occurs with the Antelope Ground Squirrel *(Ammospermophilus leucurus)*, which is small and conspicuously striped, and the Round-tailed Ground Squirrel *(S. tereticaudus)*, which is also small and has a scantily furred tail. TL

434–510 mm, T 198–235 mm, HF 53–60 mm, E 15–19 mm. Weight: 580–795 g.

DISTRIBUTION: Usually occurs on hillsides near large rocks or rocky outcrops but is sometimes seen in burrows in open country. In California it is found only in the Providence Mountains in the Mojave Desert. Its range extends from Nevada to extreme western Oklahoma and south into central Mexico.

FOOD: This species subsists largely on leafy material in spring and seeds and berries in fall, also consuming some insects.

REPRODUCTION: It may possibly have two litters per year, each with five to seven young.

COMMENTS: This squirrel readily climbs trees.

Flying Squirrels (Glaucomys)

These rather small, dull-colored squirrels are modified for gliding: a broad membrane connects the legs and forms a plane when the legs are stretched out. The tail is conspicuously flattened dorsoventrally and probably guides the animal during its flight. The fur is very soft, and the eyes are large.

In North America and Asia one or more species of flying squirrels inhabit mature forests. These species occur from the tropics to the Arctic and are always nocturnal. Like other tree squirrels, they do not hibernate but are active at all seasons.

NORTHERN FLYING SQUIRREL   *Glaucomys sabrinus*
Pl. 10, Fig. 100

DESCRIPTION: A medium-sized, brownish gray squirrel modified for gliding and nocturnal activity. A flap of skin joins the forelimbs and hind limbs and forms a "gliding sail" when the limbs are outstretched. The flattened tail has long hairs on the sides and very short hairs on the top and bottom. The postorbital processes are small, acute, and posterolaterally directed (see fig. 100). The upper premolar is peglike (fig. 100a). TL 250–310 mm, T 115–150 mm, HF 34–40 mm, E 20–29 mm. Dentition: 1/1, 0/0, 2/1, 3/3.

DISTRIBUTION: Occurs in coniferous forests, especially in mature

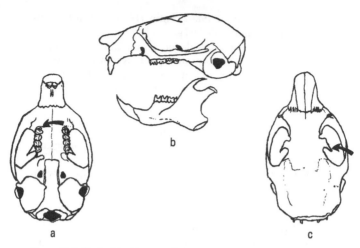

Figure 100. Skull of the Northern Flying Squirrel *(Glaucomys sabrinus):* (a) ventral view (arrow to peglike premolar); (b) lateral view; (c) dorsal view (arrow to small postorbital process).

stands of fir with an open understory. Its gliding behavior requires tall trees with unobstructed spaces between them, and in such a habitat this squirrel may abound. It is found through the North Coast, Klamath Mountains, Cascade Range, and Sierra Nevada from Washington south to Sequoia National Park, as well as in the San Jacinto and San Bernardino Mountains.

FOOD: This squirrel consumes various nuts and berries as well as some leaves, birds' eggs, and insects. In northern California its winter food may consist largely of the lichens growing on large conifers: it may feed heavily on hair moss *(Alectoria fremontii)* and less commonly on a staghorn lichen *(Usnea ceratina).* Hypogeous fungi are an important summer food.

REPRODUCTION: This species nests in hollow trees. Autumnal breeding has been recorded. One litter of two to five young is born between May and June.

STATUS: The San Bernardino Flying Squirrel *(G. s. californicus)* is a California subspecies of special concern.

COMMENTS: This is the only nocturnal squirrel in California. Although seldom observed in the daytime unless disturbed, it is occasionally seen by campers after dark and by entomologists pursuing moths. Its inexplicable attraction to feathers was discovered by early trappers who baited Marten *(Martes americana)* traps with feathers.

The flying squirrel does not actually fly. By gliding from one tree to another, however, losing a little height each time, it may quickly go well over 30 m.

YELLOW-BELLIED MARMOT   *Marmota flaviventris*
Pl. 9

DESCRIPTION: A large, heavy-bodied squirrel with a grizzled russet dorsum, a yellow venter, and a distinctive white patch on the nose. The tail is moderately bushy and rather short, and the eyes are small. The Yellow-bellied Marmot is not likely to be confused with any other California mammal. TL 470–700 mm, T 130–220 mm, HF 70–90 mm, E 18–21 mm. Weight: 1.5–4.0 kg.

DISTRIBUTION: Common about rocky outcrops and along rocky embankments of mountain roads at 2,000 m and above. It occurs widely through the eastern half of Washington and Oregon east to Montana, south to the Cascade Range and Sierra Nevada of California, and on to New Mexico.

FOOD: This marmot eats many green plants, grasses, and forbs. It forages very close to its den.

REPRODUCTION: Mating takes place after the marmot emerges from hibernation (in April or later, depending on the elevation and snow cover). After a 30-day gestation, three to eight young are born. Young emerge from the nest at four or five weeks of age. Some females mate after the first hibernation, when they are nearly one year old.

COMMENTS: Marmots are frequently seen basking in the early morning sun along roadside rock exposures, which afford den sites under rocks and boulders. When fed by tourists, marmots become very tame. They hibernate from late summer or early fall until spring.

Tree Squirrels *(Sciurus* and *Tamiasciurus)*

These are slender, bushy-tailed squirrels without alternating stripes. They have small but erect ears, a short snout, a dome-shaped skull, and postorbital processes that are short and directed posterolaterally.

Tree squirrels are genuine forest dwellers from sea level to the tree line. Conspicuous and loquacious, they are familiar to everyone who walks through wooded areas. They are active on the ground, foraging for fallen seeds and underground fungi, and high in the trees, cutting cones of pines and firs. Tree squirrels do not hibernate.

EASTERN GRAY SQUIRREL *Sciurus carolinensis*
Pl. 10

DESCRIPTION: A large, grayish squirrel with a suffusion of yellow or reddish brown dorsally and a tawny or whitish venter. The tail is long and bushy. The upper premolar is peglike. This species is similar to the Western Gray Squirrel *(S. griseus)* and also resembles the Fox Squirrel *(S. niger)*. Both these species, however, have longer tails. In addition, the Western Gray Squirrel is more silver gray, sometimes slightly russet on the dorsum, lacking the yellow or rusty color; the Fox Squirrel usually has an essentially reddish or russet dorsum and lacks the peglike premolar. TL 445–500 mm, T 184–231 mm, HF 61–70 mm, E 28–35 mm. Weight: 500–625 g.

DISTRIBUTION: Found in parks and wooded streets in many urban areas—including Chico, Sacramento, Stockton, and Palo Alto—and adjacent woodlands in California, Oregon, and Washington. It is sometimes found together with the Fox Squirrel. This species was introduced from the eastern United States, where it occurs in the Mississippi Valley and along the coast from southern Canada to the Gulf of Mexico.

FOOD: This squirrel eats leaves, flowers, and nuts of woody vegetation.

REPRODUCTION: Two to six young are born in early spring after a gestation of 40 to 45 days. Two broods are born per year when food is abundant.

WESTERN GRAY SQUIRREL *Sciurus griseus*
Pl. 10, Fig. 101

DESCRIPTION: A large, silver gray squirrel with a grizzled salt-and-pepper dorsum and white venter. The tail is very large and bushy and occasionally has a yellowish hue. The upper premolar is small and peglike (see fig. 101a). This squirrel is somewhat similar to the Eastern Gray Squirrel *(S. carolinensis)* and the Fox Squirrel *(S. niger)*. The Eastern Gray Squirrel, however, consistently has some yellow or brown on both the back and the belly, is much smaller, and has a smaller tail, whereas the Fox Squirrel lacks the peglike upper premolar and is usually (in California) very russet, frequently with an irregularly colored head. TL 500–575 mm, T 240–280 mm, HF 72–80 mm, E 28–36 mm. Weight: 735–900 g.

DISTRIBUTION: Occurs abundantly in woodlands from sea level to approximately 1,500 m in the central Sierra Nevada. It is found in most of the state except extremely high mountains and the deserts of the southeast. In recent years it has spread out, along wooded streambeds, into a number of localities in the Sacramento Valley. At low elevations it is common in groves of native walnuts , and at higher elevations it is associated with California black oak *(Quercus kelloggii)*. It is less likely than the Eastern Gray Squirrel to be observed in most urban areas. Its range extends from southern Washington to Baja California.

FOOD: Hypogeous fungi (truffles) seem to constitute this squirrel's main fare in most of the state. In contrast to mushrooms,

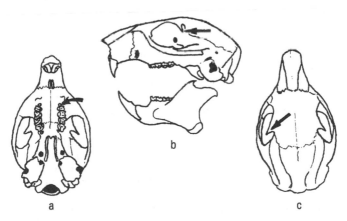

Figure 101. Skull of the Western Gray Squirrel *(Sciurus griseus):* (a) ventral view (arrow to peglike premolar); (b) lateral view; (c) dorsal view (arrows on b and c to postorbital process).

which form a minor part of its diet, these fungi are heavy, solid, and highly nutritious. Foraging both in trees and on the ground, this squirrel also eats a broad variety of fruits and green foliage, as well as the seeds of pines, oaks, and California bay *(Umbellularia californica)*.

REPRODUCTION: One or probably two litters are born each year, depending on the food supply. Litters vary from two to six young.

COMMENTS: These squirrels, among the most beautiful of California mammals, are diurnal and active all year. Unfortunately, they become serious pests in almond, walnut, and filbert orchards.

FOX SQUIRREL   *Sciurus niger*
Pl. 10

DESCRIPTION: A rather large, russet-colored tree squirrel with a long, bushy tail. Its color pattern is extremely variable, but most California individuals are very russet and lack the extensive black areas common in eastern specimens. In California this species also frequently has an irregularly colored head, sometimes with light or dark patches. The belly is often bright yellowish cream. The Fox Squirrel lacks the peglike upper premolar typical of both the Eastern Gray Squirrel *(S. carolinensis)* and Western Gray Squirrel *(S. griseus)*. For further comparisons, see the accounts of those species. TL 475–580 mm, T 220–280 mm, HF 51–80 mm, E 26–33 mm. Weight: 590–700 g.

DISTRIBUTION: Found in urban parks and woodlands of the Central Valley and Coast Ranges. It was introduced from the eastern United States, where it occurs from the northern Mississippi Valley south to the Gulf of Mexico and east to the Atlantic states, from extreme western New York and Pennsylvania south.

FOOD: This squirrel largely eats the nuts of oaks and walnuts and the leaves and buds of large trees.

REPRODUCTION: Probably two broods of one to six, but usually two to four, are born annually. The first litter may be born in March, and a second breeding may occur in late summer. Birth takes place in a nest of leaves and sticks built high in the branches of a tree.

COMMENTS: Present in a number of California cities, this is likely to be the squirrel one sees running about lawns collecting tidbits from admirers; it may also gnaw its way into the attics of homes. It can be destructive in walnut and almond orchards.

DOUGLAS'S SQUIRREL　　　*Tamiasciurus douglasii*
Pl. 10, Fig. 102

DESCRIPTION: A medium-sized tree squirrel with a chestnut brown or olive dorsum, a tawny venter, and a black stripe between the dorsal and ventral pelage in summer. Its color varies seasonally and geographically. The ears are tufted in winter. This species lacks the peglike premolar (see fig. 102a) characteristic of the two species of gray squirrels. TL 330–370 mm, T 110–140 mm, HF 48–55 mm, E 20–28 mm. Weight: 200–300 g.

DISTRIBUTION: Found in coniferous forests from the North Coast Ranges east to the Klamath Mountains, Cascade Range, and Sierra Nevada south to Kern County. Its range also extends from southwestern British Columbia (not including Vancouver Island) south through Washington and Oregon.

FOOD: This squirrel subsists on various conifer seeds, including those of hemlock, pines, firs, and Douglas-fir *(Pseudotsuga menziesii)*. It also eats berries, mushrooms, birds' eggs in season, and, in early spring, the buds and leaves of broad-leaved trees.

REPRODUCTION: One or two litters of four to seven young are born annually, the first in June. The nest is usually in a hollow tree but

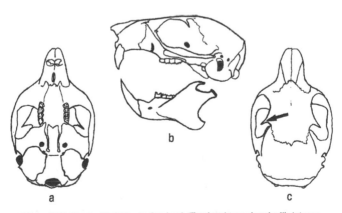

Figure 102. Skull of Douglas's Squirrel *(Tamiasciurus douglasii):* (a) ventral view; (b) lateral view; (c) dorsal view (arrow to postorbital process).

is sometimes a tightly built structure of leaves and twigs wedged high in the branches of a tall tree.

Chipmunks *(Neotamias)*

Chipmunks are small, ground-dwelling squirrels whose tail makes up more than 40 percent of their total length. They are conspicuously striped: five darkly pigmented stripes alternating with four rather lightly pigmented ones run from the snout nearly to the rump, sometimes interrupted behind the head. The skull is delicate, with small postorbital processes directed posterolaterally (fig. 103). Dentition: 1/1, 0/0, 1/1, 3/3.

These beautiful little squirrels are familiar to campers and picnickers. For the most part they dwell in the mountains, but coastal species live at low elevations, and the Least Chipmunk *(N. minimus)* occurs among the scattered and open sagebrush of the Great Basin. The 13 species in California are characterized mostly by slight differences in size, color, and color proportions—features that cannot easily be contrasted in a conventional key. For this reason we have provided comparative comments for these species (table 3); more extensive descriptions are given in the species accounts.

Although chipmunks eat many sorts of nuts, conifer seeds, and berries, they also dig up large numbers of highly nutritious underground fungi (truffles). They may also feed heavily on insects: many years ago the naturalist Vernon Bailey noted that the Least Chipmunk fed on caterpillars of butterflies and moths on sage and other chaparral species. Their seed-eating tendencies frequently make them a nuisance: when foresters reseed conifers on burned areas, chipmunks are prone to dig up the seeds and carry them to their nests.

Chipmunks carry both plague and relapsing fever and should be kept out of mountain cabins.

In North America chipmunks are known from the late Oligocene or earliest Miocene and have been referred to a primitive genus, *Nototamias,* fossils of which are known from the middle Miocene of South Dakota, Nevada, Oregon, and Colorado. Chipmunks also occurred from the Miocene onward in Greece, Pakistan, and China. Interestingly, there is a 4 to 9 million year hiatus (from the Miocene until the Pleistocene) during which chipmunk fossils are unknown in North America. Probably *Neotamias* entered the continent in the Pleistocene, and the Eastern

TABLE 3. Distribution and Diagnostic Features of California Chipmunks (Refer to Map 1 for Localities)

Locality	Species	Features
Regions north of a line from San Francisco Bay to Lake Tahoe		
Extreme northwestern part of state		
	siskiyou	Light stripes with many gray and brown hairs; lateral and median light stripes similar
	senex	Light dorsal stripes very dark, about equally pigmented
Cascades, Klamath Mountians, and North Coast Ranges		
	amoenus	Median light stripes broader than outer light stripes
	sonomae	Dorsal stripes all of about equal width
Between Van Duzen River and Bodega	ochrogenys	Not separable from *senex* in the field
Northern Sierra Nevada (north of Lake Tahoe)		
Mostly above 2,200 m	amoenus	Median light stripes broader than outer light stripes, outer stripes nearly white; sides russet; ear 13–17 mm; dark head stripes brown
Mostly above 2,200 m	speciosus	Medium-sized, larger than *alpinus*; dark head stripes black; outer dorsal stripes white; sides of body reddish; ear medium (18–20 mm) with distinct light patch behind ear
Mostly above 2,200 m	quadrimaculatus	Colors bright and ears long (18–24 mm); conspicuous white stripe from snout and below eye to rear of ear, showing as bright white patch behind ear; lateral light stripes nearly white
Great Basin (east of northern Sierra Nevada)		
	minimus	Very small; grayish with mantle of gray over shoulders; lateral light stripes conspicuous
Mostly above 1,500 m	senex	Rather large and stocky, darkly colored; lateral and median light stripes with many brown and gray hairs; gray behind ear
Great Basin (east of southern Sierra Nevada)		
	panamintinus	Reddish except for top of head, which is gray; lateral light stripes not white; ears moderate (16–18 mm)

Continued ➤

TABLE 3. *Continued*

Locality	Species	Features
Great Basin (east of southern Sierra Nevada)		
	minimus	Very small; grayish with mantle of gray over shoulders; lateral light stripes conspicuous
Mostly above 2,000 m	*senex*	Rather large and stocky, dark colors; lateral and median light stripes with many brown and gray hairs; gray behind ear

Regions south of a line from San Francisco Bay to Lake Tahoe

Locality	Species	Features
South Coast Ranges and Sierra Foothills, 250–1,500 m		
	merriami	Large, dull-colored; all light stripes grayish and not white; grayish spot behind ear
Transverse Ranges		
	merriami	Characters as for *merriami* above
	speciosus	Similar to *merriami* but sides russet and lateral light stripes white
Peninsular Ranges		
	merriami	Characters as for *merriami* above
	obscurus	Much like *merriami;* color paler; dark dorsal stripes reddish; may not always be separable from *merriami* in the field
Southern Sierra Nevada (south of Lake Tahoe)		
Mostly above 3,000 m	*alpinus*	Size very small; dark head stripes brown; dorsal stripes indistinct, light stripes not white
Mostly above 2,500 m	*umbrinus*	Medium-sized with grayish tones to pelage; lateral light stripes white, outer dark stripes faint; ear not conspicuously long (16–19 mm)
Mostly above 2,200 m	*amoenus*	Median light stripes broader than outer light stripes, outer light stripes nearly white; sides russet; ear 13–17 mm
Mostly above 2,000 m	*quadrimaculatus*	Colors bright and ears long (18–24 mm); conspicuous white stripe from snout and below eye to rear of ear, showing as bright white patch behind ear; lateral light stripes nearly white

Chipmunk *(Tamias striatus)* may be related to an earlier stock that crossed the Bering connection from Asia, perhaps as early as the Oligocene. (For details, see Jameson 1999.)

Recent studies have drastically altered the taxonomic relationships of some Pacific Coast chipmunks. These studies involve arrangement and number of chromosomes and also details of the baculum or os penis, a bone that lies within the penis in certain species. Because these structures cannot be employed in field identification, they are omitted from our discussions. Fortunately, species that differ only in these features usually do not occur together.

ALPINE CHIPMUNK   *Neotamias alpinus*

Pl. 11

DESCRIPTION: A small, drab-colored, yellowish chipmunk. Its light and dark body stripes do not contrast strongly, as the lateral light stripes are grayish, not white. The stripes on the head are brown. The ears are short. This species is smaller and darker than others with which it occurs. In addition, in the Lodgepole Chipmunk *(N. speciosus),* the lateral light stripes are white and the sides reddish or russet. TL 166–195 mm, T 70–85 mm, HF 28–31 mm, E 12–14 mm. Weight: 30–42 g.

DISTRIBUTION: Found in relatively open coniferous forests in the southern Sierra Nevada, California, at very high elevations, generally above 3,000 m. This chipmunk ranges higher than any other in California and can be seen well above timberline.

FOOD: This species eats the seeds of many kinds of forbs and grasses and is also known to gather fungi. Perhaps because it dwells at elevations where conifer growth is thin, pine seeds are not a major source of food.

REPRODUCTION: From four to five young are born in June.

COMMENTS: This species is probably only active from late spring to early fall. Very little is known of its biology, however.

YELLOW-PINE CHIPMUNK   *Neotamias amoenus*
Pl. 11

DESCRIPTION: A rather small chipmunk, but larger than the Least Chipmunk *(N. minimus)*, and with a shorter tail. The brown sides and the underside of the tail are distinctly more reddish than in the Least Chipmunk or Alpine Chipmunk *(N. alpinus)*. The lateral light stripes are white. TL 188–202 mm, T 73–85 mm, HF 29–31 mm, E 10–12 mm. Weight: 36–50 g.

DISTRIBUTION: Inhabits yellow pine *(Pinus ponderosa)* and Jeffrey pine *(P. jeffreyi)* forests from the Yosemite area north in the Sierra; found also in the Klamath, Shasta, and Warner Mountains in the north. It ranges east to Colorado and Wyoming and north through Oregon to the Olympic Mountains, Washington, and British Columbia.

FOOD: This species consumes the seeds of many forbs and shrubs, especially manzanita and ceanothus, and of conifers. It also eats substantial numbers of insects, especially in spring, and large quantities of fungi in fall.

REPRODUCTION: A single litter of four to six young is born in early June.

COMMENTS: This species accumulates fat in fall, as do other California chipmunks, but apparently it does not always hibernate. In captivity it remains active in winter under conditions that allow torpidity in the Lodgepole Chipmunk *(N. speciosus)*, Long-eared Chipmunk *(N. quadrimaculatus)*, and Shadow Chipmunk *(N. senex)*.

MERRIAM'S CHIPMUNK   *Neotamias merriami*
Pl. 11

DESCRIPTION: One of the larger California chipmunks. It is rather dull and drab in color: its light stripes and the light patches behind the ears are grayish, not white, and its head is grayish, not brownish. Its tail is relatively long. TL 233–277 mm, T 110–124 mm, HF 30–39 mm, E 16–18 mm. Weight: 60–82 g.

DISTRIBUTION: Found on the lower western

slopes of the southern Sierra and Coast Ranges and in wooded areas (generally above chaparral) south to Baja California.

FOOD: This species eats the seeds of various forbs and shrubs, as well as insects.

REPRODUCTION: Breeding occurs rather early, perhaps because this chipmunk lives at low elevations and is active most of the year. Probably a single litter of three to five young is born in late April.

COMMENTS: Like the Lodgepole Chipmunk *(N. speciosus)*, Merriam's Chipmunk may climb trees and is at home on rail fences.

LEAST CHIPMUNK   *Neotamias minimus*

Pl. 11

DESCRIPTION: The smallest chipmunk in California. It has less distinct dorsal stripes than most species; the dark stripes are wider than the adjacent light stripes. The shoulders and adjacent part of the back are gray. The underside of the tail is yellowish; in the Yellow-pine Chipmunk *(N. amoenus)*, with which the Least Chipmunk could be confused, it is reddish. TL 184–203 mm, T 71–90 mm, HF 28–32 mm, E 12–14 mm. Weight: 27–38 g.

DISTRIBUTION: Found in sagebrush areas, generally east of the Sierra Nevada, north through Oregon to central and eastern Washington. Its range extends south into Arizona and New Mexico, north through the Rocky Mountains and Yukon, and east to northern Minnesota, northern Wisconsin, and the southern margin of Hudson Bay into western Quebec.

FOOD: This species eats the seeds of desert grasses and forbs. Like other chipmunks, it sometimes feeds on insects, especially caterpillars.

REPRODUCTION: A litter of four to six young is born early in the year (April in California).

CHAPARRAL CHIPMUNK   *Neotamias obscurus*

DESCRIPTION: A large, dull-colored chipmunk, probably not separable from Merriam's Chipmunk *(N. merriami)* in the field. Its light body stripes are grayish, and the patches behind its ears are gray, not white. It is distinguished by bony parts of the genitalia of both sexes. TL 208–240 mm, T 75–120 mm, HF 30–37 mm, E 13–40 mm. Weight: 70–90 g.

DISTRIBUTION: Found in the Peninsular Ranges in extreme southern California, as well as in northern Baja California.

FOOD: Unknown.

REPRODUCTION: Breeding begins in February; there are probably more than two litters of three or four young.

COMMENTS: This chipmunk is active throughout the year.

REDWOOD CHIPMUNK   *Neotamias ochrogenys*

DESCRIPTION: A large, dark chipmunk with rather indistinct body stripes. Its general tone is olive in winter and tawny in summer; its colors are dull, never bright. Its light stripes are of about equal pigmentation. TL 252–277 mm, T 107–126 mm, HF 37–39 mm, E 15–18 mm. Weight: 65–92 g.

DISTRIBUTION: Found in the coastal conifer region from Van Duzen River south to Bodega, California, within approximately 40 km of the ocean.

FOOD: This chipmunk consumes a broad variety of small seeds, berries, and flowers of shrubs.

REPRODUCTION: It bears a single litter of three to five young.

PANAMINT CHIPMUNK   *Neotamias panamintinus*
Pl. 11

DESCRIPTION: A medium-sized, brightly colored chipmunk with a gray rump and a ventrally reddish tail. It is larger than the Least Chipmunk (*N. minimus*), in which, in addition, the shoulders and adjacent part of the back are gray and the tail is yellowish ventrally. It is smaller than the Long-eared Chipmunk (*N. quadrimaculatus*) and lacks the long ears and white postauricular patches of that species. TL 190–214 mm, T 80–95 mm, HF 28–31 mm, E 17–18 mm. Weight: 55–70 g.

DISTRIBUTION: Found in the open piñon-juniper forest of the east-

ern slope of the Sierra Nevada, California, as well as in the Kingston Range and on the northern edge of the Mojave Desert.
FOOD: This chipmunk subsists on the seeds of the piñon pine, spring forbs, and insects.

REPRODUCTION: It bears a single litter of three to six young.

LONG-EARED CHIPMUNK   *Neotamias quadrimaculatus*
Pl. 11, Fig. 103

DESCRIPTION: A medium-sized, rather brightly colored species. The lateral light stripes are white or nearly so, neither grayish nor brownish; the underside of the tail is russet. The ears are conspicuously long and slender, and the patches behind them are bright white. The colors, ear length and shape, and white postauricular patches distinguish this species from all others with which it occurs. The postorbital processes are directed laterally. (See fig. 103c.) TL 200–250 mm, T 85–118 mm, HF 34–37 mm, E 18–24 mm. Weight: 52–100 g.
DISTRIBUTION: Found in open mature coniferous forests and logged areas in the Cascade Range and Sierra Nevada from Su-

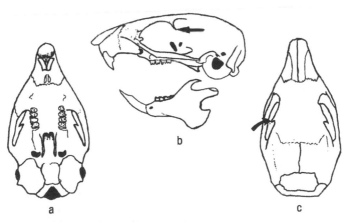

Figure 103. Skull of the Long-eared Chipmunk *(Neotamias quadrimaculatus):* (a) ventral view; (b) lateral view; (c) dorsal view (arrows on b and c to postorbital process).

sanville south to the Yosemite area. It also occurs immediately east of Lake Tahoe in Nevada.

FOOD: This species eats the seeds and leaves of many plants. It takes fresh leaves and sprouting seeds in spring, along with insects and fungi. It feeds heavily on seeds of pines and Douglas-fir (*Pseudotsuga menziesii*) when available. At other times it may subsist on hypogeous fungi.

REPRODUCTION: A single litter of four to six young is born in June.

SHADOW CHIPMUNK   *Neotamias senex*
Pl. 11

DESCRIPTION: A large, rather dull-colored chipmunk similar to both the Siskiyou Chipmunk (*N. siskiyou*) and Redwood Chipmunk (*N. ochrogenys*), with which it does not occur. Its light stripes are grayish or brownish, not white. Its ears are relatively short and lack a bright white postauricular patch. It is larger than the Lodgepole Chipmunk (*N. speciosus*), which, in addition, has reddish or russet sides and white lateral light stripes. TL 229–258 mm, T 95–112 mm, HF 35–38 mm, E 18–24 mm. Weight: 70–98 g.

DISTRIBUTION: Most commonly found in dense growth and streamside thickets in coniferous forests. Its range extends from the Klamath Mountains east to the Warner Range, south through the Cascade Range and Sierra Nevada to the Yosemite area, and north into central Oregon.

FOOD: This species consumes the seeds of forbs and various woody plants, including manzanita, ceanothus, and conifers. It also eats fresh leaves in springtime. At times, especially in fall, it may feed heavily on fungi.

REPRODUCTION: From two to five young are born in May or June; there is probably a single litter annually.

COMMENTS: This species was previously known as *N. townsendii*, which is now known not to occur in California. The common name Shadow Chipmunk refers not only to its dark coloration but also to its tendency to live and forage in forest thickets and other heavily shaded areas. It is a profound hibernator and in the northern Sierra Nevada emerges in early March, when there is still much snow on the ground.

SISKIYOU CHIPMUNK *Neotamias siskiyou*

DESCRIPTION: A large, dull-colored chipmunk similar to both the Shadow Chipmunk (*N. senex*) and Redwood Chipmunk (*N. ochrogenys*), with which it does not occur. Its outer lateral light stripes are grayish or brownish, not white, but lighter than the inner light stripes. It may occur with the Yellow-pine Chipmunk (*N. amoenus*), which is much smaller and has reddish sides and white lateral light stripes. TL 250–268 mm, T 98–117 mm, HF 35–38 mm, E 16–19 mm. Weight: 65–85 g.

DISTRIBUTION: Usually found in heavy coniferous forests in extreme northwestern California up to about 1,200 m. Its range extends north into central Oregon.

FOOD: This species eats wild berries and nuts, including pine nuts and acorns.

REPRODUCTION: One litter of four to six young is born in June.

SONOMA CHIPMUNK *Neotamias sonomae*
Pl. 11

DESCRIPTION: A rather large, light-hued chipmunk. Its lateral light stripes are grayish, not white. The head stripes are brownish or reddish, not black. The underside of the tail is reddish. This species may occur together with Redwood Chipmunk (*N. ochrogenys*), which is much darker. TL 220–227 mm, T 93–103 mm, HF 33–39 mm, E 15–22 mm. Weight: 42–65 g.

DISTRIBUTION: Found in dry, open chaparral at low elevations up to about 800 m in the North Coast Ranges of California and the southern Klamath Mountains.

FOOD: This chipmunk probably subsists on the seeds and leaves of chaparral plants, but little is actually known of its diet.

REPRODUCTION: Breeding takes place from February to the end of August. At least two broods of three to five (usually four) young are born each year.

LODGEPOLE CHIPMUNK
Pl. 11

Neotamias speciosus

DESCRIPTION: A medium-sized chipmunk with white outer dorsal stripes and russet brown sides; the light area behind the ear is not white. Among the species with which it may occur, it most nearly resembles the Long-eared Chipmunk *(N. quadrimaculatus)*; that species, however, has considerably longer ears set off by bright white postauricular patches. The Lodgepole Chipmunk *(N. speciosus)* may also be readily confused with the Yellow-pine Chipmunk *(N. amoenus)*, with which it overlaps in habitat. TL 197–218 mm, T 80–95 mm, HF 30–36 mm, E 18–20 mm. Weight: 30–64 g

DISTRIBUTION: Found in the Sierra Nevada south to the Yosemite area in forests of yellow pine Jeffrey pine, or lodgepole pine. It also occurs in Nevada in the Lake Tahoe area.

FOOD: This species eats the seeds of many grasses, forbs, and trees, especially pines. Among insects it favors caterpillars.

REPRODUCTION: A litter of three to six young is born in early July.

COMMENTS: The Lodgepole Chipmunk is distinctive in spending a large amount of time in trees, and it sometimes climbs when frightened.

UINTA CHIPMUNK
Pl. 11

Neotamias umbrinus

DESCRIPTION: A medium-sized chipmunk of rather dark, mostly grayish hue. The top of the head is gray. The lateral light stripes are white, and the lateral dark stripes are obscure. TL 210–225 mm, T 86–103 mm, HF 30–34 mm, E 16–19 mm. Weight: 52–71 g.

DISTRIBUTION: Found in high-elevation open forests on the eastern slopes of the southern Sierra Nevada up to 3,000 m. Its range extends from northern Arizona to Wyoming, Montana, and Colorado.

FOOD: This chipmunk eats the seeds of various montane trees and shrubs. Piñon nuts are a favorite.

REPRODUCTION: From four to six young are born in late June or early July; young may continue to nurse until mid-August.

Beavers (Castoridae)

These large, stocky rodents are highly modified for an aquatic life. Their fur is dense and water repellent. The legs are short, and the hind feet are completely webbed. The tail is dorsoventrally flattened, muscular, and naked. The eyes are small, the external ears are reduced, and the nostrils are valved. The skull is heavy and deep with no postorbital projections (fig. 105).

Beavers are at home in water and are seldom seen far from a creek or pond. They dive skillfully and have been reported to remain submerged for up to 15 minutes, during which time a very slow heartbeat (bradycardia) reduces blood flow and blood pressure.

One species of beaver is found in North America today, and a separate but similar species is found in Eurasia. About 10 to 14 million years ago giant beavers evolved in North America, and some survived until the end of the glacial epochs, perhaps 10,000 years ago. One of these ancient rodents approximated the Black Bear *(Ursus americanus)* in size.

BEAVER   *Castor canadensis*

Figs. 104, 105

DESCRIPTION: A large rodent, superficially squirrel-like, with the family characteristics described above. It is dark brown dorsally and brownish ventrally. The second claw of each hind foot is divided; the significance of this is not known. TL 1.0–1.2 m, T 260–330 mm, HF 150–200 mm, E 23–29 mm. Weight: 11–26 kg. Dentition: 1/1, 0/0, 1/1, 3/3 (see fig. 105).

DISTRIBUTION: Found in streams and small lakes throughout the northern two-thirds of the state, north to watercourses in Oregon and Washington. It is frequently encountered in the Sacramento–San Joaquin Delta; it occurs up to 2,000 m in the mountains. A pale-colored population exists along the Colorado River. Its range covers most of North Amer-

Figure 104. Beaver *(Castor canadensis).*

ica, from northern Mexico north to Newfoundland and Alaska and east to the Atlantic Ocean.

FOOD: The Beaver eats willows, aspens, poplars, and other broad-leaved trees, as well as grasses and forbs in spring.

REPRODUCTION: Mating takes place in winter (February), and two to six (usually three or four) kits are born 106–110 days later. The young are weaned by two months and disperse at about two years, as they approach sexual maturity.

COMMENTS: The Beaver's gut contains microbiota (bacteria and protozoa) for digesting cellulose, an estimated 30 percent of which is absorbed. Reingestion of feces also aids in absorption of food material and vitamins. The Beaver stores food in winter.

In the past beavers were widely trapped for their fur and their musk glands, which presumably serve to mark territories and

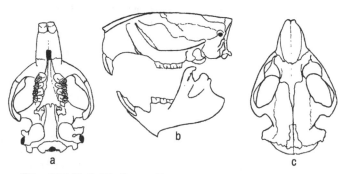

Figure 105. Skull of the Beaver *(Castor canadensis):* (a) ventral view; (b) lateral view; (c) dorsal view.

were once considered to have medicinal value. Today their value as a furbearer is low, about $20 for a good skin, and not more than about 2,000 are taken annually in California.

Beavers are common in many parts of North America and can become a pest. Their presence is frequently indicated by their typical gnawing of tree trunks and construction of dams and lodges (dome-shaped stick houses). In the Delta they typically live in burrows in levees. Some levee breaks in the area are probably at least partly due to their burrowing.

Jumping Mice (Dipodidae)

These small mice are clearly modified for jumping: the hind legs and hind feet are enlarged, and the tail is much longer than the head and body combined. They have yellow sides and a wide brown middorsal stripe. The hind feet have five toes. The anterior orbital foramina are large (fig. 106b). The upper incisors are grooved (fig. 106a); the upper premolar is peglike (fig. 106a).

Jumping mice occur in moist meadows and forests, especially along small creeks, in contrast to pocket mice, which occur in dry areas. In addition, jumping mice have cheek pouches that open within the mouth; the cheek pouches of pocket mice open outside the mouth and are fur lined.

Meadow Jumping Mice *(Zapus)* in California

Two extremely similar species of meadow jumping mice occupying separate ranges are currently recognized in our state. Dentition: 1/1, 0/0, 1/0, 3/3.

Key to Species of Jumping Mice *(Zapus)*

1a   Upper premolar 0.70–0.75 mm in diameter; ear of even pigmentation; coastal redwood forests. . . . . *trinotatus*

1b   Upper premolar 0.70–0.75 mm in diameter; ear with yellow margin; montane forests . . . . *princeps*

WESTERN JUMPING MOUSE   *Zapus princeps*

Fig. 106

DESCRIPTION: A jumping mouse with a dorsum conspicuously darker than its straw yellow sides. The tail is bicolored, and the head is grayish; the ear has a lightly colored or yellow margin.

Upper premolar 0.55 × 0.50 mm (see fig. 106a). TL 215–255 mm, T 121–155 mm, HF 30–37 mm, E 14–18 mm. Weight: 17–28 g. Baculum spatulate at tip.

DISTRIBUTION: Found in meadows and along creeksides in mountains from the Klamath Mountains through the Warner Mountains south to the Sierra Nevada.

FOOD: This mouse largely eats the seeds of grasses and forbs. It also consumes pulpy berries and sometimes insects.

REPRODUCTION: A single litter of four to eight young is born annually.

COMMENTS: These dainty little mice are both diurnal and nocturnal and may occasionally be startled as they forage in tall grass. They escape by making leaps of several feet. They are profound hibernators and in late summer accumulate quantities of both subcutaneous and visceral fat.

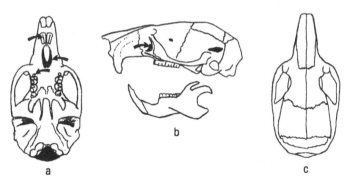

Figure 106. Skull of the Western Jumping Mouse *(Zapus princeps):* (a) ventral view (top arrow to grooved incisors, middle arrow to incisive foramina, bottom arrow to peglike premolar); (b) lateral view (arrow to anterior orbital foramen); (c) dorsal view.

PACIFIC JUMPING MOUSE *Zapus trinotatus*
Pl. 16

DESCRIPTION: A jumping mouse with dark yellowish sides and dark wide band the length of the back and evenly dark ears. Upper premolars 0.70 × 0.75 mm. TL 211–242 mm, T 112–155 mm, HF 30–36 mm, E 13–17 mm. Weight: 16–24 g. Baculum lanceotate at tip.

DISTRIBUTION: Found in coastal meadows from Marin County north to Oregon and Washington and in coastal forests north to southern British Columbia, not including Vancouver Island.

FOOD: This mouse consumes the seeds of forbs and grasses and some leafy material.

REPRODUCTION: One litter of four to eight young is produced annually.

STATUS: The Point Reyes Jumping Mouse (*Z. t. orarius*) is a California subspecies of special concern.

Pocket Gophers, Kangaroo Rats, Kangaroo Mice, and Pocket Mice (Geomyidae)

The pocket gophers are today in the subfamily Geomyinae, and the kangaroo rats, kangaroo mice, and pocket mice are in the subfamily Heteromyinae. They share certain characteristics, such as fur-lined cheek pouches and a transverse infraorbital foramen. They are unlikely to be confused, however. Pocket gophers have a short tail and greatly enlarged forelegs with long claws; kangaroo rats and kangaroo mice have a long tail, short forelegs, and large hind legs; and pocket mice have nearly equal forelegs and hind legs. The groups also have different markings.

The earliest fossils are from the early Oligocene of North America, at which time gophers and kangaroo rats were already distinctive.

Pocket Gophers (Geomyinae)

Pocket gophers are medium-sized, rather stocky rodents greatly modified for burrowing. The tail is nearly nude, the forefeet are enlarged, the eyes are very small, and the ear pinnae are reduced. The skull is broad, flat, and heavy, with small ear capsules. Dentition: 1/1, 0/0, 1/1, 3/3.

These rodents range widely from central Mexico north to the edges of the continental coniferous forests, though they are absent from northeastern North America. Their diggings are apparent on many light soils. Although pocket gophers and moles may occur on the same soils, moles favor moist ground, and pocket gophers favor somewhat drier regions where the ground allows burrowing. (See fig. 107.)

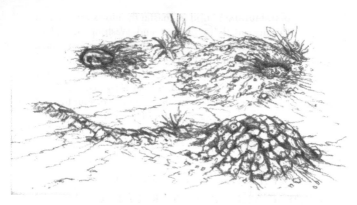

Figure 107. Upper: Fan shaped-mound of Botta's Pocket Gopher *(Thomomys bottae)*, with plugged hole and scratched soil. Lower: Dome mound of a mole *(Scapanus spp.)* and raised tunnel, both of small clods of earth pushed up; no hole.

Pocket gophers build elaborate, branching tunnels, placing the dirt in old or abandoned tunnels or pushing it out onto the surface. In winter they come to the surface and burrow through the snow; later they fill these tunnels with dirt from tunnels through the ground. When the snow melts, these earth cores remain in testimony to their activity during winter.

Pocket gophers are virtually always solitary, socializing only during the breeding season. One individual occupies a system of burrows and alone may account for the destruction of many plants and vegetables in a garden. The fortunate aspect to this unpleasant situation is that the offender can be removed. Gopher traps are best for this purpose. Poison can also be effective, especially when applied on fresh baits, such as carrots, and placed within a tunnel. Commercially prepared baits are much less attractive to pocket gophers.

These little rodents are notoriously variable. Taxonomists have named innumerable populations with minor morphological distinctions and sometimes rather major differences. In the past many of these geographic variants were described as separate species, but careful study reveals that the vast majority of them intergrade with one another. Very few distinct species are recognized today. Generally, they are mutually exclusive and do not share the same habitat. In certain regions, such as northern California, there may be several species, each in a different habitat.

Key to Species of Pocket Gophers (Thomomys) in California (modified from Hall, 1981)

1a Sphenoidal fissure absent (figs. 109b, 110b, 111b)........2
1b Sphenoidal fissure present (figs. 109b, 112b) 4
    2a Ear relatively long, more than 6.9 mm from notch....
     3
    2b Ear relatively short, less than 6.9 mm from notch.....
      ....*talpoides*
3a Reddish brown to black; ear from notch 7–8.5 mm........
    ....*mazama*
3b Dark brown; ear from notch 8–9 mm ....*monticola*
    4a Hind foot 30–40 mm; northeastern part of state .....
      ....*townsendii*
    4b Hind foot 22–35 mm; widely distributed ......*bottae*

BOTTA'S POCKET GOPHER   *Thomomys bottae*
Pl. 12, Figs. 107, 108

DESCRIPTION: A highly variable species in both size and color. Its color tends to match the soil, usually dull brown but sometimes yellow, buff, or black. The skull has a distinct and deep sphenoidal fissure (see fig. 108b). It also has incisive foramina, mostly behind the rear margin of the infraorbital canal (see fig. 108a). TL 190–300 mm, HF 22–35 mm, E 5–8 mm. Weight: 102–209 g.
DISTRIBUTION: Prefers light soils but is occasionally found on clay. It occurs virtually statewide, though it is absent from higher elevations in the Sierra Nevada and Cascade Range. It may occur in some areas with the Mountain Pocket Gopher (*T. monticola*) and, in northwestern California, with the Mazama Pocket Gopher (*T. mazama*). Its range extends north to Oregon and Colorado and south to northern Mexico, including Baja California.
FOOD: Attracted to a continuously growing root system, it eats roots, bulbs, and the tender bases of growing plants.
REPRODUCTION: Breeding takes place from late winter to summer,

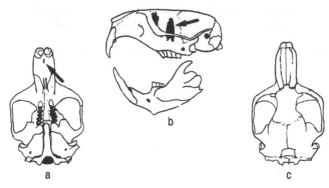

Figure 108. Skull of Botta's Pocket Gopher *(Thomomys bottae):* (a) ventral view (arrow to incisive foramina); (b) lateral view (arrow to sphenoidal fissure); (c) dorsal view.

but the reproductive season is prolonged where the land is irrigated. There may be from one to four litters of two to 12 young.

COMMENTS: This gopher is a pest on agricultural lands and in home gardens and lawns. It is especially fond of alfalfa and can also be very destructive to potatoes, sugar beets, and carrots. On serpentine soils, where there is a very restricted flora, it feeds almost exclusively on the corms of small lilies (*Brodiaea* spp.) and may indeed depend upon them.

MAZAMA POCKET GOPHER   *Thomomys mazama*
Pl. 12, Fig. 109

DESCRIPTION: A small, reddish brown pocket gopher, especially reddish dorsally. There is black on the head area, including black patches about the eyes. The feet are gray, scarcely lighter than the forearms. The ear is shorter than in the Mountain Pocket Gopher *(T. monticola)*. The premaxillary bones extend far beyond the nasal bones (see fig. 109c). The skull has no sphenoidal fissure but does have incisive foramina anterior to the infraorbital canal (see fig. 109a). TL 175–262 mm, T 45–79 mm, HF 21–32 mm, E 7–8. mm5. Weight: 90–120 g.

DISTRIBUTION: Found on light soils and in cultivated areas in the

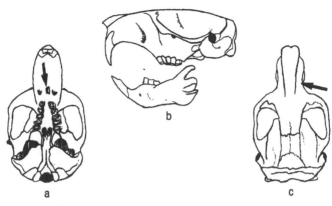

Figure 109. Skull of the Mazama Pocket Gopher *(Thomomys mazama):*
(a) ventral view (arrow to incisive foramina); (b) lateral view; (c) dorsal
view (arrow to premaxillary bone).

northern parts of California. It occurs in the Cascade Range west
to the mountains of northwestern California, east to the west side
of Mount Shasta, and north through western Oregon.

FOOD: This species eats the underground parts of plants and some
leaves. It is known to feed on both the leaves and roots or tubers
of potatoes, clover, alfalfa, and carrots.

REPRODUCTION: Breeding takes place through spring and summer. Litters vary from five to seven, and females probably produce at least two litters per year.

MOUNTAIN POCKET GOPHER   *Thomomys monticola*
Pl. 12, Fig. 110

DESCRIPTION: A brown pocket gopher of medium size.
The fur on the back of the forefeet is white, the fur
about the ears black. The tail has more hairs than in
other species. The ears are pointed and relatively
long, in contrast to those of the Mazama
Pocket Gopher *(T. mazama).* The posterior margins of the nasal bones form a V;
the premaxillary bones extend approximately as far back as the nasals (see fig.

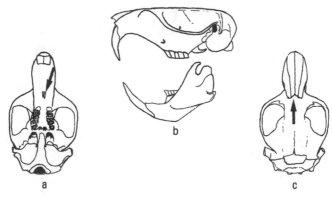

Figure 110. Skull of the Mountain Pocket Gopher *(Thomomys monticola)*: (a) ventral view (arrow to incisive foramina); (b) lateral view; (c) dorsal view (arrow to nasal bones).

110c). The skull lacks a sphenoidal fissure; the incisive foramina are at least partly anterior to the rear margin of the infraorbital foramina (see fig. 110a). TL 190–220 mm, T 55-80 mm, HF 26–30 mm, E 8–9 mm. Weight: 50–105 g.

DISTRIBUTION: Found on light sandy or occasionally on gravelly soils in the Cascade Range and Sierra Nevada from the west side of Mount Shasta to central Modoc County. A separate population lives south of Pit River from central Shasta County to Fresno County, from 1,000 to 2,700 m. This species also occurs in extreme western Nevada.

FOOD: This pocket gopher eats leaves and underground parts of plants. In winter, it feeds on leaves as it burrows aboveground under a protective cover of snow.

REPRODUCTION: Breeding takes place in spring and early summer. Probably one litter of three to four young is born each year. Young become sexually mature at one year.

NORTHERN POCKET GOPHER   *Thomomys talpoides*
Fig. 111
DESCRIPTION: A rather small, brown or yellow brown pocket gopher with gray feet and black ear patches. The fur is grayer in win-

ter. The posterior ends of the nasal bones are square ended, not V shaped as in *monticola,* and the premaxillary bones extend slightly behind the nasals (see fig. 111c). There is no sphenoidal fissure; the incisive foramina are anterior to the rear margin of the infraorbital canal (see fig. 111a). TL 170–225 mm, T 47–70 mm, HF 24–30 mm, E 5–6 mm. Weight: 74–110 g.

DISTRIBUTION: Generally found on sandy or loose soils in Modoc Plateau, south to Honey Lake; also occurs in the Mono Lake area and in Big Valley (Bieber) in Lassen County. Its local range extends north through eastern Oregon and eastern Washington, with isolated populations in the higher Olympic Mountains, Vail Prairie, Rochester Prairie, and the lower Columbia River Valley. This is a widely distributed species, found from southern Canada to northern Arizona and New Mexico and east to the Dakotas.

FOOD: This pocket gopher largely consumes underground parts of forbs and woody species. It is known to eat such species as dandelion, yarrow, sagebrush, and penstemon.

REPRODUCTION: Breeding takes place in June and July; the female probably bears one litter of four to six young each year.

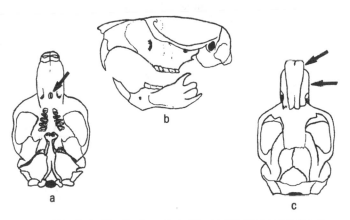

Figure 111. Skull of the Northern Pocket Gopher *(Thomomys talpoides):* (a) ventral view (arrow to incisive foramina); (b) lateral view; (c) dorsal view (top arrow to nasal bones, bottom arrow to premaxillary bones).

TOWNSEND'S POCKET GOPHER *Thomomys townsendii*
Fig. 112

DESCRIPTION: The largest pocket gopher in California. It occurs in two color phases: the gray phase has black ears, whereas the black phase has small white patches on the chin and/or feet. The ears are relatively small. A sphenoidal fissure is present (see fig. 112b). TL 240–305 mm, T 72–100 mm, HF 35–40 mm, E 5–8 mm. Weight: 220–280 g.
DISTRIBUTION: Found in northeastern California, usually on deep soils; frequently abundant on irrigated land. It does not occur with any other species of pocket gopher. Its range extends east to Nevada and north to southeastern Oregon and southwestern Idaho.

FOOD: This pocket gopher consumes roots and aboveground parts of many forbs and feeds heavily on grasses.

REPRODUCTION: Breeding takes place in late winter or early spring in undisturbed, nonirrigated environments. A litter usually comprises six to eight young. Females experience postpartum estrus and sometimes are both pregnant and lactating.

COMMENTS: This species is sometimes destructive on irrigated land, where it may damage such crops as potatoes and alfalfa.

This pocket gopher is notable for the size of its burrows. The

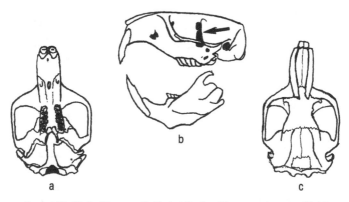

Figure 112. Skull of Townsend's Pocket Gopher *(Thomomys townsendii):* (a) ventral view; (b) lateral view (arrow to sphenoidal fissure); (c) dorsal view.

mounds of dirt it brings to the surface may measure up to 1 m across. Because it seems to favor moist soils, it is sometimes destructive to irrigation levees.

Kangaroo Rats, Kangaroo Mice, and Pocket Mice (Heteromyinae)

These small rodents are variously modified, morphologically, physiologically, and behaviorally, for a nonfossorial life in a semiarid or arid environment. The tail is elongated. There are well-developed, external, fur-lined cheek pouches and greatly enlarged ear capsules. Dentition: 1/1, 0/0, 1/1, 3/3.

These rodents are typical of rather sparsely vegetated, seasonally arid or desert lands, grasslands, and chaparral, and frequently constitute a substantial part of the total small rodent population. Like many animals of open areas, kangaroo rats and kangaroo mice are adapted for saltatory, or jumping, locomotion. They use their cheek pouches to carry seeds, which they take to their underground nests or hide in surface caches.

Kangaroo rats (*Dipodomys* spp.) are conspicuously marked and have a dorsal skin gland. Pocket mice (*Chaetodipus* and *Perognathus* spp.) and kangaroo mice (*Microdipodops* spp.) are usually more uniform in color and have no dorsal gland. In many areas pocket mice and kangaroo rats occur together; in some areas all three genera may live on the same ground.

Key to Genera of Heteromyinae in California

1a   Tail without conspicuous lateral white stripe; pelage rather evenly colored dorsally (sometimes grizzled but not spotted or striped); no dermal gland on back 2

1b   Tail with conspicuous lateral white stripe; pelage of contrasting light and dark spots or stripes above eyes and behind ears, with white stripe on thigh; clearly marked dermal gland on neck between shoulder blades .... *Dipodomys*

    2a   Tail tapering in width; sole of hind foot sometimes hairy; auditory bullae only slightly enlarged ....
    .... *Perognathus* and *Chaetodipus*

    2b   Tail wider at middle than at base; sole of hind foot well furred; auditory bullae greatly enlarged ....
    .... *Microdipodops*

Kangaroo Rats *(Dipodomys)*

These rats are conspicuously spotted and striped dorsally and have a dorsal skin gland. There are dark spots above the eyes and ears and a distinctive white stripe on the thigh. The tail has a dorsal (and usually a ventral) pigmented stripe, is usually crested, and nearly always has a dark brown or black tip with a conspicuous tuft of hairs. The fifth toe on the hind foot is vestigial or lacking. The very large ear capsules make the skull triangular (or nearly heart shaped when viewed from above). The ear pinnae are usually rather small. The molars are flat crowned; in adults, the enamel forms a circle or oval not interrupted by a groove. The teeth grow from a persistent pulp; the roots do not close. All species have unique karyotypes.

Kangaroo rats occur commonly on rather loose, sandy soils and are well adapted for a desert life. They survive water scarcity by releasing metabolic wastes in very concentrated urine and very dry feces. This ability to secrete high concentrations of salts is so well developed that a kangaroo rat can drink seawater and still maintain proper water balance in its tissues. It can subsist on so-called metabolic water bound chemically and released by digestion of its dry food. Some species can obtain water from leaves, which are very salty.

In the laboratory these rodents can adjust their metabolism to an air temperature of 37 degrees C. Behavioral adaptations provide escape from higher desert temperatures. The air may reach 45 degrees C just above the surface of the ground, and the soil surface may reach an intolerable 70 degrees C, but the kangaroo rat's burrow is deep enough to remain a warm but comfortable 30 to 34 degrees C. In areas of creosote bush *(Larrea tridentata)* or other large desert shrubs, kangaroo rat burrows are easily located. Each rat owns a castlelike mound of soil or sand surrounding the base of a bush, and there are usually several entrances, the most frequently used made obvious by a profusion of radiating tracks and tail marks.

The conspicuous dorsal gland seems to have a role in scent marking. This may replace urine marking, which would be metabolically expensive for a mammal that must conserve water. Sand bathing is a characteristic activity of kangaroo rats; among captives, odor in sand seems to attract them to habitual sand-bathing sites. Another means of communication is thumping with the hind feet. (See fig. 114.)

The reproductive season of at least some kinds of kangaroo rats seems to be correlated with rainfall and the subsequent growth of forbs and grasses. Recently ecologists have pointed out the increased intake of fresh leaves by breeding kangaroo rats, a phenomenon noted by Joseph Grinnell in 1932.

In the desert these animals are the most commonly seen mammals crossing the road at night, especially roads not bordered by ditches. To the uninitiated, they are a startling sight, appearing like small ghosts hopping about the pavement. In campgrounds, kangaroo rats quickly learn to approach humans and accept proffered tidbits, not in the least disturbed by flashlights.

Key to Species of Kangaroo Rats *(Dipodomys)* in California (modified from Hall 1981)

1a Hind foot with four toes.................................. 2
1b Hind foot with five toes (fifth toe represented by claw above inside of foot) 5
    2a Tail length 160 mm or less 3
    2b Tail length 161 mm or more........................ 4
3a North of Tehachapi Mountains; San Joaquin Valley, California.. *nitratoides*
3b South and east of Tehachapi Mountains, California .... *merriami*
    4a Dorsal color dark brown .... *californicus*
    4b Dorsal color pale buff or whitish .... *deserti*
5a Head and body usually less than 130 mm 6
5b Head and body usually more than 130 mm; Merced and Kern Counties south to Santa Barbara County, California .... *ingens*
    6a East of Sierra Nevada, Transverse Ranges (Tehachapi Mountains), and South Coast Ranges.............. 7
    6b West of the Sierra Nevada, Transverse Ranges (Tehachapi Mountains), and South Coast Ranges 9
7a Lower incisors rounded on anterior surface 8
7b Lower incisors flat on anterior surface (fig. 115c) .... *microps*
    8a Hind foot usually less than 44 mm; total length less than 280 mm................................ *ordii*
    8b Hind foot usually more than 44 mm; total length more than 280 mm .... *panamintinus*

9a   South of Coast Ranges, Point Conception south 10

9b   North from Point Conception; Coast Ranges and Central Valley. 12

   10a   Ear pinna completely yellow brown; dorsal and ventral tail stripes with many white hairs . . . . *stephensi*

   10b   Ear pinna with white (or light) spot at base; dorsal and ventral tail stripes with very few white hairs 11

11a Southern California, Los Angeles Basin to foothills of southern Sierra Nevada; skull relatively narrow; auditory bullae oval when viewed from above . . . . *agilis*

11b Southern California, south through most of Baja California; auditory bullae kidney shaped when viewed from above. . . . . *simulans*

   12a   Zygomatic arch with posterior margin rounded, shallowly concave (fig. 116a); ear 15–18 mm 13

   12b   Zygomatic arch with posterior margin straight, distinctly set off from a ventral concavity (fig. 114a); ear 12–15 mm. . . . . *heermanni*

13a Ear mostly brown, 16–18 mm from crown; ventral dark tail stripe narrower than lateral white stripe at midpoint of tail . . . . *elephantinus*

13b Ear mostly black, 15–16 mm from crown; ventral dark tail stripe wider than lateral white stripe at midpoint of tail . . . . *venustus*

PACIFIC KANGAROO RAT   *Dipodomys agilis*

DESCRIPTION: A medium to large-sized kangaroo rat. Its tail has a brown or black tuft and dorsal and ventral dark stripes with very few white hairs. The hind foot has five toes. This species is similar to Stephen's Kangaroo Rat *(D. stephensi)*, but the Pacific Kangaroo Rat *(D. agilis)* has a white or light spot at the base of the ear pinna, whereas Stephen's Kangaroo Rat does not. The Pacific Kangaroo Rat is also similar to the Baja California Kangaroo Rat *(D. simulans)*, but when viewed from above, the auditory bullae are oval in the Pacific Kangaroo Rat and kidney shaped in the Baja California Kangaroo Rat. TL 265—319 mm, T 155–197 mm, HF 40–46 mm. Weight: 67–76 g.

DISTRIBUTION: Found in sagebrush and chaparral on sandy soils of the South Coast Ranges, up to 2,200 m in the San Gabriel Mountains. Its range extends south into Baja California.

FOOD: This species eats the seeds of forbs, grasses, and shrubs and is known to store seeds of laurel-sumac *(Rhus laurina)* and chamise *(Adenostoma fasciculatum)*. It also eats some insects.

REPRODUCTION: Two to four young are born in June or early July.

CALIFORNIA KANGAROO RAT · *Dipodomys californicus*

DESCRIPTION: A rather large, dark kangaroo rat with a white-tipped tail. The hind foot has four toes. TL 270–340 mm, T 174–194 mm, HF 39–47 mm. Weight: 57–78 g.

DISTRIBUTION: Found in brushy areas and chaparral in the western North Coast Ranges, up to 1,300 m in El Dorado County. It ranges north into extreme south-central Oregon.

FOOD: This species eats berries and seeds of grasses, forbs, and shrubs, such as manzanita, buckthorn *(Ceanothus)*, and rabbitbrush *(Chrysothamnus)*. It also eats green leaves.

REPRODUCTION: Breeding takes place from late winter to summer. There are two to four young in a litter.

STATUS: The Marysville California Kangaroo Rat *(D. c. eximus)* is a California subspecies of special concern.

DESERT KANGAROO RAT · *Dipodomys deserti*
Pl. 13

DESCRIPTION: A very large kangaroo rat, very pale buff in color. The tail has a white tip and usually also lacks a ventral dark stripe. The hind foot has four toes. TL 305–377 mm, T 180–215 mm, HF 50–58 mm. Weight: 83–138 g.

DISTRIBUTION: Lives in areas of extremely loose, dry sand in the deserts of southeastern California, southern Arizona, and northern Mexico.

FOOD: The Desert Kangaroo Rat feeds on seeds and leaves of desert forbs; it may consume green vegetation in quantity in winter and spring.

REPRODUCTION: Breeding occurs from midwinter to summer and throughout the year in some areas. There are three to four young in a litter.

COMMENTS: The tracks of this kangaroo rat are frequently noticed in the loose sand before the animal itself is seen. Burrow entrances are often found at the base of the creosote bush (*Larrea tridentata*).

BIG-EARED KANGAROO RAT   *Dipodomys elephantinus*

DESCRIPTION: A large, dark kangaroo rat with large ears and five toes on the hind foot. In the terminal half of the tail, the dark ventral stripe is narrower than the lateral white stripe (in the Narrow-faced Kangaroo Rat [*D. venustus*], the reverse is true). TL 305–336 mm, T 183–210 mm, HF 44–50 mm, E 16–18 mm (from crown). Weight: 81–97 g.

DISTRIBUTION: Occurs in sagebrush and chaparral on the eastern slope of the Coast Ranges in San Benito County, California.

FOOD: This species eats the seeds and leaves of forbs and shrubs.

REPRODUCTION: One or two litters of two to three young are born in late winter and spring.

STATUS: This is a California species of special concern.

HEERMANN'S KANGAROO RAT   *Dipodomys heermanni*

Pl. 13, Fig. 113

DESCRIPTION: A medium to large, dark kangaroo rat. The hind foot has five toes. The upper rear margin of the zygomatic arch is straight and sharply set off from a distinctly lower concavity (see fig. 113a). TL 250–340 mm, T 160–200 mm, HF 38–47 mm. Weight: 56–74 g.

DISTRIBUTION: Inhabits brushy and grassy slopes and flats of chaparral-covered hillsides in the South Coast Ranges. It is also found in most of the Central Valley, north to the San Francisco Bay Area and the foothills of El Dorado County, California.

FOOD: This species is known to gather seeds from a broad spectrum of forbs and grasses, including red brome (*Bromus rubens*),

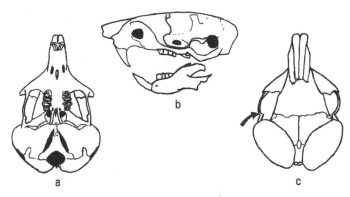

Figure 113. Skull of Heermann's Kangaroo Rat *(Dipodomys heermanni):* (a) ventral view; (b) lateral view; (c) dorsal view (arrow to zygomatic arch).

lupines, and red-stem filaree *(Erodium cicutarium)*, as well as green leaves and occasionally insects.

REPRODUCTION: A variable number of litters, each containing two to five (usually four) young, are born from February to August. Young born early in the year breed in the same year.

STATUS: The Morro Bay Kangaroo Rat *(D. h. morroensis)* is federally listed as endangered, listed by the state as endangered, and fully protected in California.

COMMENTS: One student noted that this kangaroo rat gleaned seeds from the entrances to dens of the Black Harvester Ant *(Veromessor pergandei)*.

GIANT KANGAROO RAT   *Dipodomys ingens*

DESCRIPTION: A very large, brownish kangaroo rat with a light brown tip to the tail. The tail is relatively short (usually less than 130 percent of the length of the head and body). The hind foot has five toes. TL 311–348 mm, T 157–198 mm, HF 46–55 mm. Weight: 131–180 g.

DISTRIBUTION: Found in open areas on fine soils on the southwest side of the San Joaquin Valley. It occurs in Merced and Kern Counties south to Santa Barbara County, California.

FOOD: The Giant Kangaroo Rat eats the seeds and sometimes leaves of such plants as pepper-grass *(Lepidium nitidum)*, red-stem filaree *(Erodium* *cicutarium)*, saltbush *(Atriplex)*, and cudweed *(Gnaphalium)*. It is known to store seeds.

REPRODUCTION: Breeding takes place from February to June or later. Litters commonly contain four to five or even six young, a rather large number for a kangaroo rat.

STATUS: This species is federally listed and listed by the state as endangered.

COMMENTS: The Giant Kangaroo Rat has declined with the increase in cultivation of its range. Also, cattle probably destroy the burrow system close to the surface. Continuous grazing by cattle, however, reduces the introduced annual grasses, which, if allowed to increase, would destroy the habitat of the Giant Kangaroo Rat. On conservation lands, such as Bureau of Land Management and Nature Conservancy lands of the Elkhorn and Carrizo Plains, grazing continues only because of its benefit to this endangered species.

MERRIAM'S KANGAROO RAT *Dipodomys merriami*
Pl. 13, Fig. 114

DESCRIPTION: A rather small, buff-colored kangaroo rat with four toes on its hind foot. It is rather similar to Ord's Kangaroo Rat *(D. ordii)*, from which it can be distinguished by the absence of a minute fifth toe on the hind foot. TL 220–260 mm, T 123–160 mm, HF 34–39 mm. Weight: 39–52 g.

DISTRIBUTION: Lives on light, sandy soils across much of the southern half of the California Coast Ranges. It is found in the Mojave and Colorado Deserts and also in a small area in extreme northeastern California. Its range extends north to northern Nevada, east to Texas, and south into Mexico, including Baja California. When found in sand dunes together with the Desert Kangaroo Rat *(D. deserti)*, Merriam's Kangaroo Rat tends to occur on rocky patches, though the two species' burrows may be separated by only about 1 m.

FOOD: This species eats the seeds of grasses and forbs, probably exploiting what species are available. It is known also to feed on leaves of *Franseria* and other forbs, including winter annuals.

REPRODUCTION: Breeding may take place from January to June or later. A litter contains one to five young. Reproduction seems to

Figure 114. Two Merriam's Kangaroo Rats *(Dipodomys merriami)* kick-fighting in a territorial dispute.

depend on the growth of winter annuals, which is much greater following falls with heavy rain. Thus an extended breeding season follows heavy fall rains, and reproduction may not occur in dry years.

STATUS: The San Bernardino Kangaroo Rat *(D. m. parvus)* is federally listed as endangered and is a California subspecies of special concern.

CHISEL-TOOTHED KANGAROO RAT   *Dipodomys microps*
Fig. 115

DESCRIPTION: A medium-sized kangaroo rat of dark color. The hind foot has five toes. The anterior face of the lower incisors is flat (see fig. 115c). This species is similar to Ord's Kangaroo Rat *(D. ordii)*, which has lower incisors with rounded or awl-shaped anterior surfaces. It may also be found with the Panamint Kangaroo Rat *(D. panamintinus)*, which has a larger hind foot and rounded lower incisors. TL 244–290 mm, T 140–173 mm, HF 38–44 mm. Weight: 55–75 g.

DISTRIBUTION: Found in piñon-juniper associations on rather light soils in southeastern California, mostly the Mojave Desert.

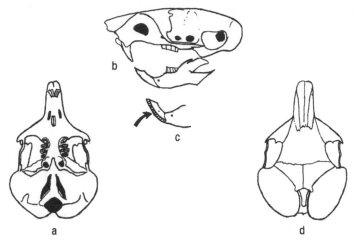

Figure 115. Skull of the Chisel-toothed Kangaroo Rat *(Dipodomys microps)*: (a) ventral view; (b) lateral view; (c) lower incisor enlarged (arrow to anterior surface); (d) dorsal view.

It also lives in sagebrush flats. Its range extends through Nevada north to southeastern Oregon, east to Salt Lake, and south to northern Arizona.

FOOD: This kangaroo rat eats the seeds and especially the leaves of desert grasses and forbs. It is known to gather the seeds of saltbush *(Atriplex)*, pepperwort *(Lepidium)*, and clover and the leaves of sagebrush. Fresh leaves constitute the bulk of its diet.

REPRODUCTION: Breeding takes place from late winter through spring and, in wet years, summer. There are one to four (usually two) young to a litter.

COMMENTS: This species is known to climb bushes to gather food. The flattened lower incisors are used to shave off the outer, saline layers of the leaves of saltbush, exposing the succulent inner tissue, which is eaten.

SAN JOAQUIN KANGAROO RAT *Dipodomys nitratoides*
Pl. 13

DESCRIPTION: A somewhat small, rather dark kangaroo rat with four toes on its hind foot. It has a relatively short, buff-tipped tail.

TL 211–253 mm, T 120–152 mm, HF 33–37 mm. Weight: 39–47 g.

DISTRIBUTION: Found in alkali sink communities of western Fresno and Tulare Counties south to Kern County, San Luis Obispo County, and Ventura County, California. It most closely resembles Merriam's Kangaroo Rat *(D. merriami)*, which does not occur in the San Joaquin Valley.

FOOD: Though its diet is apparently not known, this species presumably feeds on the seeds of such annuals as filaree, shepherd's purse *(Capsella bursa-pastoris)*, and perennials such as saltbush *(Atriplex)*, which are abundant in its habitat.

REPRODUCTION: Breeding takes place throughout the year. Known litters have consisted of two young.

STATUS: The northern subspecies, the Fresno Kangaroo Rat *(D. n. exilis)*, is federally listed and listed by the state as endangered but is likely extinct. The Tipton Kangaroo Rat *(D. n. nitratoides)* is federally listed and listed by the state as endangered. The Short-nosed Kangaroo Rat *(D. n. brevinasus)* is a California subspecies of special concern.

ORD'S KANGAROO RAT   *Dipodomys ordii*

DESCRIPTION: A medium-sized, brownish kangaroo rat. The hind foot has five toes and (in California) is shorter than 44 mm. The anterior edge of the lower incisors is rounded. TL 208–281 mm, T 100–163 mm, HF 33–41 mm. Weight: 50–61 g.

DISTRIBUTION: Inhabits the sagebrush deserts of the Great Basin. Its range extends north through eastern Oregon to south-central Washington and southern Canada, east to central Kansas, and south to central Mexico.

FOOD: This species eats the seeds of grasses, forbs such as *Helianthus,* and shrubs such as mesquite and *Ambrosia,* as well as the tender growing leaves of forbs, and some insects.

REPRODUCTION: From one to six young are born from late winter to early summer. There may be more than one brood per year.

PANAMINT KANGAROO RAT   *Dipodomys panamintinus*
Pl. 13

DESCRIPTION: A rather large kangaroo rat whose color varies from ashy gray to dark brown or cinnamon brown. It is distinguished from the Desert Kangaroo Rat *(D. deserti)*, which is very pale and with which it may occur, by its darker color and by the five toes on its hind foot (the Desert Kangaroo Rat has four). Its lower incisors have a rounded anterior surface. TL 285–334 mm, T 156–202 mm, HF 42–48 mm (usually more than 44). Weight: 64–81 g.

DISTRIBUTION: Found in Joshua Tree and piñon-juniper associations in the eastern California deserts (Mojave Desert and Great Basin), north to Plumas County.

FOOD: This kangaroo rat is known to feed on green leaves of forbs and on juniper berries.

REPRODUCTION: Breeding takes place from March through July; more than a single brood a season is the rule. A second mating occurs shortly after the first birth, so not infrequently females are both pregnant and lactating. Litters vary from three to five young (usually four).

BAJA CALIFORNIA KANGAROO RAT   *Dipodomys simulans*
DESCRIPTION: A medium-sized kangaroo rat with five toes on the hind foot. It is similar to the Pacific Kangaroo Rat *(D. agilis)*, but the skull is relatively broader and the auditory bullae are kidney shaped rather than oval.

DISTRIBUTION: Found in southern California (Los Angeles Basin) south through Baja California.

FOOD: Unknown.

REPRODUCTION: Unknown.

STEPHENS' KANGAROO RAT   *Dipodomys stephensi*
DESCRIPTION: A medium-sized, rather dark kangaroo rat with five toes on the hind foot. The tail is crested and conspicuously striped,

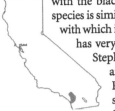

with the black stripes having many white hairs. This species is similar to the Pacific Kangaroo Rat *(D. agilis)*, with which it may occur, but the Pacific Kangaroo Rat has very few white hairs in the black tail stripes. Stephens' Kangaroo Rat also has ear pinnae of an even yellow brown color, whereas in the Pacific Kangaroo Rat, the pinnae have light spots at the base. TL 277–300 mm, T 164–180 mm, HF 39–43 mm. Weight: 60–74 g.

DISTRIBUTION: Lives in sagebrush areas and grassy patches on sandy or gravelly soils. It is known from the San Jacinto Valley of San Diego, Riverside, and San Bernardino Counties.

FOOD: Species of sage, buckwheat, horehound, dock, and goldenbush are common in this rat's habitat; presumably their seeds are its staple diet.

REPRODUCTION: A litter of two or three young is born in late spring.

STATUS: This species is federally listed as endangered and listed by the state as threatened.

NARROW-FACED KANGAROO RAT *Dipodomys venustus*
Pl. 13, Fig. 116

DESCRIPTION: A large, dark kangaroo rat. The dark ventral stripe is wider than the lateral white stripe in the terminal half of the tail. The hind foot has five toes. The ears are larger than in most members of the genus except for the Big-eared Kangaroo Rat *(D. elephantinus)*. The zygomatic arch has a rounded posterior margin. TL 293–332 mm, T 175–203 mm, HF 44–47 mm, E 15–16 mm (from crown). Weight: 66–74 g. (See fig. 116a.)

DISTRIBUTION: Occurs in open sandy areas or dense chaparral in the South Coast Ranges from San Francisco Bay to Point Conception.

FOOD: This species eats the seeds of grasses, forbs such as bur clover *(Medicago)*, and shrubs.

REPRODUCTION: One or two litters of two to four young are born each year.

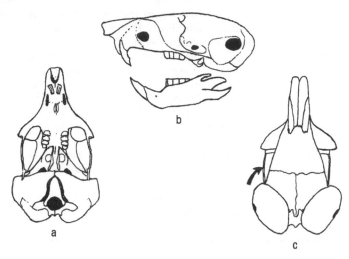

Figure 116. Skull of the Narrow-faced Kangaroo Rat *(Dipodomys venustus)*: (a) ventral view; (b) lateral view; (c) dorsal view (arrow to zygomatic arch).

COMMENTS: This species may be a geographic race of the Pacific Kangaroo Rat *(D. agilis),* but the two can be separated on the basis of their distributions. Alternatively, it may be conspecific with the Big-eared Kangaroo Rat *(D. elephantinus).*

Kangaroo Mice *(Microdipodops)*

These large-headed rodents have small forelegs and much larger hind legs. Their long, untufted tails are constricted at the base, slightly swollen in the middle, and tapered toward the tip. Like the pocket mice *(Chaetodipus* and *Perognathus* spp.) but unlike the kangaroo rats *(Dipodomys* spp.), they lack a dorsal skin gland. The soles of the hind feet are furred, and the cheek pouches are fur lined. The ear capsules are extremely large (see fig. 117a).

The two species of kangaroo mice are unique to the sagebrush deserts of the Great Basin. Like the kangaroo rats, they are bipedal, jumping rodents, but they may also scurry about on all fours like pocket mice. Species of the three genera frequently occur together.

Like other heteromyid rodents in arid habitats, kangaroo mice release highly concentrated urine and can subsist on the water in such dry foods as seeds, which may contain from 5 to 8 percent free water. In nature they undoubtedly obtain a great deal of water from the insects and fresh leaves they eat. Kangaroo mice hibernate.

Key to Species of Kangaroo Mice *(Microdipodops)* in California

1a   Dorsum blackish brown or dark gray with hairs dark at base (next to skin); incisive foramina wider posteriorly; tip of tail black (fig. 117d) . . . . *megacephalus*

1b   Dorsum pink cinnamon with hairs white at base (next to skin); incisive foramina with outer margins parallel; tail evenly colored (fig. 117e) . . . . *pallidus*

DARK KANGAROO MOUSE *Microdipodops megacephalus*
Pl. 13, Fig. 117

DESCRIPTION: The Dark Kangaroo Mouse is brownish, blackish, or gray dorsally; its venter is paler, but the base of the hairs is dark or smoky. The upper part of the tail is clearly darker than the body. This species is distinguished from the Pale Kangaroo Mouse *(M. pallidus)* by color and by a smaller hind foot. The incisive foramina are pointed, converging anteriorly and wider posteriorly (see fig. 117d). TL 140–177 mm, T 68–103 mm, HF 23–25 mm. Weight: 10–17 g.

DISTRIBUTION: Found in sagebrush deserts east of the Sierra Nevada, characteristically on somewhat coarse gravelly and sandy soils, where it may occasionally be very common. Its range extends from southeastern Oregon through Nevada east to Utah.

FOOD: This mouse takes a variety of seeds of desert forbs and grasses as well as insects.

REPRODUCTION: Breeding takes place from April to September; litters consist of two to seven young. Young may breed in the year of their birth.

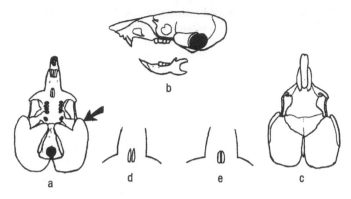

Figure 117. Skull of the Dark Kangaroo Mouse *(Microdipodops mega-cephalus):* (a) ventral view (arrow to auditory bullae); (b) lateral view; (c) dorsal view; (d) incisive foramina of the Dark Kangaroo Mouse; (e) incisive foramina of the Pale Kangaroo Mouse *(M. pallidus).*

COMMENTS: Insects are found in its stomach, although apparently they are not carried in the cheek pouches.

PALE KANGAROO MOUSE   *Microdipodops pallidus*
Pl. 13, Fig. 117

DESCRIPTION: This kangaroo mouse has a rather pale pink dorsum and a white venter; both dorsal and ventral hairs are white at the base. The tail is about the same color as the body, not blackish or dark at the tip. The hind foot is 25 mm or longer (longer than in the Dark Kangaroo Mouse *[M. megacephalus]*). The incisive foramina have parallel outer margins (see fig. 117e). TL 85–90 mm, T 25–26 mm. Weight: 10–17 g.

DISTRIBUTION: Found in Mono and Inyo Counties, generally in regions of very fine sand; at times it is extremely common. It inhabits the extreme eastern margin of the state, east of Yosemite National Park, and adjacent western Nevada.

FOOD: The diet seems similar to that of the Dark Kangaroo Mouse: a mixture of seeds and insects.

REPRODUCTION: Breeding takes place from late winter until June or July. There are two to six young in a litter and more than one litter a season.

Pocket Mice *(Perognathus* and *Chaetodipus)*

These are small, long-faced mice with nearly equal forelegs and hind legs and moderately long tails. The pelage is rather stiff and often spiny in *Chaetodipus* and soft or silky in *Perognathus*. The hind foot has five toes. These mice have external, fur-lined cheek pouches. The snout is relatively longer than in either kangaroo rats (*Dipodomys* spp.) or kangaroo mice (*Microdipodops* spp.). The ear capsules are small to moderately sized, and the external ear is rather short. The molars are tuberculate and have closed roots and a transverse groove.

Both genera of pocket mice occur together in many parts of arid and seasonally arid western North America. Both have behavioral and physiological modifications for living in hot, dry habitats. Some species have been carefully studied in the laboratory.

Seeds and insects are both important in the diet of these mice and provide the water they need. The free water in fresh seeds can be substantial, and chemically bound water is also released from seeds during digestion. As these mice forage, they place seeds in their cheek pouches for later storage near or in their homes (fig. 118). They seem not to place insects in their cheek pouches, which has led to the misconception that they are strictly granivorous. Their stomachs, however, may be crammed with grasshoppers or cutworms. Pocket mice tend to forage under the cover of shrubs; kangaroo rats, in contrast, readily move much greater distances when gathering food. At least some kinds of pocket mice (e.g., the Desert Pocket Mouse *[Chaetodipus penicillatus]*) have been known to climb up at least 25 cm in bushes, possibly in search of insects.

Like most desert rodents, pocket mice are nocturnal. Some species become dormant and spend long periods of winter underground. Some experience daily torpor. In some species the cue for torpidity is temperature decline, in others it is deprivation of food; in still others it is not obvious. While torpid, the pocket mouse has reduced breathing and heartbeat. Such an energy-saving suspension of bodily activity is valuable in the dry, harsh deserts.

Key to Genera of Pocket Mice
(*Chaetodipus* and *Perognathus*) in California

1a Tail crested and with terminal tuft (pencil); interparietal bone wider than long; heel of hind foot unfurred ....
....*Chaetodipus*

1b Tail usually short haired (sometimes with slight apical crest), no terminal tuft; interparietal bone longer than wide ....*Perognathus*

Key to Species of *Perognathus* in California

1a Hind foot longer than 20 mm; skull longer than 24 mm ...
....2

1b Hind foot shorter than 20 mm; skull less than 24 mm. Tail short haired except for some longer hairs projecting beyond tip; pelage soft....................................*parvus*
(also *alticola* and *xanthonotus*, see *parvus* species account)

    2a Auditory bullae large, in contact anteriorly (see fig. 121a) ....*inornatus*

    2b Auditory bullae smaller, not in contact anteriorly (fig. 122a)................................*longimembris*

Key to Species of *Chaetodipus* in California

1a Tail with or without crest, hairs of terminal tuft less than 15 mm; pelage coarse 2

1b Tail with crest and tuft, with long hairs more than 15 mm long ....*formosus*

    2a Pelage on rump or hips with spines 3

    2b Pelage on rump or hips without spines.............5

3a Ears tall, usually longer than 9–10 mm ....*californicus*

3b Ears rounded, usually shorter than 9–10 mm............4

    4a With yellowish lateral line ....*fallax*

    4b Without yellowish lateral line, with spines sometimes from shoulders to rump....................*spinatus*

5a Dorsum grayish; total length usually more than 210 mm; extreme southern California......................*baileyi*

5b Dorsum yellow brown; total length usually less than 210 mm; desert areas of southern and eastern California ....*penicillatus*

Figure 118. Pocket mouse (*Perognathus* spp.) burrow.

BAILEY'S POCKET MOUSE — *Chaetodipus baileyi*
Pl. 14

DESCRIPTION: A rather large, gray buff pocket mouse. The tail is bicolored with a distinct tuft. The heel of the hind foot is naked. The auditory bullae do not touch anteriorly. TL 201–230 mm, T 110–125 mm, HF 26–28 mm. Weight: 15–20 g.

DISTRIBUTION: Found in the rocky deserts of extreme southern California; often associated with mesquite. Its range extends south throughout Baja California, east to New Mexico, and along the eastern coast of the Gulf of California.

FOOD: A genuine dietary generalist, this pocket mouse eats seeds from a broad spectrum of desert grasses, forbs, and woody plants. It favors the seeds of the jojoba (*Simmondsia chinensis*). It also feeds on insects.

REPRODUCTION: Breeding takes place from June to October; there are one or more litters of two to five young.

COMMENTS: This pocket mouse is the only mammal known to be able to metabolize the oil of jojoba seeds.

CALIFORNIA POCKET MOUSE   *Chaetodipus californicus*
Pl. 14, Fig. 119

DESCRIPTION: A rather large pocket mouse with a mixture of yellow and black hairs dorsally and strong, spiny hairs on its sides and rump. The heel of the hind foot is naked. The ears are unusually long (9–14 mm). The auditory bullae are separated anteriorly (see fig. 119a). The interparietal bone is nearly twice as wide as it is long (see fig. 119c). TL 190–235 mm, T 103–143 mm, HF 24–29 mm. Weight: 16–21 g.

DISTRIBUTION: Occurs around the margins of the San Joaquin Valley and in coastal and inland areas south of San Francisco Bay. It has been found in a variety of habitats from open scrub oak to sagebrush and may occur up to 2,800 m in the southern part of the state. It also appears in northern Baja California.

FOOD: This mouse eats the seeds of grasses and shrubs, such as sage.

REPRODUCTION: Breeding takes place from April through June; two to five young are in a litter.

STATUS: The Dulzura (California) Pocket Mouse *(C. c. femoralis)*, is a California subspecies of special concern.

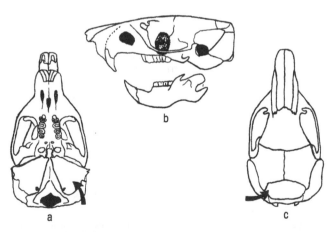

Figure 119. Skull of the California Pocket Mouse *(Chaetodipus californicus):* (a) ventral view (arrow to auditory bulla); (b) lateral view; (c) dorsal view (arrow to interparietal bone).

COMMENTS: This species spends brief periods in torpor with lowered breathing, heartbeat, and body temperature. It does not appear aboveground in winter.

SAN DIEGO POCKET MOUSE *Chaetodipus fallax*
Pl. 14

DESCRIPTION: A medium-sized, brownish pocket mouse. There are conspicuous white, spiny hairs on the dorsum and sides and black spines on the rump. A buff stripe, or lateral line, separates the venter and dorsum. The tail is bicolored and crested. The heel of the hind foot is naked. The auditory bullae are well separated anteriorly. TL 176–200 mm, T 88–118 mm, HF 21–26 mm. Weight: 15–18 g.

DISTRIBUTION: Restricted to the southwestern region of the state; common on compact soils in open desert areas up to 2,000 m. It is frequently associated with species of yucca.

FOOD: This mouse eats the seeds of plants with which it occurs. It is known to collect and store seeds of yucca, sage, ryegrass, and other grasses.

REPRODUCTION: Breeding takes place in fall; a litter contains two to four young.

STATUS: This is a California species of special concern.

COMMENTS: Although this species may not enter prolonged periods of dormancy, it does become inactive in cold weather.

LONG-TAILED POCKET MOUSE *Chaetodipus formosus*
Pl. 14, Fig. 120

DESCRIPTION: A rather large, dark pocket mouse with a bicolored tail, conspicuously tufted at the tip. TL 172–211 mm, T 86–118 mm, HF 22–26 mm. Weight: 19–25 g.

DISTRIBUTION: Occurs on mixed sandy and rocky soils in relatively open arid areas of eastern California. It typically lives where there are many small rocks that resemble it in size and color. Its range extends east to Utah and northern Arizona and south to Baja California.

Figure 120. Long-tailed Pocket Mouse *(Chaetodipus formosus)* sand bathing and perhaps simultaneously scent marking.

FOOD: This pocket mouse consumes the seeds of grasses and forbs; insects; and green leaves in springtime.

REPRODUCTION: Breeding follows the emergence of grasses and forbs in spring. There are commonly five to six young in a litter.

COMMENTS: This species is active in midwinter and presumably does not hibernate. Some students, however, reported that it becomes torpid in the cold months. Apparently winter activity varies with the locality.

DESERT POCKET MOUSE · *Chaetodipus penicillatus*
Pl. 14

DESCRIPTION: A medium-sized pocket mouse with rather coarse, but not spiny, yellow gray pelage. Its tail is faintly bicolored, often annulated (ringed), with a conspicuous tuft and crest. The heel of the hind foot is naked. The ears are relatively pointed; the auditory bullae are distinctly separated anteriorly. TL 153–221 mm, T 91–121 mm, HF 21–27 mm. Weight: 14–20 g.

DISTRIBUTION: Occurs on sandy soils, sometimes on sand and small stones; common in open areas with scattered, low bushes. It is found in the southern California deserts, chiefly the Colorado

Desert. Its range extends east to New Mexico and south to central Mexico, including Baja California.

FOOD: This mouse eats the seeds of a wide variety of desert plants, and possibly insects. It is known to store seeds of mesquite, creosote bush *(Larrea tridentata),* and snakeweed.

REPRODUCTION: Breeding takes place from late winter to spring and sometimes again in fall. In moist, streamside habitats reproduction may continue all summer. Litters vary in size from two to six.

COMMENTS: This species has been captured in live traps placed 24 cm aboveground in bushes, which it may climb in search of insects. In contrast to many pocket mice, it is active on the coldest winter nights. Apparently it does not hibernate.

SPINY POCKET MOUSE   *Chaetodipus spinatus*
Pl. 14

DESCRIPTION: A rather large, yellow brown pocket mouse with a bicolored, crested tail and whitish, spiny hairs on the rump. The heel of the hind foot is naked. The ears are rather small (5–7 mm). The auditory bullae are well separated anteriorly. TL 164–225 mm, T 89–128 mm, HF 20–28 mm. Weight: 19–29 g.

DISTRIBUTION: Lives in the southeastern deserts of the state, ranging south through Baja California.

FOOD: Its diet is not definitely known, but it presumably eats the seeds of grasses and forbs.

REPRODUCTION: This species breeds from April to July and bears at least two broods per year. There are usually two to three young in a litter, rarely as many as five.

SAN JOAQUIN POCKET MOUSE   *Perognathus inornatus*
Pl. 14, Fig. 121

DESCRIPTION: A small, brown or buff orange pocket mouse with a sprinkling of dark guard hairs (but no spiny hairs) on the dorsum. The body has an indistinct lateral line. There is hair on the heel of the hind foot. The auditory bullae are rather large and in contact anteriorly. The interparietal bone is about 1.5 times as wide as it is long (see fig. 121c). This species is similar to the Little

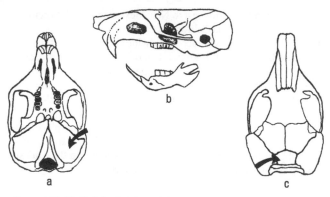

Figure 121. Skull of the San Joaquin Pocket Mouse *(Perognathus inornatus):* (a) ventral view (arrow to auditory bulla); (b) lateral view; (c) dorsal view (arrow to interparietal bone).

Pocket Mouse *(P. longimembris),* but the adults are larger, on average, and in the Little Pocket Mouse the auditory bullae are not in contact. TL 128–163 mm, T 63–78 mm, HF 18–21 mm. Weight: 15–18 g.

DISTRIBUTION: Found on flat ground and low hills in the Central Valley north to Marysville Buttes and south to Carrizo Plain; also occurs in the Salinas Valley. This species is not known outside California.

FOOD: It eats the seeds of grasses, forbs, and shrubs such as *Artemisia* and *Atriplex,* as well as soft-bodied insects, such as cutworms.

REPRODUCTION: Breeding takes place from March to July. Females have at least two litters a season, with four to six young in a litter.

STATUS: The Salinas Pocket Mouse *(P. i. psammophilus)* is a California subspecies of special concern.

COMMENTS: This species experiences daily torpor independent of obvious cues.

LITTLE POCKET MOUSE   *Perognathus longimembris*
Pl. 14, Fig. 122

DESCRIPTION: A small, pink buff pocket mouse. Its tail is either bi-colored or evenly pale; it is more heavily furred on the distal third and is tufted on the distal 3–7 mm. The posterior third of the sole of the hind foot is hairy. The auditory bullae are not in contact anteriorly (see fig. 122a). The interparietal bone is about 1.5 times as wide as it is long (see fig. 122c). TL 112–138 mm, T 50–76 mm, HF 17–20 mm. Weight: 7–10 g.

DISTRIBUTION: Lives on fine, sandy soils in many parts of the southern half of California. It is widely distributed in arid regions from southern Oregon to western Utah and Arizona.

FOOD: This species eats the seeds of many desert plants, including grasses, goosefoot *(Chenopodium)*, and the desert trumpet *(Eriogonum inflatum)*. It also takes soil-dwelling insects.

REPRODUCTION: Pregnancies occur in spring and fall with a summer lull. This species is rather prolific; litter size ranges from two to eight.

STATUS: The Pacific Pocket Mouse *(P. l. pacificus)* is federally listed as endangered and is a California subspecies of special concern. The Palm Springs Pocket Mouse *(P. l. bangsi)*, the Los Angeles Pocket Mouse *(P. l. brevinasus)*, and the Jacumba Pocket Mouse *(P. l. internationalis)* are California subspecies of special concern.

COMMENTS: The Little Pocket Mouse is a profound hibernator and tends to become torpid with a decline in temperature at any season. It remains underground in winter. It seems not to accu-

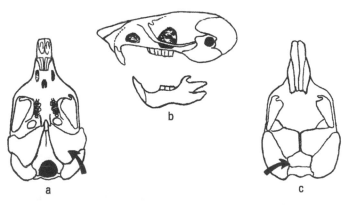

Figure 122. Skull of the Little Pocket Mouse *(Perognathus longimembris)*: (a) ventral view (arrow to auditory bulla); (b) lateral view; (c) dorsal view (arrow to interparietal bone, which is only slightly broader than long).

mulate fat for winter and probably subsists on food stored during summer and autumn.

GREAT BASIN POCKET MOUSE   *Perognathus parvus*
Pl. 14

DESCRIPTION: A buff or ashy pocket mouse with soft, silky fur. The long, bicolored tail has a slight apical crest. The heel of the hind foot is furred. The ears are small with an inner lobe at the base. TL 160–195 mm, T 85–100 mm, HF 22–27 mm. Weight: 16–28 g.

DISTRIBUTION: Occurs in open sagebrush east of the Sierra Nevada, north through much of eastern Oregon and eastern Washington. It may sometimes abound on heavily grazed land with sandy soils and also along creeksides. Its range extends through the Great Basin north to Canada.

FOOD: This species takes the seeds of many forbs and shrubs, especially in the ripening stage, when their water content is very high. It is also known to gorge itself on caterpillars.

REPRODUCTION: Breeding seems to occur irregularly throughout the year, except in winter. The number of broods is not known; each litter contains two to eight young.

COMMENTS: This pocket mouse becomes torpid, at least sporadically, in winter; it is not active aboveground in the coldest weather.

Two described species, the White-eared Pocket Mouse (*P. alticola*) and the Yellow-eared Pocket Mouse (*P. xanthonotus*), are rather similar to *P. parvus* and are apparently extremely rare. *P. alticola* is known from Kern and San Bernardino Counties; *P. xanthonotus* is known from Walker Pass. Both may be subspecies or synonyms of the Great Basin Pocket Mouse.

Rats, Mice, Voles, and Allies (Muridae)

This large, diverse family includes most native mice and rats of North America. The species differ in ear development, eye size, tail length, limbs, and color. The various species also have a variety of adaptations to different habitats. Members of the following subfamilies occur in California: deer mice, wood rats, and allies (Sigmodontinae); voles (Arvicolinae); and Old World rats and mice (Murinae). All of these rodents are small. They vary in color and form, but all have a broad zygomatic plate (see fig. 127a) and narrow infraorbital foramen (see fig. 127c). They lack postorbital processes (which are well developed in squirrels [Sciuridae]). In some genera the front surface of each upper incisor is deeply grooved. Slight grooves sometimes can be seen in other genera. Dentition (in California species): 1/1, 0/0, 0/0, 3/3.

Various species of deer mice, wood rats, and voles are among the most familiar of California rodents. They are frequently the most commonly encountered in the field and are among those most likely to enter mountain cabins.

Deer Mice, Wood Rats, and Allies (Sigmodontinae)

Members of the many genera of Sigmodontinae tend to be earthy or sandy brown. The tail is usually more than one-third of the animal's total length. The appendages are modified for neither digging nor jumping. Cheek pouches, if present, are membranous, not fur lined. The molar surfaces are flat or have two longitudinal rows of tubercles.

Key to Genera of Sigmodontinae in California

1a   Cheek teeth flat crowned with enamel patterns of irregular loops (fig. 125a); mostly rat sized........................ 2
1b   Cheek teeth tuberculate (fig. 127a); mostly mouse sized
     3
     2a   Venter white or buff, dorsum usually even shade of brown or gray; ears rather nude; skull weakly ridged (fig. 125b).............................. *Neotoma*
     2b   Venter light gray, dorsum grizzled gray; ears well covered with fur; skull ridged above eyes (fig. 128b) .....
       .... *Sigmodon*
3a   Upper incisors not grooved............................ 4
3b   Upper incisors deeply grooved........... *Reithrodontomys*
     4a   Tail slender, at least one-third total length; hind foot with five or six plantar tubercles; coronoid process short (fig. 127b) .... *Peromyscus*
     4b   Tail thickened, less than one third total length; hind foot with four plantar tubercles; coronoid process long (fig. 126b) .... *Onychomys*

Wood Rats *(Neotoma)*

Wood rats are rat sized, long-tailed rodents with large, usually naked ears and large black eyes. They are an even gray or brown (not grizzled or agouti) dorsally. With its large eyes and ears, soft pelage and color, and long tail, a wood rat resembles a giant deer mouse (*Peromyscus* spp.). The skull has weak dorsal (supraorbital) ridges (fig. 125c). The molars are flat crowned (fig. 125b).

Most areas in California have at least one species of wood rat, and some have two (usually in different habitats). Wood rats build large houses, usually on the ground, in a tree, or partly beneath a rock outcrop, using twigs, small sticks, cactus joints, cow pies, and other detritus.

Wood rats are commonly known as packrats or trade rats from their habit of stealing small objects about campsites and leaving others in exchange. This behavior has been traced to their habit of carrying small items about in their mouths. If a wood rat encounters something of special interest, it apparently drops what it is carrying and picks up the new object. Thus, it appears to be exchanging one object for another. This also accounts for pieces of campsite garbage, such as bottle caps, around houses of wood rats.

Key to Species of Wood Rats *(Neotoma)* in California

1a Tail with hairs few or moderate but never bushy 2
1b Tail bushy; Sierra Nevada and northern California ....
.... *cinerea*

    2a Tail conspicuously bicolored 3
    2b Tail very faintly bicolored; statewide except desert regions and high Sierra Nevada .... *fuscipes*

3a Fur on chin and throat completely white; extreme southwest desert regions .... *albigula*
3b Fur on throat gray at base, with white patch on upper chest; southern half of state except San Joaquin Valley ..... *lepida*

WHITE-THROATED WOOD RAT   *Neotoma albigula*
Pl. 15

DESCRIPTION: A grayish wood rat with a pure white throat (the throat hairs are white to the base). The tail is bicolored. This species is larger than the Desert Wood Rat *(N. lepida)*. TL 282–400 mm, T 76–185 mm, HF 30–39 mm, E 28–30 mm. Weight: 145–200 g.

DISTRIBUTION: Characteristically occurs in regions of piñon-juniper and at lower life zones in deserts of mesquite and prickly pear. It is seldom found far from chollas. It occurs in the deserts of extreme southeastern California, east of the Tehachapi Mountains. Its range extends from southern Utah and southern Colorado east to Texas and south to central Mexico.

FOOD: This rat feeds heavily on succulent sections of cacti and other desert plants. It also takes grasses, forbs, and yucca leaves in lesser amounts. In some regions it browses heavily on juniper branches.

REPRODUCTION: A small litter, possibly of one to three young, is born in spring. Studies do not indicate more than one litter per year.

COMMENTS: As this wood rat eats cacti, spines accumulate about its pathways. With time, they become hard and form a protective layer that may well deter predators from entering its house. The wood rat, however, runs over these trails with no injury. It also climbs chollas with impunity.

BUSHY-TAILED WOOD RAT   *Neotoma cinerea*
Pl. 15

DESCRIPTION: A large wood rat with a pale gray dorsum and a long, bushy tail. A large ventral skin gland produces a musky odor. The palatine foramina are long. TL 335–425 mm, T 140–185 mm, HF 38–44 mm, E 31–33 mm. Weight: 220–435 g.

DISTRIBUTION: Found in rock outcrops of coniferous forests and piñon-juniper woodland and even sage in the northern and eastern parts of California, up to 2,800 m or above. Its range extends from British Columbia through the Rocky Mountain states south to northern Arizona and New Mexico.

FOOD: It eats a broad variety of leaves and berries, as well as insects and small lizards.

REPRODUCTION: From three to five young are born in spring.

COMMENTS: The presence of this species is made apparent by the white urine stains and amber discoloration from glandular secretions on rock ledges in its habitats.

DUSKY-FOOTED WOOD RAT   *Neotoma fuscipes*
Fig. 123

DESCRIPTION: A large wood rat with a grayish brown dorsum and a pale or white venter. The feet are brown at the base, with the distal half white. The tail is faintly bicolored and scantily haired. The ear pinnae are broad and moderately hairy. TL 260–439 mm, T 130–212 mm, HF 37–44 mm, E 31–34 mm. Weight: 184–358 g.

DISTRIBUTION: Found in hardwood forests and brushlands, along the western edge of the Bay and along the coast north to Oregon, and in the northeastern part of the state south to Lake Tahoe.

FOOD: This species consumes many sorts of leaves, flowers, nuts, and berries. Its favorites are the leaves and berries of coffeeberry (*Rhamnus californica*), poison-oak (*Toxicodendron diversilobum*), blackberry, and roses. It eats hypogeous fungi during spring.

REPRODUCTION: This species breeds chiefly in winter and spring and apparently is sexually inactive in late fall and early winter. In exceptionally wet years, however, it may breed nearly year-round. It usually has more than one, and rarely up to five, litters per year. There are one to three young in a litter, most commonly two.

STATUS: The Riparian Wood Rat (*N. f. riparia*) is federally listed as endangered and is a California subspecies of special concern. The San Francisco Dusky-footed Wood Rat (*N. f. annectens*) and the Monterey Dusky-footed Wood Rat (*N. f. luciana*) are California subspecies of special concern.

COMMENTS: Radiotelemetry has shown that the Dusky-footed Wood Rat frequently forages high above the ground.

This wood rat builds large houses of twigs, leaves, and other debris. These are usually on the ground, frequently in thickets of poison-oak, but they sometimes occur in trees.

This rat is known to rattle its tail in a circular fashion when excited or under duress. The sound so produced lasts three or four seconds and can be heard by human ears from 15 m. The behavior usually occurs when the animal is in its house; an approaching human can sometimes hear the rattling from inside. It may also

Figure 123. Dusky-footed Wood Rat *(Neotoma fuscipes)*: note dark marking on top of hind feet only, not including toes.

occur when the rat is in a tree. On occasion, tail rattling by one animal may elicit the same response from another. Its purpose may be to announce a strange object. Tail rattling is discussed in great detail by Linsdale and Tevis in *Dusky-footed Wood Rat: A record of observations made on the Hastings Natural History Reservation* (1951).

The Dusky-footed Wood Rat has been divided into two species, the new one being called *Neotoma macrotis*. They are separated on the basis of mitochondrial DNA, details of the phallus, and details of the skull. *N. macrotis* occurs south of San Francisco Bay along the coast and in the mountains south of Lake Tahoe. Aside from their distribution, they are probably not separable in the field. The two species are not known to differ in habits.

DESERT WOOD RAT *Neotoma lepida*
Pl. 15, Figs. 124, 125

DESCRIPTION: A relatively small, pale gray wood rat with a distinctly bicolored tail. The underparts are pale or white, but the hairs are gray at the base. There is a small white throat patch. TL 282–305 mm, T 113–128 mm, HF 29–34 mm, E 23–25 mm. Weight: 100–190 g. (See fig. 125.)

DISTRIBUTION: Found in a variety of habitats in the southern half of California, including chaparral in the inner Coast Ranges, often in the vicinity of rocky outcrops. It is a typical inhabitant of the piñon-juniper-covered hillsides of southeastern California and is the common wood rat in most Cal-

Figure 124. Desert Wood Rat *(Neotoma lepida)* sleeping in its nest in hot weather.

ifornia deserts. In the mountains it is generally replaced by the Bushy-tailed Wood Rat *(N. cinerea)*. Its range extends from southeastern Oregon to Colorado and south into Baja California.

FOOD: Like other wood rats, this species eats the leaves and seeds of many sorts of forbs; it also browses on the leaves of shrubs and eats berries.

REPRODUCTION: Breeding takes place in fall, winter, or spring. Two to four young are born within the protection of a stick house. Some females may possibly have two litters per year.

STATUS: The San Diego Desert Wood Rat *(N. l. intermedia)* is a California subspecies of special concern.

COMMENTS: This wood rat is often associated with cholla, yucca,

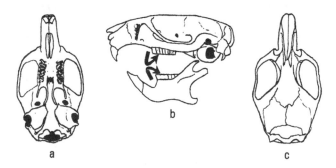

Figure 125. Skull of the Desert Wood Rat *(Neotoma lepida):* (a) ventral view; (b) lateral view (arrows to cheek teeth); (c) dorsal view.

and other desert succulents, from which it obtains necessary water, food, and probably protection. It often builds stick houses beneath the cover of small rock ledges or at the base of yuccas. The remains of partly eaten cacti are often strewed about the entrances to its houses. Radiocarbon dating has established that these houses may persist among rocks for thousands of years. This species is known also to occupy old burrows of ground squirrels or kangaroo rats.

Grasshopper Mice (Onychomys)

These are pale, stocky mice with short, thick tails. The sole of the hind foot is mostly furred (in comparison, it is usually nude in deer mice). The distinctive nasal bones come to a point posteriorly (see fig. 126a); in related genera they are rounded or truncate posteriorly. The lower jaw is unusual in having a long, slender coronoid process (see fig. 126b).

Grasshopper mice seem to favor compact soils with a sparse growth of perennial grasses. These stout little mice are generally a small part of the rodent fauna in the western United States. They are highly predatory and feed extensively on insects and sometimes on other small mice. They possess delightful vocal talents, and the pleasant "song" often heard from captives includes both sonic and ultrasonic frequencies.

Key to Species of Grasshopper Mice (Onychomys) in California

1a   Tail usually more than 50 percent of body length; desert and arid areas in southern half of state . . . . *torridus*

1b   Tail usually less than 50 percent of body length; Great Basin, roughly northeastern section of state . . . . *leucogaster*

NORTHERN GRASSHOPPER MOUSE   *Onychomys leucogaster*

Pl. 16, Fig. 126

DESCRIPTION: A heavy-bodied mouse with a sandy gray brown or rufous dorsum and white venter. The tail is relatively thick and less than 50 percent of the animal's total length. TL 120–190 mm, T 29–62 mm, HF 17–25 mm, E 12–17 mm. Weight: 24–38 g. (See fig. 126.)

DISTRIBUTION: Found in sagebrush regions of the Great Basin in

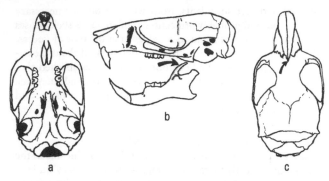

Figure 126. Skull of the Northern Grasshopper Mouse *(Onychomys leucogaster)*: (a) ventral view; (b) lateral view (arrow to coronoid process, which is much longer than in species of deer mice; see fig. 110); (c) dorsal view (arrow to nasal bones).

northeastern California, north through eastern Oregon to central Washington. Its range extends from southern Canada to northeastern Mexico.

FOOD: This mouse eats many insects, especially large species such as fleshy larvae of moths and large beetles, as well as orthopterans. It also takes some small mice, which it easily subdues and kills. It eats little vegetable food.

REPRODUCTION: The breeding season is prolonged, lasting from February to August or even later. At least two broods of one to six young are born each year.

SOUTHERN GRASSHOPPER MOUSE   *Onychomys torridus*
Pl. 16

DESCRIPTION: A heavy-bodied mouse with a buff to cinnamon back and white belly. The tail is thick and usually more than 50 percent of the animal's total length. TL 120–165 mm, T 39–52 mm, HF 18–20 mm, E 11–17 mm. Weight: 20–26 g.

DISTRIBUTION: Found in desert regions of the southern half of California, especially sandy areas of the Mojave and Sonora Deserts and parts of the San Joaquin Valley. Its range extends from central Nevada and southern Utah south through central Mexico, including Baja California.

FOOD: This mouse consumes large insects and small mice; it takes little plant material. Captives, however, can be maintained on unsalted sunflower seeds.

REPRODUCTION: Breeding takes place from May to July. Litters usually comprise two to three young, sometimes up to five or six. There are usually two litters per year. Young probably do not breed in the year of their birth.

STATUS: The Southern Grasshopper Mouse *(O. t. ramona)* and the Tulare Grasshopper Mouse *(O. t. tularensis)* are California subspecies of special concern.

Deer Mice *(Peromyscus)*

These small rodents have a well-developed, nearly nude tail. In adults, the dorsum can be various shades of brown, buff, or brownish gray; the feet and venter are white. The young tend to be rather gray. The ears are moderately developed and project out from the fur. The coronoid process is rather short (see fig. 127b). The upper incisors are not grooved.

This is a large genus in North America, and most regions have at least one species. In some California habitats up to four may occur close together. Many species are rather responsive to environmental conditions and form subspecies or geographic races. They are also adaptable, and the habits and habitat of a given species may vary from time to time and place to place.

Because of their abundance and diversity, several species of deer mice are favorite subjects for both field and laboratory studies. They reproduce easily and are no trouble to care for. They thrive on a broad variety of foods, including insects, seeds, leaves, and fungi.

The following key to the *Peromyscus* in California is designed to assist in the identification of specimens found in the field. We have omitted details of dental, cranial, and other internal characteristics that require dissection and microscopic study. Because individual and age variation increases the difficulty of identification, this key should be used together with the descriptions of species, the color plates, and the geographic ranges of the species. Immature specimens differ not only in size and color but also in body proportions, and the ratios in the key may not be correct for them.

Key to Species of *Peromyscus* in California

1a   Tail (not including terminal tufts of hair) 55 percent or more of total length..................................2

1b   Tail (not including terminal tufts of hair) 51 percent or less of total length 4

     2a   Hind foot usually 17–25 mm; total length usually less than 235 mm; deserts 3

     2b   Hind foot usually 25–29 mm; total length usually 220–266 mm; southern half of state, low elevations and foothills, except deserts. .... *californicus*

3a   Tail with long terminal hairs (4–10 mm from tip); Great Basin and deserts from northeastern corner of California to Mojave Desert and eastern half of Oregon .... *crinitus*

3b   Tail with short terminal hairs (2–4 mm from tip); deserts of southern California .... *eremicus*

     4a   Tail usually about 50–51 percent of total length (sometimes more); hind foot usually 20–27 mm 5

     4b   Tail about 45 percent of total length; hind foot usually 18–22 mm, longer than ear (measured from notch) ..
        .... *maniculatus*

5a   Ear quite large, usually as long as or longer than hind foot; ear 18–27 mm (measured from notch); Sierra Nevada and Great Basin .... *truei*

5b   Ear moderate, usually less than length of hind foot; ear 15–20 mm (measured from notch)..................... 6

     6a   Tail more than half total length................ *boylii*

     6b   Tail less than half total length .... *truei*

BRUSH MOUSE   *Peromyscus boylii*
Pl. 16

DESCRIPTION: A moderately large deer mouse. The dorsum is light chocolate brown, often with a pinkish area on the forearm, and the venter is white. This mouse is distinguished by its rather long tail (more than 55 percent of its total length) and medium-sized ears (shorter than the hind foot). TL 180–238 mm, T 91–123 mm, HF 20–26 mm, E 15–20 mm. Weight: 25–35 g.

DISTRIBUTION: Widely distributed in the northern two-thirds of California, usually in brushy areas from sea level to 2,000 m and rarely to 3,000 m. It is common on hillsides of manzanita, ceanothus, and associated shrubs, and in oak-bay woodlands along the coast. Its range extends east to Texas and south through Mexico.

FOOD: This mouse eats a great variety of forb and shrub seeds, leaves, and insects; it sometimes feeds heavily on hypogeous fungi. Insects are usually a major food. This arboreal species commonly forages in shrubs for such items as manzanita berries and fruits and leaves of the silk-tassel *(Garrya)*.

REPRODUCTION: Breeding takes place from April to October; litters contain four to six young. Reproduction is dependent upon the abundance of food.

PARASITIC MOUSE · *Peromyscus californicus*
Pl. 16, Fig. 127

DESCRIPTION: The largest deer mouse in California, if not in the United States. It is distinguished by its large size, relatively long tail (more than 55 percent of its total length), and dark brown coloration. TL 220–285 mm, T 117–156 mm, HF 24–31 mm, E 20–26 mm. Weight: 33–55 g. (See fig. 127.)

DISTRIBUTION: Mostly found in South Coast Range woodlands and East Bay hills, especially in oaks, buckeyes, bay, and other hardwoods. It also lives in the foothills of the Sierra Nevada south into Baja California.

FOOD: This mouse eats a variety of leafy material and some insects. It is known to feed heavily on seeds of the California bay *(Umbellularia californica)* and also on fresh green foliage. In addition, it consumes fruits, flowers, and seeds of shrubs.

REPRODUCTION: From one to four young are born after a gestation of 21 to 25 days. Lactation lasts for 35 to 45 days. This species is unusual among sigmodontine mice in forming long-term or permanent pair bonds. The female releases a chemical inducer in her urine that causes the male to defend the nest and share in care of the young.

COMMENTS: This mouse climbs well and probably does much of its foraging in bushes or trees.

This species has long been called the Parasitic Mouse, referring to its proclivity for making its home within the houses of wood rats, a habit also seen in the Piñon Mouse *(P. truei)*. The Parasitic Mouse may actually depend to some extent on these stick houses, for one student observed a decline in its abundance following the systematic removal of wood rat houses. The species

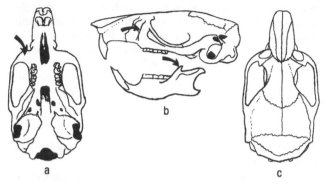

Figure 127. Skull of the Parasitic Mouse *(Peromyscus californicus):*
(a) ventral view; (b) lateral view; (c) dorsal view.

has also been called the California Mouse, a name that has no
special significance and could be applied with equal logic to a
number of other mouselike rodents in our state.

CAÑON MOUSE   *Peromyscus crinitus*
Pl. 16

DESCRIPTION: A rather small deer mouse with soft,
loose fur. The well-furred tail is long (more than half
of its total length) and bicolored and has a tuft of hairs
projecting some 4–10 mm beyond the tip. The ear
is sometimes shorter than the hind foot but
usually about the same length. TL 161–192
mm, T 80–118 mm, HF 17–23 mm, E
17–21 mm. Weight: 13–18 g.

DISTRIBUTION: Found in desert areas east
of the Sierra Nevada, usually not above 1,800 m. It prefers sandy
habitat with rocky outcrops. Its range extends from western Oregon to western Colorado south to northern Mexico, including
Baja California.

FOOD: This mouse eats the seeds of a great variety of forbs, shrubs,
and trees, and presumably also takes insects.

REPRODUCTION: This species apparently breeds throughout much
of the year. Reproductive activity may depend on sporadic rain-

fall, as it does with many desert rodents. Litters of three to five young have been found from February to July.

COMMENTS: Like some species of kangaroo rats *(Dipodomys)* and pocket mice *(Chaetodipus* and *Perognathus),* the Cañon Mouse produces extremely concentrated urine and is thus able to survive with very little free water in its food. In nature, insects probably provide ample water.

CACTUS MOUSE   *Peromyscus eremicus*
Pl. 16

DESCRIPTION: A rather small, buff-colored deer mouse with loose, silky fur. The tail is usually more than 55 percent of its total length and has a terminal tuft 1–4 mm long. The hind foot is larger than the ear. In contrast, the Cañon Mouse *(P. crinitus)* has a longer tuft on the tail and a slightly longer hind foot. Also, the tail is well furred in the Cañon Mouse and sparsely furred in the Cactus Mouse except for the apical tuft. TL 160–200 mm, T 84–120 mm, HF 18–22 mm, E 13–20 mm. Weight: 18–30 g.

DISTRIBUTION: Found in the Mojave and Sonora Deserts and west to the coast, generally in sandy areas with some shrubby growth. Its range extends from southern Nevada and southern Utah south to Texas and Mexico, including Baja California.

FOOD: This mouse eats a varied diet of seeds, leaves, and many insects.

REPRODUCTION: Breeding extends from February to June and sometimes occurs in fall; a litter contains one to five young. Reproductive activity probably varies according to the duration of winter rains and consequent growth of green plants and insects.

COMMENTS: The Cactus Mouse is sometimes difficult to find in summer because it is known to become torpid or estivate in warm weather. This behavior might be induced by a lack of moisture, for in streamside environments it remains active all summer.

DEER MOUSE   *Peromyscus maniculatus*
Pl. 16

DESCRIPTION: One of the smaller members of the genus, distinctive in having a relatively short, bicolored tail and moderate ears.

TL 150–200 mm, T 60–91 mm, HF 18–22 mm, E 14–20 mm. Weight: 14–25 g.

DISTRIBUTION: Found in the coastal states from sea level to above 3,000 m. It frequents all kinds of habitat, including forests, brush, grassland, and chaparral. Its range extends from northern Canada and Newfoundland south to Mexico and includes the Channel Islands.

FOOD: The Deer Mouse eats many kinds of seeds and sometimes leaves. It may feed heavily on insects, such as orthopterans and soil-dwelling insect larvae, and may eat hypogeous fungi.

REPRODUCTION: Breeding takes place from April through November or even December; the breeding season is determined largely by the abundance of food. A litter contains two to eight young.

STATUS: The Anacapa Island Deer Mouse *(P. m. anacapa)* and the San Clemente Deer Mouse *(P. m. clementis)* are California subspecies of special concern.

COMMENTS: This is the commonest member of the genus and is likely to be encountered almost anywhere. This native mouse most commonly enters mountain cabins. In winter it invades these shelters and makes nests of such materials as pillows, mattresses, toilet paper, and tampons; it nibbles on virtually anything but glass and metal.

PIÑON MOUSE   *Peromyscus truei*
Pl. 16

DESCRIPTION: A moderately large deer mouse. The Sierran and Great Basin populations usually have very large ears (longer than the hind foot), but in coastal populations the ears are approximately the same length as the hind foot. The tail is about 50 percent of the animal's total length. In contrast, the ears of the Brush Mouse *(P. boylii)* are shorter than its hind foot, and its tail is more than 55 percent of its total length. The two species are easily separated when they occur in the Sierra Nevada, but in coastal woodlands they look more similar and may be difficult to distinguish. TL 177–195 mm, T 87–98 mm, HF 22–24 mm, E 18–27 mm. Weight: 20–29 g.

DISTRIBUTION: Found in open woodland and brushy areas, especially piñon- and juniper-covered hillsides, frequently near rock outcrops. It avoids the Central Valley, high mountains, and extremely arid regions. Although it is commonly associated with the piñon pine *(Pinus monophylla)*, in California it may occur hundreds of miles from this tree. Its range extends from central Oregon and northern Colorado south through Mexico, including many regions where the piñon pine is absent.

FOOD: This mouse eats a variety of seeds, berries (including juniper berries), and insects. It is fond of pine nuts; the shells can sometimes be found on the ground under the cover of a fallen log or boulder.

REPRODUCTION: Litters of three to six young are born from April to June or July; there is time for two or three litters per year.

COMMENTS: This species sometimes nests within wood rat stick houses. It is also at home in trees; when released, a captive frequently climbs the nearest tree or shrub.

Harvest Mice *(Reithrodontomys)*

These are small, delicate mice, somewhat like the House Mouse *(Mus musculus)* but smaller. Adults are yellow brown, and immature individuals are gray. They have long, nearly nude tails and large, almost hairless ear pinnae. The upper incisor has a pronounced groove. Jumping mice, which also have grooved upper incisors, are distinguishable by their long hind foot (more than twice as long as the ear) and large infraorbital foramen.

These little mice are frequently found in brushy areas with a dense ground cover of long grass. They are usually smaller than deer mice *(Peromyscus* spp.); the two may occur together.

HARVEST MOUSE   *Reithrodontomys megalotis*
Pl. 16

DESCRIPTION: A small, delicate, buff or brownish mouse with a white venter and sometimes a buff spot on the chest. The tail is bicolored and rather long. TL 114–145 mm, T 50–70 mm, HF 15–18 mm, E 12–15 mm. Weight: 9–14 g.

DISTRIBUTION: Found throughout California in lowland and midelevations in grassy areas and open oak woodlands, north to southeastern Washington. It may occur along margins of cultivated areas if grass and weeds are present. It is rather adaptable but avoids forest. Its range extends from south-western Canada across much of the western United States to Mexico.

FOOD: The Harvest Mouse eats the seeds of grasses and weeds as well as some insects, especially cutworms.

REPRODUCTION: Breeding occurs in spring and sometimes again in fall. A litter usually has three to five young (sometimes up to nine). Young are weaned in three and a half weeks and may breed in the year of birth, providing for occasional increases to high densities.

COMMENTS: The nest is a distinctive ball-shaped structure of fine grass, usually on the surface of the ground but sometimes in a dense bush .3 m or more aboveground.

SALT MARSH HARVEST MOUSE   *Reithrodontomys raviventris*

DESCRIPTION: This species is similar to the Harvest Mouse *(R. megalotis)*, but its tail is thicker near the base (or in the basal third) and is one color or only faintly bicolored. In addition, the venter is brownish or buff, whereas in the Harvest Mouse it is white or nearly so. TL 118–175 mm, T 56–95 mm, HF 15–21 mm, E 12–14 mm. Weight: 8–12 g.

DISTRIBUTION: Confined to the salt marshes about San Francisco Bay and the Napa, Petaluma, and Suisun salt marshes. It is commonly associated with dense growth of pickleweed *(Salicornia virginica)*. As its habitat is being obliterated, this mouse is disappearing.

FOOD: The Salt Marsh Harvest Mouse presumably feeds on seeds of grasses and forbs as well as insects.

REPRODUCTION: Breeding takes place from spring to fall; litters contain one to seven young (usually three or four).

STATUS: This species is federally listed and listed by the state as endangered and is fully protected in California.

HISPID COTTON RAT   *Sigmodon hispidus*
Pl. 15, Fig. 128

DESCRIPTION: A small, blackish or dark brown rat with grizzled fur. Its grizzled color and shorter tail distinguish it from the Roof Rat *(Rattus rattus)*, which has a tail more than half its total length. The Brown Rat *(R. norvegicus)* is larger and browner than the Hispid Cotton Rat. *Rattus* species are also clearly distinguishable by the three rows of tubercles on the molar teeth; in *Sigmodon,*

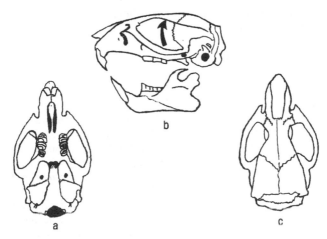

Figure 128. Skull of the Hispid Cotton Rat *(Sigmodon hispidus):* (a) ventral view; (b) lateral view (arrow to ridge over eyes); (c) dorsal view.

there are two rows, and in older specimens these are worn to reveal transverse ridges of enamel. TL 224–365 mm, T 81–166 mm, HF 28–41 mm, E 16–24 mm. Weight: 100–225 g. (See fig. 128.)

DISTRIBUTION: Found along margins of watercourses in the region of the Colorado River and near the Salton Sea. It occurs throughout much of the southern half of the United States into South America, usually near water.

FOOD: This species is more vegetarian than most native mice. It eats many grasses and forbs, taking few insects.

REPRODUCTION: The breeding season is extended and may be dependent on the presence of fresh vegetation. After a gestation of 26 to 28 days, three to seven young are born.

STATUS: The Yuma Hispid Cotton Rat *(S. h. eremicus)* is a California subspecies of special concern.

COMMENTS: The Hispid Cotton Rat resembles voles in some ways: it is active both day and night and makes runways through grassy areas.

Another Cotton Rat *(Sigmodon arizonae)* is known from the extreme southeastern part of California, along the Colorado River, and may occur with or near the Hispid Cotton Rat. The two species resemble each other, but when the occiput is viewed from the rear, it is curved or bowed in the Cotton Rat and angular in the Hispid Cotton Rat.

Voles (Arvicolinae)

Voles are stocky, mostly small rodents with short legs, small eyes, and ears partly hidden in fur. They are grizzled brown or blackish, sometimes reddish or buff, but never have a white belly. The tail is medium or short, less than 50 percent of the animal's total length. The skull is flat and broad. The cheek teeth have flat surfaces with distinctive loops and triangles that are important in generic and specific identification (fig. 137b). The molars have closed roots in some genera (*Clethrionomys, Phenacomys*, and *Arborimus*). Dentition: 1/1, 0/0, 0/0, 3/3.

Most species feed on vegetative parts of grasses and forbs, but they sometimes eat the cambium of trees and the underground parts of both woody and herbaceous plants. Insects are not generally a major item in their diet.

Voles are active both day and night and make runways through grassy areas.

Voles tend to be prolific. They mature at an early age—three weeks in some species—and may have several large litters per year. Sporadically they become very abundant. At such times they attract many predatory birds and mammals and are also capable of doing substantial damage to field crops and orchards. The largest California vole is the Muskrat (*Ondatra zibethicus*), the most important furbearer in the United States.

There are many genera and species of voles, and their identification is sometimes difficult. Because an area usually supports only a few different kinds, the locality of capture greatly aids in their recognition.

Key to Genera of *Arvicolinae* (voles) in California

1a   Tail round in cross section; hind feet without lateral fringe of stiff hairs 2

1b   Tail compressed with dorsal and ventral ridges; hind toes and feet with lateral fringes of stiff hairs. . . . .
. . . . *Ondatra zibethicus*

    2a   Cheek teeth with roots closed in adults 3

    2b   Cheek teeth with roots remaining open in adults. . . . .
    5

3a   Lower cheek teeth with inner angles deeper than outer (fig.
      138b)..4
3b   Lower cheek teeth with inner angles not conspicuously
      deeper than outer (fig. 129b)........................
      ....*Clethrionomys californicus.*

   4a   Tail less than 20 percent of total length; four pairs of
         nipples; hip glands present ....*Phenacomys*
   4b   Tail more than 30 percent of total length; two (rarely
         three) pairs of nipples; hip glands absent ....
         ....*Arborimus*

5a   Soles of feet with dense hairs; tail scarcely longer than hind
      foot; external ears mostly concealed in fur......*Lemmiscus*
5b   Soles of feet hairless; tail considerably longer than hind foot
      in most species; external ears projecting slightly from fur
      ....*Microtus*

Climbing Voles (Arborimus)

In these voles the tail is more than 30 percent of the animal's total
length. Females usually have two pairs of nipples (indicating two
pairs of mammae). The molar roots are closed. This genus is restricted to the coastal forests of northwestern California and Oregon. Species of *Arborimus* are sometimes placed in *Phenacomys*,
but the hip glands characteristic of *Phenacomys* are absent in *Arborimus.*

Key to Species of Arborimus (Climbing Voles) in California

1a   Color reddish; tail hairy............*longicaudus* and *pomo*
1b   Color brown or gray; tail scantily haired ....*albipes*

WHITE-FOOTED VOLE   *Arborimus albipes*

DESCRIPTION: A small, dark brown vole with a long,
clearly bicolored tail. The venter is gray, sometimes
with a pinkish cast. The snout is darker than the rest
of the head. The claws are straight. Most females
have four nipples, but some have up to six. TL
165–181 mm, T 62–71 mm, HF 19–20
mm, E 10–12 mm. Weight: 17–28 g.
DISTRIBUTION: Known from streamside

thickets in redwood forests in northwestern California and the coastal forests of Oregon.

FOOD: This vole browses on the leaves of the red alder *(Alnus rubra)* and willows. It also feeds on the leaves of many forbs.

REPRODUCTION: There are usually two or three young in a litter.

STATUS: This is a California species of special concern.

COMMENTS: This vole climbs well. It and the Red Tree Vole seem to be the only voles that climb more than several inches above the ground.

RED TREE VOLES   *Arborimus longicaudus, A. pomo*
Pl. 12

DESCRIPTION: Bright chestnut red or brick red voles with a long, well-furred tail, curved claws, and ears partly concealed in fur. The species *A. pomo* and *A. longicaudus* have chromosomal differences but apparently cannot be separated in the field. Both might possibly be confused with the California Red-backed Vole *(Clethrionomys californicus)*, which is a much darker red and has a shorter tail. TL 158–186 mm, T 60–76 mm, HF 19–21 mm, E 10–11 mm. Weight: 24–27 g.

DISTRIBUTION: *A. pomo* is found in coastal forests in the humid fog belt north of San Francisco Bay (Marin County) north to the Klamath Mountains. *A. longicaudus* is found along the coast of Oregon, north of the Klamath Mountains. The point at which the two ranges meet may be the Klamath River. Both species are associated with open stands of Douglas-fir *(Pseudotsuga menziesii)* and are strictly tree dwelling.

FOOD: Red Tree Voles feed only on conifer leaves. In California they browse almost solely on needles of Douglas-fir, but in Oregon they are known to eat needles of the Coast Hemlock and Tideland Spruce. They shun grasses, forbs, seeds, and insects.

REPRODUCTION: Litters of one to three young are generally distributed throughout the year.

STATUS: *A. pomo* is a California species of special concern.

COMMENTS: Red Tree Voles are among the most unusual of North American mammals. They are perhaps the most specialized feeders. They are confined to a narrow range; they do not venture into less humid areas of the interior, even where Douglas-fir may abound. Because of their specialized diet, they are not attracted to bait. They can sometimes be captured within their nests, however, which are built of fir needles and are usually located 3 to 20 m above the ground in the outer branches of conifers, usually Douglas-firs.

CALIFORNIA RED-BACKED VOLE

Clethrionomys californicus

Pl. 12, Fig. 129

DESCRIPTION: A dark chestnut-colored vole with many blackish dorsal hairs giving a dark aspect to the reddish color. The venter is buff or cream colored, the tail faintly bicolored. The cheek teeth have equal inner and outer angles (fig. 129b); the molar roots are closed. TL 155–165 mm, T 46–55 mm, HF 17–21 mm, E 10–12 mm. Weight: 17–33 g. (See fig. 129.)

DISTRIBUTION: Found in coniferous forests north of San Francisco Bay along the coast and north of Lake Tahoe in the Sierra Nevada and Cascade Range. In years of abun-

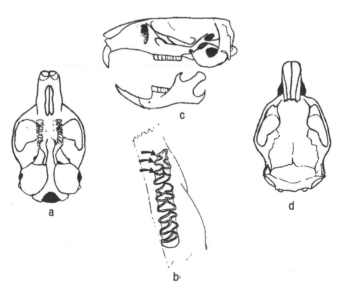

Figure 129. Skull of the California Red-backed Vole *(Clethrionomys californicus):* (a) ventral view; (b) right lower tooth row (arrows to inner angles); (c) lateral view; (d) dorsal view.

dance, it can be found in manzanita and silk-tassel brushlands at around 1,500 m in Plumas County, California. It also occurs in western Oregon north to the Columbia River and at lower elevations in western Washington.

FOOD: This vole eats seeds and insects of the forest floor, as well as quantities of fungi. The green plant material found in its stomach includes lichens.

REPRODUCTION: A rather long breeding season, from March to October, provides for several litters. This vole is not very prolific, however, and usually bears two or three young per litter.

COMMENTS: This little vole seems not to be common, but population densities do vary.

SAGEBRUSH VOLE — *Lemmiscus curtatus*
Pl. 12, Fig. 130

DESCRIPTION: A light grayish vole with a rather short tail. The soles of the feet are furred. The auditory bullae are clearly enlarged (fig. 130a). TL 108–140 mm, T 15–26 mm, HF 15–18 mm. Weight: 20–30 g.

DISTRIBUTION: Typically found in the high, arid sagebrush areas of the Great Basin. In California, it occurs east of the Sierra Nevada both north and south of Lake Tahoe. It is

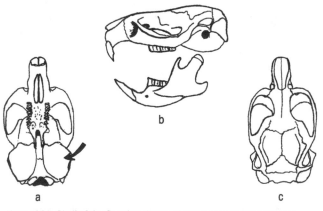

a   c

Figure 130. Skull of the Sagebrush Vole *(Lemmiscus curtatus):* (a) ventral view (arrow to auditory bulla); (b) lateral view; (c) dorsal view.

said to favor rather open areas where the sage is sufficiently mature that some branches touch the ground. Its range extends north to southeastern Washington.

FOOD: This little vole eats sage leaves at all seasons and also eats the cambium layer of the woody growth of sage. Apparently it does not eat seeds, and it is not known to eat insects.

REPRODUCTION: Some desert rodents that feed on seasonally available plants are rather seasonal in their breeding activity, but the food of the Sagebrush Vole is available year-round, and this species seems not to have a distinct reproductive season. From three to six young are born in a concealed nest of grass and shredded bark.

Meadow Voles *(Microtus)*

These are dark blackish or brownish rodents with grizzled fur. The tail is usually less than 50 percent of the animal's total length. The incisors have open roots, and the outer and inner reentrant angles of the molar surfaces are about equal (see fig. 132a).

These little rodents, variously known as voles, meadow mice, or field mice, feed very largely on vegetative parts of green plants but may also eat cambium layers of stems and sometimes seeds. They are therefore capable of doing tremendous damage to cultivated crops, such as alfalfa and sugar beets, and sometimes have been known to girdle orchard trees. They are generally prolific and attain very high densities.

Five species occur in the Pacific States. They all resemble each other rather closely but can usually be identified on the basis of (1) length of tail relative to either hind foot or total length, (2) shape of incisive foramina, (3) plantar tubercles, and (4) geographic range. The incisive foramina can easily be seen by peeling back the soft flesh in the roof of the mouth from just behind the base of the upper incisors (see figs. 132a–136a–135a).

Key to Species of *Microtus* (Meadow Voles) In California

1a   Six plantar tubercles 2
1b   Five plantar tubercles; hind foot less than 22 mm; side glands (on females) reduced or absent; incisive foramina not constricted (fig. 135a)........................ *oregoni*
    2a   Incisive foramina abruptly constricted and narrower posteriorly (figs. 134a, 136a) 3
    2b   Incisive foramina tapered gradually posteriorly or not at all (figs. 131a, 132a) 4

3a   Tail less than twice length of hind foot; Great Basin ....
.... *montanus*
3b   Tail more than twice length of hind foot; coastal marshes of
extreme northern California, coastal lowlands of Oregon
and Washington .... *townsendii*
    4a   Tail more than 33 percent of total length; montane
forests................................. *longicaudus*
    4b   Tail less than 32 percent of total length; usually low-
land meadows.......................... *californicus*

CALIFORNIA MEADOW VOLE   *Microtus californicus*
Pl. 12, Figs. 131, 132

DESCRIPTION: A medium-sized vole with a faintly bicolored tail more than twice the length of the hind foot. It
has six plantar tubercles. The incisive foramina are
not constricted posteriorly (see fig. 132a). TL
157–211 mm, T 39–68 mm, HF 20–25 mm, E
13–16 mm. Weight: 35–72 g.

DISTRIBUTION: Found in lowlands and
foothills of California up to 1,500 m in the
northern Sierra Nevada. It prefers wet
meadows but is also common in alfalfa plantings and irrigated
pastures, where it sometimes reaches high densities. Its range extends from northern Baja California to extreme southern Oregon.

FOOD: The California Meadow Vole consumes many kinds of
forbs and grasses, favoring
fresh, tender leaves and developing seeds. Mature
seeds are not preferred and
insects are shunned.

REPRODUCTION: This vole
may breed throughout the
year if fresh green food is
abundant. Otherwise it
breeds when grasses and
forbs are sprouting.
There are usually
three to eight young in
a litter. Females be-

Figure 131. California Meadow Vole
(*Microtus californicus*).

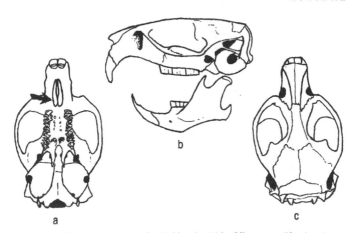

Figure 132. Skull of the California Meadow Vole *(Microtus californicus):*
(a) ventral view (arrow to rounded posterior ends of palatine foramina);
(b) lateral view; (c) dorsal view.

come sexually mature between three and four weeks of age. Populations experience multiannual cycles.

STATUS: The Amargosa Vole *(M. c. scirpensis)* is federally listed and listed by the state as endangered. The Mohave River Vole *(M. c. mohavensis),* the San Pablo Vole *(M. c. sanpabloensis),* the Stephens' California Vole *(M. c. stephensi),* and the Owens Valley Vole *(M. c. vallicola)* are California subspecies of special concern.
COMMENTS: This is the common meadow vole found throughout most of the state except the higher elevations. It is the only vole in the Central Valley, except for the large aquatic muskrat.

LONG-TAILED VOLE   *Microtus longicaudus*
Pl. 12, Fig. 133

DESCRIPTION: A moderately large vole with an indistinctly bicolored, rather long tail (more than twice the length of the hind foot and more than 33 percent of the animal's total length). It has six plantar tubercles. The incisive foramina are of even width from front to back (see fig. 133a). TL 155–221 mm, T 50–93 mm, HF 20–25 mm, E 11–15 mm. Weight: 21–56 g.

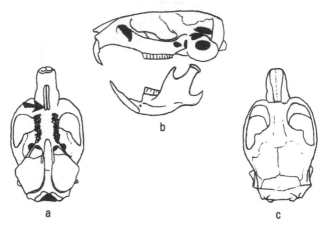

Figure 133. Skull of the Long-tailed Vole *(Microtus longicaudus):* (a) ventral view (arrow to palatine foramina, which are of even width throughout their lengths); (b) lateral view; (c) dorsal view.

DISTRIBUTION: Found in open woodlands in the Sierra Nevada and Cascade Range, the San Bernardino Mountains, and northwestern California. Its range extends throughout the western United States, including much of Oregon and Washington, and north to Alaska.

FOOD: This vole eats grasses, forbs, and the seeds of many forest shrubs and trees.

REPRODUCTION: Two or more litters of three to eight young are born each year.

COMMENTS: Because this vole is not characteristically found in meadows, it does not make runways, as other *Microtus* species do.

MONTANE VOLE   *Microtus montanus*
Fig. 134

DESCRIPTION: A vole with a short, bicolored tail (usually less than twice the length of the hind foot). It has six plantar tubercles. The incisive foramina are conspicuously narrow posteriorly (see fig. 134a). TL 169–189 mm, T 39–54 mm, HF 20–27 mm, E 11–14 mm. Weight: 30–65 g.

DISTRIBUTION: Found in the meadows of the High Sierra and

Great Basin. Its range extends through the western United States north through much of eastern Oregon and eastern Washington to British Columbia.

FOOD: This vole consumes almost exclusively green grasses, sedges, and forbs.

REPRODUCTION: Breeding takes place in spring and summer, but seasonality varies from place to place and year to year. There are successive litters of three to eight young.

COMMENTS: Like several other *Microtus* species, this vole commonly makes discrete runways through grasses and sedges. This widespread vole sometimes attains populations estimated at several thousand per acre; at other times it may become very scarce.

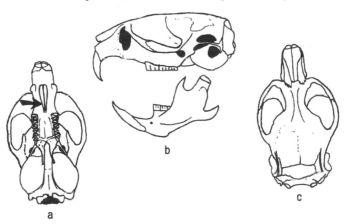

Figure 134. Skull of the Montane Vole *(Microtus montanus):* (a) ventral view (arrow to the palatine foramina, which are constricted posteriorly); (b) lateral view; (c) dorsal view.

CREEPING VOLE *Microtus oregoni*

Pl. 12, Fig. 135

DESCRIPTION: A medium-sized vole with a short tail (less than twice the length of the hind foot). It has five plantar tubercles. The incisive foramina are not constricted but rounded posteriorly (see fig. 135a). TL 129–154 mm, T 32–42 mm, HF 16–19 mm, E 9–10 mm. Weight: 18–22 g.

DISTRIBUTION: Found in forests with deep soil, from lowlands to high mountains; generally avoids meadows. It occurs in northwestern California and ranges north through western Oregon and western Washington to southwestern British Columbia.

FOOD: Its diet is not well known; presumably it consumes roots, bulbs, and stems of forest forbs and grasses.

REPRODUCTION: Breeding may take place throughout the year. Litters of three to six young have been recorded.

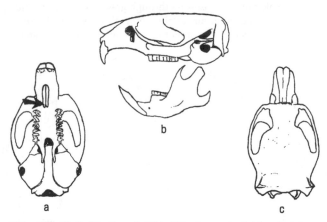

Figure 135. Skull of the Creeping Vole *(Microtus oregoni):* (a) ventral view (arrow to palatine foramina, which are of even width throughout their lengths); (b) lateral view; (c) dorsal view.

TOWNSEND'S VOLE   *Microtus townsendii*
Fig. 136

DESCRIPTION: A large, dark vole with a faintly bicolored tail more than twice the length of the hind foot. It has six plantar tubercles. The incisive foramina are constricted posteriorly (see fig. 136a). TL 169–222 mm, T 48–70 mm, HF 20–26 mm, E 15–17 mm. Weight: 75–82 g.

DISTRIBUTION: Found in marshes in northwestern California and in the lowlands of western Oregon and Washington. Its range extends north to Vancouver Island.

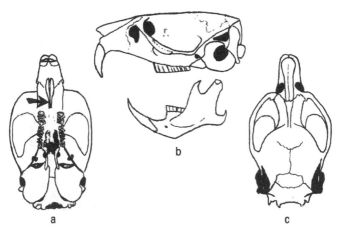

Figure 1.36. Skull of Townsend's Vole *(Microtus townsendii):* (a) ventral view (arrow to palatine foramina, which are constricted posteriorly); (b) lateral view; (c) dorsal view.

FOOD: This vole eats grasses and forbs.

REPRODUCTION: Breeding may take place throughout the year on irrigated land but occurs mostly in spring and summer in natural environments. Litter size varies from five to eight.

COMMENTS: This species is sometimes a serious pest in fields of clover and alfalfa.

MUSKRAT *Ondatra zibethicus*

Pl. 15, Fig. 137

DESCRIPTION: A rat-sized vole modified for an aquatic life. Its fur is chocolate brown, with long, glossy guard hairs and dense, water-repellent underfur. The hind feet are partly webbed with a fringe of hairs; the tail is laterally compressed. The eyes are small. Conspicuous scent glands near the anal opening give this mammal its name; scent is deposited with feces. The skull is heavy and angular, and the incisive foramina are narrowed at each end. (See fig. 137a.) TL 456–553 mm, T 200–250 mm, HF 65–75 mm, E 20–22 mm. Weight: .7–1.8 kg.

DISTRIBUTION: Introduced in the Sacramento and San Joaquin Valleys, where it is now rather common along watercourses, artificial and natural. It also occurs in many low-lying wetlands

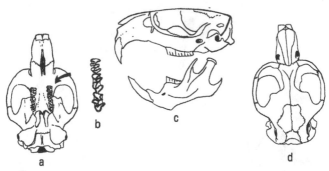

Figure 137. Skull of the Muskrat *(Ondatra zibethicus):* (a) ventral view (arrow to upper tooth row); (b) the upper tooth row shows the typical "loops and triangles" of the cheek teeth of voles; (c) lateral view; (d) dorsal view.

in Oregon and Washington. Its range extends from Alaska through most of Canada and the northern two-thirds of the United States. It has also been introduced in Eurasia.

FOOD: The Muskrat feeds on water plants, exhibiting an obvious preference for cattails. It may also take some animal food, such as freshwater clams and crayfish.

REPRODUCTION: Breeding takes place from late winter through spring and sometimes summer. Two to 10 young are born after a gestation of 25 to 29 days. They can swim in two weeks, are weaned in four weeks, and may breed in the year of birth.

COMMENTS: The Muskrat may nest on a platform of dead cattails or in a burrow in the bank of a levee. Thus, it can be destructive to irrigation canals. It may also be a host to such diseases as tularemia and, through its urine, can introduce tularemia into the water in which it swims. Usually common and easily trapped, it is an important furbearer and sometimes serves as food for humans.

HEATHER VOLE · *Phenacomys Intermedius*
Fig. 138

DESCRIPTION: A grayish brown vole similar to *Arborimus* and some kinds of *Clethrionomys*. It has a short, sharply bicolored

tail. Hip glands are present. Females have four pairs of nipples. This species is distinguishable from the Red-backed Vole *(Clethrionomys californicus)* and the Long-tailed Vole *(Microtus longicaudus)* by its grayish brown (not reddish or blackish) dorsum and sharply bicolored tail. The roots of the cheek teeth are closed in the adult, and the lower cheek teeth have inner angles deeper than the outer (see fig. 138b).
TL 130–153 mm, T 26–41 mm, HF 16–18 mm, E 13–16 mm. Weight: 21–40 g.

DISTRIBUTION: Found in the high (2,500 m and above) central Sierra Nevada, such as the forests above Lake Tahoe or around Mount Shasta. It is sometimes associated with patches of red heather *(Phyllodoce breweri)*. It is not confined to these areas, however, but occurs in a variety of high-elevation woodlands. It is widely distributed in coniferous forests across Canada and the western United States, including high elevations in Oregon and Washington.

FOOD: Its diet is not well known but apparently consists of green

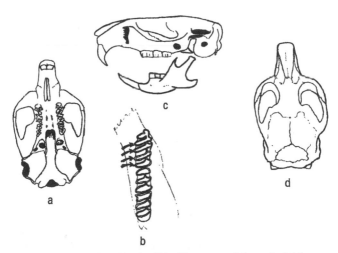

Figure 138. Skull of the Heather Vole *(Phenacomys intermedius)*: (a) ventral view; (b) lower right tooth row (arrows to longer inner loops of the first lower cheek tooth); (c) lateral view; (d) dorsal view.

plants. This vole is readily taken in mouse traps baited with walnut meats.

REPRODUCTION: A litter of four to six young is born in spring or summer. There are probably two broods per year.

COMMENTS: This is the only known species of *Phenacomys*. The Heather Vole seems to be rather infrequently captured, so enterprising students have an opportunity to learn the basic facets of this little-known vole.

Old World Rats and Mice (Murinae)

These small to medium-sized rodents, diverse in form and color, are represented in North America only by the familiar commensal species. They are distinguishable from the native Sigmodontinae by the presence of three longitudinal rows of tubercles on the cheek teeth (fig. 139b); in Sigmodontinae the molar surfaces are either flat or have two rows of tubercles. The Murinae in North America are also characterized by conspicuous ears and a rather long, nearly hairless, scaly tail. Dentition: 1/1, 0/0, 0/0, 3/3 (fig. 139c).

This is the largest subfamily of rats and mice in the Old World, and three commensal species are common in most parts of the United States. They have been carried by ships to all parts of the Earth.

Key to Commensal Rats and Mice (Murinae)

1a	Adult total length more than 320 mm 2	
1b	Adult total length less than 225 mm . . . . *Mus musculus*	
	2a	Tail more than half total length . . . . *Rattus rattus*
	2b	Tail less than half total length . . . . *Rattus norvegicus*

HOUSE MOUSE   *Mus musculus*

Pl. 16

DESCRIPTION: A small, variously colored mouse. The dorsum may be brown, dark brown, grayish, or buff; the venter is buff to dark gray brown. The tail is long, uncolored, and almost hairless. This species is similar to but larger than the harvest mice (*Reithrodontomys* spp.), which, in addition, have a light dorsum, a white venter, and deeply grooved upper incisors. Species of *Peromyscus* usually also have a lighter dorsum than the House Mouse and a white venter. Moreover, native mice have two longitudinal rows of tubercles on cheek teeth, whereas the House Mouse has three. Finally, the House Mouse has an unpleasant odor unlike that of any native species. TL 155–204 mm, T 70–95 mm, HF 17–20 mm, E 11–16 mm. Weight: 12–24 g.

DISTRIBUTION: Found statewide, about human habitations and in fields and brushy areas up to at least 2,200 m. It occurs virtually worldwide.

FOOD: This mouse eats almost any edible material: seeds, leaves, insects, and foods stored in warehouses and pantries.

REPRODUCTION: Breeding takes place more or less throughout the year. A litter of four to eight (sometimes more) is born after a gestation of three weeks, and as many as five litters are born each year. Young become sexually mature between seven and eight weeks. This pattern allows rapid population increases.

COMMENTS: These dirty and destructive little mice soil what they do not eat. They destroy stored foods everywhere and pollute what remains. They are among the most universal mammalian pests. They are among the most common laboratory animals, however, and are the mammal with which much biological research is conducted.

BROWN RAT or NORWAY RAT   *Rattus norvegicus*
Pl. 15, Fig. 139

DESCRIPTION: A brownish rat with a gray venter and rather rough or coarse fur. The naked, scaly tail is less than 50 percent of its total length. The ears are rather large and naked, and the nose is somewhat blunt (Roman). The Brown Rat is distinguishable from wood rats (*Neotoma* spp.) by the bare, scaly tail, the coarse fur, and the tuberculate molar surfaces (see fig. 139b); wood rats have well-haired tails, soft, lax fur, and flat-surfaced molars. The rather similar Roof Rat (*R. rattus*) has a long tail that is more than 50 percent of its total length. TL 300–475 mm, T 120–215 mm, HF 32–44 mm, E 19–24 mm. Weight: 300–525 g.

DISTRIBUTION: Found statewide about buildings and in wild environments up to about 1,000 m. It sometimes occurs along the coast. This species is present worldwide.

FOOD: The Brown Rat is omnivorous.

REPRODUCTION: This prolific species breeds more or less continuously from about 12 weeks of age. There are four to 10 young in a

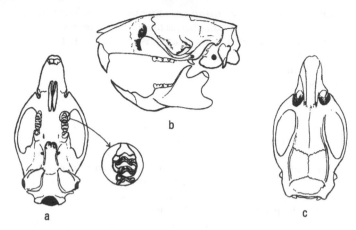

Figure 139. Skull of the Brown Rat, or Norway Rat, *(Rattus norvegicus)*: (a) ventral view (the cheek teeth, in circle, are characterized by tubercles forming three indistinct longitudinal rows); (b) lateral view; (c) dorsal view.

litter. Females may mate immediately after giving birth and thus may be simultaneously pregnant and nursing.

COMMENTS: The Brown Rat is extremely destructive to food products of all sorts. It may be common in rice fields and sometimes occurs in plantings of row crops. It destroys food both in the field and in storage and damages warehouses and homes when gaining entry. It is a reservoir of plague, some forms of hepatitis, trichina worms, and typhus. It is also the white rat of scientific laboratories and, like the House Mouse *(Mus musculus)*, a very valuable experimental mammal. Its specific name *(norvegicus)* is misleading, for it is probably a native of China. The name "Brown Rat" is preferable and is the name used in other English-speaking countries where this rat occurs.

ROOF RAT, or BLACK RAT   *Rattus rattus*
Pl. 15

DESCRIPTION: A slender rodent of variable color. It may be entirely black or nearly so, in which case it is called the Black Rat, or it may be various shades of brown, with a pale buff venter; brownish forms seem to be more common in California. It is distinguishable from the Brown Rat *(R. norvegicus)* by its long tail, more than half its total length, and from wood rats *(Neotoma* spp.) by its naked tail and the tuberculate surface of its molars (they are flat-surfaced in wood rats). TL 320–435 mm, T 170–240 mm, HF 32–39 mm, E 19–26 mm. Weight: 160–205 g.

DISTRIBUTION: Widely distributed except above 800 m. Its range extends north to Washington, including the San Juan Islands. Probably native to the tropical Orient, it now occurs worldwide in warm and temperate regions.

FOOD: The Roof Rat largely consumes plant material, in California most commonly fresh vegetables and fruits.

REPRODUCTION: This species is less prolific than the Brown Rat. It breeds throughout the year, however, and has three to five litters per year of five to eight young (sometimes more). Young become sexually mature between three and four months.

COMMENTS: This rat's name refers to its tendency to climb. It climbs trees, runs along telephone wires, and commonly enters attics, where it is more common than the Brown Rat. In foothill areas it may become common in dense blackberry thickets. It is also known to invade orchards and eat almonds. In southern California it may live in groves of avocados, where it is called the tree rat. Although less destructive to supplies of stored food than the Brown Rat, the Roof Rat is an important reservoir of both plague and typhus fever and has been implicated in serious outbreaks of the latter in the United States.

LAGOMORPHA

These small, superficially rodentlike herbivores have very short, usually fluffy tails. Their eyes are lateral, providing for a large field of vision but limiting binocular vision. The incisive foramina are very large (see fig. 142a). Unlike rodents, lagomorphs have two pairs of upper incisors; a small incisor lies behind each large, deeply grooved upper incisor (see figs. 140a, 142a–144a). The upper rows of cheek teeth are farther apart than the lower; this requires some lateral movement of the lower jaw so that an upper row can meet the corresponding lower row. Between the incisors and the cheek teeth is a large diastema, and there are no canine teeth.

This order includes rabbits, hares (*Lepus* spp.), and pikas (*Ochotona* spp.). Pikas have hind limbs scarcely longer than their forelimbs, as well as vestigial tails. Rabbits and hares have hind limbs greatly enlarged for jumping and small but bushy tails.

Hares (*Lepus* spp.) can be distinguished from rabbits (*Sylvilagus* spp.) on the basis of size and cranial features, but the two groups closely resemble each other. The young of hares are precocial (born well furred and capable of walking a short time after birth); those of rabbits are altricial (born naked, blind, and helpless). Generally the larger species are characteristic of open spaces, and the smaller are more prone to live in denser cover.

The lagomorphs first appeared in the early Eocene of Asia, and the two living families have been distinct since that time. Lagomorphs and rodents share many characteristics and are frequently combined in the superorder Glires, which quite possibly arose in the earliest Cenozoic. On the other hand, molecular data suggest that lagomorphs may be closer to primates than to rodents. Many such conflicts exist between morphological and molecular data. This order has not experienced the extreme diversity seen in rodents.

Pikas (Ochotonidae)

Pikas are small herbivores with short legs, minute tails, and rather short, rounded ears. The hind legs are almost the same size as the forelegs. There are five pairs of upper cheek teeth (fig. 140a). Dentition: 2/1, 0/0, 3/2, 2/3.

In North America pikas live on talus slopes of high mountains, where they can sometimes be seen perched atop large boulders. They are diurnal and active throughout the year. In eastern Asia some species occur deep in coniferous forests, where they burrow elaborate systems of tunnels, and some live in open grasslands. "Ochotona" is the Mongolian name for the animal.

PIKA   *Ochotona princeps*

Pl. 17, Fig. 140

DESCRIPTION: A small, gray buff mammal with a minute tail. The ears are short but broad and project above the fur. Two pairs of upper incisors distinguish the skull of a Pika from that of a rodent (see fig. 140a). TL 150–210 mm (tail vestigial), HF 27–35 mm. Weight: 120–130 g.

DISTRIBUTION: Found on high mountains and talus slopes from the southern Sierra Nevada north to Mount Shasta, Oregon, and Washington, and east to the Warner Mountains. In California it seldom occurs as low as 1,700 m. In the Lava Beds National Monument, it lives in short grass over lava outcrops. Its range extends from New Mexico and Colorado north to British Columbia.

FOOD: This little mammal largely eats stems and the leafy growth of forbs and shrubs.

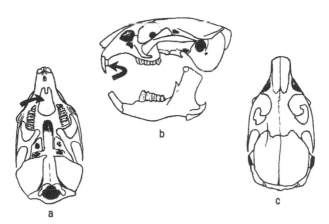

Figure 140. Skull of the Pika (Ochotona princeps): (a) ventral view (arrow to enlarged incisive foramen); (b) lateral view (arrow to inner upper incisor); (c) dorsal view.

REPRODUCTION: A single litter of two to four young is born in June or July.

COMMENTS: The Pika gathers stems and leaves, dries them in the sun, then stores them beneath rocks. These so-called hay piles do not, however, sustain it through winter, for it continues to feed outside its burrow.

Though alert and shy, Pikas are not inconspicuous and can be readily observed by the patient field zoologist. They frequently sit at the top of large rocks from which they can observe other Pikas, as well as enemies. They are sometimes heard when not seen.

Rabbits and Hares (Leporidae)

These herbivores have enlarged hind legs and well-developed external ears, long in most species. The eyes are rather large and placed so as to provide for a broad field of vision. The skull has a well-developed supraorbital process (see figs. 142c, 143c). Dentition: 2/1, 0/0, 3/2, 3/3.

The native species include the various cottontails (*Sylvilagus* spp.) and the hares (*Lepus* spp.). They tend to be abundant and prolific, and some species are conspicuously cyclic, attaining population highs approximately every 10 years. Like other herbivores, rabbits and hares are especially sensitive to the quality and amount of plant food available; their reproduction is clearly enhanced by a rich food supply.

They occur in a variety of habitats, with each species' limbs and ears adapted for its type of cover and mode of life. Those species, such as the jackrabbits, that live in the open have extremely long ears (for the detection of enemies) and long legs (for escape). In contrast, those that live in dense cover and whose defense lies in concealment have short ears and relatively short legs. The long ears of desert jackrabbits also release excess heat from the animal's body; on a very hot day a jackrabbit can sit in the shade and allow heat to escape from its well-vascularized ears.

Not included in the key is the European Rabbit (*Oryctolagus cuniculus*), which has been introduced in the Farallon Islands of California and the San Juan Islands of Washington.

Key to Genera and Species of Rabbits and Hares (Leporidae) in California

1a Size large (hind foot 110 mm or longer); interparietal bone not clearly distinguishable in skull (see fig. 142c) 2

1b   Size medium or small (hind foot 105 mm or shorter); inter-
       parietal bone outlined by distinct suture (figs.
       143c–145c–144c) 4

      2a   Postorbital projection entirely separate from skull (fig.
             142c); upper part of tail white 3

      2b   Postorbital projection fused posteriorly to lateral mar-
             gin of frontal bone; upper part of tail black . . . .
             . . . . *Lepus californicus*

3a   Hind foot more than 138 mm; ear from notch more than
       100 mm; High Sierra and eastward . . . . *Lepus townsendii*

3b   Hind foot less than 138 mm; ear from notch less than 100
       mm; coniferous forests from 1,000 to 2,500 m . . . .
       . . . . *Lepus americanus*

      4a   Tail white below; postorbital projections various but
             not long and not forming open rounded concavity
             with lateral margin of skull . . . . *Sylvilagus* 5

      4b   Tail dusky or gray below; postorbital projection short
             (less than twice as long as antorbital projection) and
             forming open rounded concavity with lateral margin
             of skull . . . . *Brachylagus idahoensis*

5a   Ears sparsely haired on inner surface; vibrissae entirely
       black or sometimes with small amount of white ventrally
       6

5b   Ears heavily furred on inner surface, black edged at tip; vib-
       rissae partly white . . . . *Sylvilagus nuttallii*

      6a   Ears relatively long, 70–75 mm, dark tipped and
             rounded . . . . *Sylvilagus auduboni*

      6b   Ears relatively short, 43–63 mm, uniformly colored
             and pointed . . . . *Sylvilagus bachmani*

SNOWSHOE HARE   *Lepus americanus*
Pl. 17

DESCRIPTION: A medium-sized hare with relatively
short ears and large feet. It is white in winter. TL
365–390 mm, T 24–28 mm, HF 112–135 mm, E
68–72 mm. Weight: .9–1.1 kg.

DISTRIBUTION: Occurs in the coniferous forests
of the Sierra Nevada and Cascade Range
up to 2,500 m, north through Oregon and
Washington. It favors dense creekside
alder or willow thickets. Its range extends

throughout the boreal coniferous forests of the United States, Canada, and Alaska.

FOOD: This species eats willows, alders, and many low shrubs.

REPRODUCTION: From three to six young are born in June. There is probably a single brood; gestation is long, and nursing may continue until late August.

STATUS: The Oregon Snowshoe Hare *(L. a. klamathensis)* and the Sierra Nevada Snowshoe Hare *(L. a. tahoensis)* are California subspecies of special concern.

COMMENTS: In the northern Sierra Nevada this hare is sometimes abundant and thrives in the dense stands of manzanita that grow up after a major forest fire.

This species causes extreme damage to young pines in montane plantations and is a major obstacle to reforestation of conifers after forest fires.

Although it is seldom seen, its huge tracks in the snow reveal it to be common. Depending on its concealing coloring for protection, it does not readily run from enemies.

BLACK-TAILED JACKRABBIT
Lepus californicus
Pl. 17, Figs. 141, 142

DESCRIPTION: A distinctive, long-legged hare with very long ears. Its tail is black (or partly black); its rump is partly black dorsally and grayish ventrally. TL 495–550 mm, T 76–95 mm, HF 117–130 mm, E 105–123 mm. Weight: 1.5–3.6 kg. (See fig. 142.)

DISTRIBUTION: Widely distributed in the state except in forested areas and eastern slopes of the high mountains; may occur up to 2,500 m. It is common in deserts, irrigated pastures, and row crops. Its range covers much of the western United States, north through arid and semiarid Oregon and south-central Washington and south to central Mexico and Baja California.

Figure 141. Black-tailed Jackrabbit *(Lepus californicus).*

FOOD: This species feeds on many herbs and grasses, including many cultivated crops.

REPRODUCTION: It may breed at almost any time of the year, probably depending on the nature of the available food. There are usually three or four young in a litter but sometimes as many as seven.

STATUS: The San Diego Black-tailed Jackrabbit *(L. c. bennetti)* is a California subspecies of special concern.

COMMENTS: This hare is frequently active in the daytime. Its nest is a depression in the ground.

Sometimes two males, in the presence of a female, can be seen boxing, a form of male-male combat. This behavior, seen also in the European hare, gave rise to the notion of the so-called mad March hare.

WHITE-TAILED JACKRABBIT   *Lopus townsendii*
Pl. 17

DESCRIPTION: A large gray jackrabbit with a hind foot more than 140 mm long. The pelage is white in winter; the tail is conspicuous and white throughout the year. It may be confused with the Black-tailed Jackrabbit *(L. californicus)*, but that species has a blackish tail, longer ears, and a smaller hind foot. TL 545–650 mm, T 66–103 mm, HF 145–165 mm, E 98–110 mm. Weight: 2.15–3.44 kg.

DISTRIBUTION: Found in rather open areas in the eastern part of California north through eastern Oregon and Washington, from 1,500 m on the western slopes eastward. Its range covers much of the northwestern United States and south-central Canada.

FOOD: This jackrabbit browses on many low-growing shrubs.

REPRODUCTION: A single litter of four to six young is born each year. As in the Snowshoe Hare *(L. americanus)*, lactation may occupy most of summer.

STATUS: This is a California species of special concern.

COMMENTS: This high-elevation hare is not frequently seen and is little known in our state. It seems not to occur with either the Black-tailed Jackrabbit (which is generally found at lower elevations) or the Snowshoe Hare (which remains close to cover).

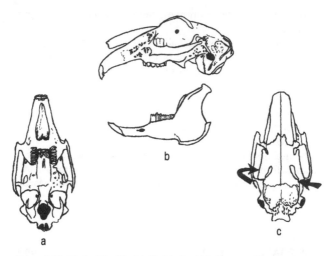

Figure 142. Skull of the Black-tailed Jackrabbit *(Lepus californicus):* (a) ventral view; (b) lateral view; (c) dorsal view (left arrow to supraorbital process, right arrow to postorbital projection).

AUDUBON'S COTTONTAIL   *Sylvilagus audubonii*
Pl. 17, Fig. 143

DESCRIPTION: A large, long-legged cottontail with rather short fur. It most nearly resembles Nuttall's Cottontail *(S. nuttallii).* Its dorsum is generally gray, however, whereas Nuttall's Cottontail has a brownish dorsum. In addition, Audubon's Cottonatail has larger, sparsely furred ears and rather slender, sparsely furred hind feet. TL 370–400 mm, T 40–56 mm, HF 80–95 mm, E 70–75 mm. Weight: 750–900 g (males), .88–1.25 kg (females). (See fig. 143.)

DISTRIBUTION: Found in most of the shrub-covered part of California, except the northern third. It inhabits much of the Great Basin, east to the Rocky Mountains and south to central Mexico, including Baja California. Where it occurs together with Nuttall's Cottontail, Audubon's Cottontail tends to occur at lower elevations.

FOOD: This rabbit eats forbs, grasses, and tender branches of shrubs.

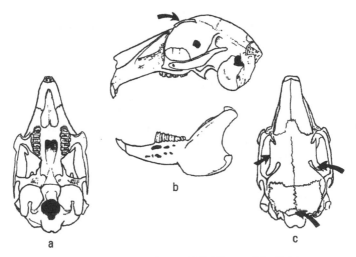

Figure 143. Skull of Audubon's Cottontail *(Sylvilagus adubonii):* (a) ventral view; (b) lateral view (arrow to supraorbital process); (c) dorsal view (left arrow to supraorbital process, right arrow to postorbital process, bottom arrow to interparietal bone).

REPRODUCTION: Breeding takes place from January to June, or throughout the year on irrigated land. Litters vary from two to six.
COMMENTS: This is the common cottontail in brushlands, deserts, and orchards at lower elevations. It may occur together with the Brush Rabbit *(Sylvilagus bachmani)* but is far more likely to forage out in the open.

This species is known to climb small trees, apparently drinking dew from the leaves.

BRUSH RABBIT   *Sylvilagus bachmani*
Pl. 17, Fig. 144
DESCRIPTION: A small, short-legged rabbit with moderately pointed (clearly not rounded) ears, a gray tail, and black vibrissae. TL 300–360 mm, T 12–28 mm, HF 70–85 mm, E 43–63 mm. Weight: 560–840 g (females usually larger than males). (See fig. 144.)
DISTRIBUTION: Occurs throughout the chaparral of the western two-thirds of the state, from the western slopes of the Sierra

Nevada and Cascade Range to the coast. It favors dense brush and seldom strays far from cover. It is also found on San Jose and Año Nuevo Islands. Its range extends from Baja California north to the western part of Oregon.

FOOD: This species feeds largely on grasses and forbs, including sow thistle *(Sonchus)* and sea lettuce *(Dudleya farinosa).*

REPRODUCTION: Breeding takes place from January to June or July; there are probably two broods a year, each with three to four young. As in other species of *Sylvilagus,* the young are born blind and helpless. They become sexually mature in about 10 months.

STATUS: The Riparian Brush Rabbit *(S. b. riparius)* is federally listed and listed by the state as endangered.

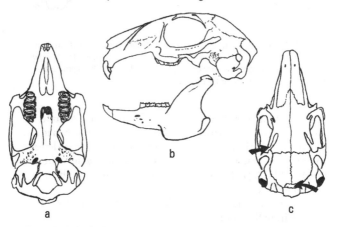

Figure 144. Skull of the Brush Rabbit *(Sylvilagus bachmani):* (a) ventral view; (b) lateral view; (c) dorsal view (left arrow to postorbital projection, bottom arrow to interparietal bone).

NUTTALL'S COTTONTAIL   *Sylvilagus nuttallii*
Pl. 17, Fig. 145

DESCRIPTION: A rather large cottontail that generally has a brownish, not gray, dorsum. The vibrissae are partly whitish. The moderately short ears have rounded tips and a furred inner surface. TL 335–392 mm, T 30–53 mm, HF 86–100 mm, E 57–85 mm. Weight: 650–980 g. (See fig. 145.)

DISTRIBUTION: Found in the canyons and creekbeds of the Great Basin, often where there is a cover of sagebrush. It occurs east of the Sierra Nevada south to the Mojave Desert and north in open areas through eastern Oregon to Washington. Its range extends from Arizona and New Mexico north to Canada, mostly in the intermountain region.

FOOD: This is a browsing species, especially fond of bitterbrush *(Purshia tridentata)* and sagebrush. It grazes on a broad spectrum of forbs and grasses in spring.

REPRODUCTION: Breeding takes place from late winter or early spring until early summer. Probably two broods of four to eight (most frequently six) are common, but in some years four broods are possible. The young are altricial, born blind and naked. The mother makes a nest in a crevice among rocks or in a depression in the ground and lines it with fine hair pulled from her belly.

COMMENTS: This cottontail appears to be unique among lagomorphs in its arboreal tendency. During arid periods in midsummer it may spend much of the early daylight period up to 3 m above the ground in junipers, apparently licking water from the foliage.

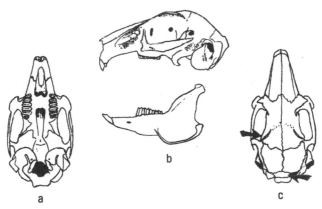

Figure 145. Skull of Nuttall's Cottontail *(Sylvilagus nuttallii):* (a) ventral view; (b) lateral view; (c) dorsal view (left arrow to postorbital projection, bottom arrow to interparietal bone).

PIGMY RABBIT
Pl. 17

Brachylagus idahoensis

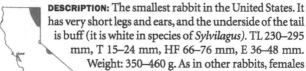

DESCRIPTION: The smallest rabbit in the United States. It has very short legs and ears, and the underside of the tail is buff (it is white in species of *Sylvilagus*). TL 230–295 mm, T 15–24 mm, HF 66–76 mm, E 36–48 mm. Weight: 350–460 g. As in other rabbits, females tend to be larger than males.

DISTRIBUTION: Found along the eastern margin of California and in the southeastern quarter of Oregon, especially in rocky areas dominated by sagebrush. It inhabits the Great Basin east to Utah and north to the central Columbian Plateau of Washington and southeast Montana.

FOOD: This rabbit eats sagebrush and grasses.

REPRODUCTION: Breeding takes place in late winter and spring; probably two litters of four to eight young are born annually. The young become sexually mature in one year.

STATUS: This is a California species of special concern.

COMMENTS: This little rabbit is sometimes common and frequently active in the daytime. It is not difficult to see but may not venture far from the protection of dense cover. It nests underground and builds burrow systems, into which it retreats when pursued. It has a distinctive gait—a mixture of hopping and walking. Sagebrush, a favorite food, is said to impart a strong flavor to its flesh.

DOMESTIC RABBIT

Oryctolagus cuniculus

DESCRIPTION: The familiar domestic rabbit. Its color is extremely variable, depending upon what domestic breed was released. TL 450–600 mm, T 66–80 mm, HF 92–110 mm.

DISTRIBUTION: Occurs in feral populations on the Farallon Islands, California, and the San Juan Islands, Washington.

COMMENTS: This species has been introduced to many parts of the world, and feral populations thrive in areas with a Mediterranean climate (Australia, South Africa, and southern Argentina). They cause extensive habitat destruction.

GLOSSARY

Adaptive radiation The evolution of diverse ecotypes from a common ancestor in the absence of competition.

Agouti A fur color with a mixture of black and brown hairs.

Alloparent An individual, frequently a sibling of a prior generation, that acts as a parent but is not reproductively active.

Alpha The leader of a group, such as the dominant individual in a herd of elephants or pack of wolves.

Altricial The condition at birth in which the newborn is hairless (or nearly so), blind (with eyes unopened), with earflap closed, and unable to walk. See *precocial.*

Alveolus A blind, enlarged end to a duct or tube, such as the alveolus into which milk is secreted (see fig. 6). The plural is *alveoli.*

Annulated Ringed, such as the rings on the tail of a raccoon.

Antero-posterior The length from head to tail.

Anterior The front, or in front of.

Antiperistaltic The reverse muscular wave of the gut, toward the stomach and away from the colon.

Apical At the top or apex.

Arteriole A minute artery, usually branching from an artery.

Atlas The first vertebra, which supports the skull.

Auditory bulla A spherical bony covering, enclosing the inner ear, at the bottom of the rear end of the skull (see fig. 4). Also called *tympanic bulla.* The plural is *auditory bullae,* or *tympanic bullae.*

Autonomic nervous system The nerves controlling the heart and other internal organs. It consists of the sympathetic and parasympathetic nervous systems.

Baculum A bone within the penis of many species of mammals.

Blastocyst An early embryonic stage consisting of a hollow ball of cells.

Boreal Northern.

Bradycardia A slowed heartbeat, occurring usually with a drop in body temperature, as in hibernation.

Brown fat, or brown adipose tissue (BAT) A type of fat with a rich supply of nerves and blood vessels that is capable of providing an infusion of heat into a hibernating mammal or a cold-stressed mammal.

Caecum A pocket or diverticulum at the junction of the small and large intestines.

Calcar A small bone extending from the heel in bats. It provides for an extension of the skin between the leg and the tail (see fig. 25).

Canine tooth The tooth between the last incisor and the first premolar. Usually conspicuously long, more or less rounded in cross section, and usually slightly curved.

Canine Doglike.

Capital breeder A species that accumulates substantial amounts of body fat before the breeding season and uses this energy in reproduction.

Carnassial The modification of the upper fourth premolar and the lower first molar so that their surfaces meet to form a sharp cutting edge.

Carnivoran A member of the order Carnivora.

Carnivore A meat-eating animal.

Carrying capacity Environmental productivity (or other environmental conditions) adequate to meet the needs of a species.

Caudal Pertaining to the tail.

Cheek teeth The molars and sometimes also the premolars; those teeth behind the canines (or behind the incisors in rodents, lago-morphs, and many artiodactyls).

Colostrum A fluid from the mammary gland, produced immediately after birth of the young.

Commensal An association of two organisms, usually unrelated phylogenetically, in which one benefits without harming the other.

Conspecific Individuals that belong to one species.

Cooperative breeding Raising of young in which their care is shared by individuals (in addition to one of the parents), usually siblings of the previous generation.

Coprophagy The consumption or reingestion of soft fecal material that is semidigested, as done by rabbits and gophers, and makes available for absorption nutrients that have been digested by intestinal microbes as well as protein from the microbes themselves.

Copulation Mating.

Coronold process A dorsal projection on the lower jaw (see fig. 4).

Coxal Of the basal segment (or coxa) of the jointed legs of an arthropod.

Crepuscular Active at dawn and at dusk with nocturnal and diurnal rest periods.

Critical thermal maximum The body temperature so high that it is lethal.

Deciduous Disposable, usually referring to body parts that are shed, as in the horny covering of the horns of the Pronghorn (*Antilocapra americana*).

Defense polygyny Aggressive activity of a male toward other males in protecting a group of females during the reproductive season.

Delayed development Embryonic development that proceeds at a very slow but continuous rate.

Delayed fertilization Fertilization that occurs weeks or months after mating, as a result of sperm having been stored in the reproductive tract of the female. Characteristic of vespertilionid bats.

Delayed implantation A condition in which the embryo does not immediately attach to the uterus but remains for some weeks or months in a state of arrested development. See *embryonic diapause*.

Density-dependent Referring to the limiting of population through its own numbers or its own density.

Dentary bone The paired bones of the lower jaw of mammals. Anteriorly they are usually fused.

Dentition The number and kind of teeth, expressed as a dental formula indicating the number of upper and lower incisors, canines, premolars, and molars.

Dermal Referring to the skin.

Dichotomous Branching into two more or less equal parts, as opposed to smaller branches from a continuous main stem.

Diaphragm A muscle at the bottom or rear of the thoracic cavity; when the diaphragm contracts, the thoracic cavity enlarges, causing the lungs to inflate and draw in air.

Digitigrade The condition in which a mammal normally walks on its toes, as does a dog or a cat.

Diphyodont The occurrence of two successive sets of teeth: milk teeth and permanent teeth.

Dispersal Movement away from the place of birth to make a home at another site, usually before sexual maturity.

Distal Away from the center.

Diurnal Active in daylight hours, as opposed to nocturnal, or active at night.

Dormancy A resting condition characterized by slowed breathing and heartbeat and a reduced body temperature, as during hibernation.

Dorsum The upper part or back of an animal.

E Abbreviation meaning "ear."

Ear ossicles Three small bones in the middle ear.

Echolocation Determination of an object's position by the echo returning from it; a bat can determine the distance, size, and flight speed of an insect by the reflected sound waves that the bat emitted. Echolocation is also used by toothed whales.

Embryonic diapause The resting state of an embryo, characterized by a delay of weeks or months, before it implants or attaches to the uterus of the mother. Characteristic of many species of the weasel family. See *delayed implantation*.

Endocrine glands Glands that secrete one or more hormones into the circulatory system.

Endothermy The condition in which metabolic heat allows the body temperature to rise above the ambient temperature.

Entrainment The process by which a regular physiological state, such as sleep or a seasonal activity, is timed by an environmental factor such as light.

Eocene The second epoch in the Cenozoic.

Estivation A period of dormancy occurring in summer.

Estrogen A hormone from the ovaries (and also the adrenal cortex); it stimulates the development of female characteristics and functions in reproduction.

Estrous cycle The periodic release of one or more eggs and preparation of the uterus to receive an embryo.

Estrus The point in the estrous cycle when the female is receptive to sexual advances of a male; a time commonly referred to as "heat."

Eutheria Mammals in which the young develop with a placental attachment; as opposed to Marsupialia, or Metatheria, in which the placental attachment is developed in only one family, and in which the very small infant develops within a pouch.

Exogenous Referring to a factor, influence, or cue that comes from the environment and affects bodily activity.

Extirpation The disappearance of a population from a specified area.

F Abbreviation meaning "forearm."

Facultative specialist An animal that concentrates in foraging on a specific food type that is temporarily abundant.

Feral The condition in which a domestic animal has reverted to a wild existence.

Fertilization The union of the genetic elements of an egg and a sperm.

Fitness The contribution of a given genotype to the next generation, relative to other genotypes in the population.

Flehmen The act in which many mammals raise their heads and simultaneously raise their upper lips and expose the mouth to air. This action apparently allows odors to reach the vomeronasal organ at the anterior part of the roof of the mouth. See *vomeronasal organ*.

Follicle-stimulating hormone (FSH) A hormone from the anterior pituitary gland that stimulates the ovary to develop one or more eggs or ova.

Fossorial Burrowing in habit, as a pocket gopher or mole.

Free-running rhythm The pattern of activity and rest in an experimental situation with no external cues, such as changes in light or noise.

Frequency modulation The change in pitch, as in an echolocation call, from beginning to the end.

Gestation The period of embryonic development.

Gonad The reproductive organ in which eggs or sperm are made; also the site of production of such hormones as estrogen, progesterone, and androgens. Gonads are testes in males and ovaries in females.

Gonadotropin A hormone (such as follicle-stimulating hormone or luteinizing hormone) that stimulates the gonads and is produced by the anterior pituitary gland.

Gonadotropin-release hormone (Gn-RH) A hormone that stimulates the anterior pituitary gland to release gonadotropins; produced by the hypothalamus.

Guard hairs The long hairs that protrude slightly from a mammal's shorter body fur.

HF Abbreviation meaning "hind foot."

Hard palate The bony part of the roof of the mouth, separating the intake of air from that of food and water.

Heterodont The occurrence of teeth of specialized types, such as incisors, canines, premolars, and molars.

Heteropteran A member of the insect order Heteroptera, which comprises the true bugs.

Hibernaculum The nest or cover a mammal uses during hibernation.

Hibernation A state of dormancy, in winter, in which most body activities are greatly slowed and body temperature drops.

Holocene The epoch after the last ice age, and the period we are now in.

Home range The area over which a mammal moves for food on a regular basis.

Homing The tendency for an animal to return to its home; generally shown by displacement experiments in which an animal is taken from its home, marked by a band or some other identifying mark, and then released.

Homeothermy The maintenance of a relatively constant temperature.

Hormono A chemical messenger. Manufactured at one site (an endocrine gland), a hormone circulates throughout the body and stimulates activity (circulation and cellular activity) in a certain area, the target organ.

Hymenopteran A member of the insect order Hymenoptera, which includes wasps, ants, and bees.

Hypogeous Growing underground.

Hypothalamus A structure at the base of the brain that receives stimuli from both outside and inside the animal and mediates many daily and seasonal activities. It is a major regulator.

Hypothermia A body temperature significantly below normal temperature.

Immunoglobulin Substances that provide resistence to disease.

Implantation The attachment of an embryo to the uterus. At this site the placenta, from both embryonic and maternal tissue, subsequently develops.

Incisive foramina A pair of slits in the roof of the mouth through which two branches of the olfactory nerve enter the mouth. See *Flehmen.*

Incisor One or more teeth, located in front of the canine.

Income breeder A species that accumulates body energy immediately prior to reproducing and uses virtually all of this energy in growth and development of young.

Induced ovulation Ovulation (the release of an egg from the ovary) as the result of mating.

Infraorbital foramen or canal An opening or canal between the snout and the eye region through the bone directly in front of the eye cavity.

Interdigital Located between the fingers or toes.

Intergrade A genetically intermediate condition, as in the offspring of a hybrid mating.

Interfemoral membrane A fold of skin connecting the hind legs and frequently enclosing the tail in bats.

Interorbital breadth On the skull, the bony space between the eyes.

Interparietal The large unpaired bone at the top and rear of the skull.

Interscapular A region between the scapula or shoulder blades.

Karyotype The chromosomal composition of an individual.

Lactation The process of giving milk; also suckling.

Lactogenesis The manufacture of milk by the mammary glands.

Lateral On the side.

Latero-caudad On the side and toward the tail.

Luteinizing hormone (LH) A hormone from the anterior pituitary gland. It stimulates the release of the egg from the ovary, and also stimulates the testes.

Mammary glands The structures that, in female mammals, produce milk.

Mammogenesis The development of mammary glands.

Mandible The jawbone, or the lower jaw; in modern mammals, the dentary bone.

Mantle The area over the shoulders and behind the neck.

Marsupium The pouch in which the nipples lie and which houses the young of marsupials.

Maxillary The upper jaw; the tooth-bearing bone of the bottom part of the skull.

Melatonin A hormone that suppresses gonadal activity. It comes from the pineal gland and becomes more common during periods of darkness.

Metatarsal One of the bones connecting the foot and the ankle.

Microbiota Microscopic plants or animals.

Midventral See *ventral.*

Migration A movement from one area to another with a subsequent return.

Miocene The epoch between the Oligocene and the Pliocene.

Molar A tooth that develops behind the premolars; molars are not preceded (in development) by milk teeth.

Monogamy A mating relationship involving a long-term pairing between one male and one female.

Monotreme An egg-laying mammal, such as the Duckbill *(Ornithorynchus anatinus)* of Australia.

Mortality The rate at which individuals die over a specified period of time.

Movement Wandering from one place to another, usually in search of food or a mate, within the home range.

Myoepithelial strands Small contractile fibers about the alveoli in mammary glands (see fig. 6).

Natality Rate of birth.

Navigation The ability of an animal to determine direction from environmental cues.

Neuroendocrine loop A physiological communication involving both nervous and hormonal messages, such as the release of milk after stimulation of the nipples.

Neotropical Referring to the flora and fauna of South America north to southern Mexico.

Nocturnal Active at night.

Nonshivering thermogenesis The production of body heat without shivering but using the breakdown of brown fat.

Noradrenalin A hormone, synthesized in the sympathetic nervous system, which stimulates the activity of brown fat.

Normothermic The range of body temperatures at which an animal is active.

Obligate specialist A kind of mammal that depends upon a specific kind of food.

Occiput The dorsal ridge at the rear of the skull, or occipital bone.

Olfaction The detection of or sensitivity to odors.

Oligocene The epoch between the Eocene and the Miocene.

Orbital foramen An orifice at the front end of the eye cavity.

Orientation The determination of an animal's geographic position from environmental cues.

Orthopteran A member of the insect order Orthoptera, which include grasshoppers, cockroaches, and crickets.

Os clitoris A small bone in the clitoris of some mammals.

Ovary The gonad of females, which produces ova or eggs.

Oviduct The small tube that carries eggs from the ovary to the uterus.

Oviposition Egg laying.

Ovulation The process in which one or more eggs are released from the ovary or ovaries.

Ovum An egg. The plural is *ova*.

Oxytocin A hormone from the posterior pituitary (made in the hypothalamus and released through the posterior pituitary) that stimulates the release of milk from the alveoli of the mammary glands and also induces maternal behavior.

Palate The roof of the mouth, separating the mouth from the nasal passage and consisting of fleshy tissue and bony (or hard) tissue.

Palatine foramen A paired orifice in the roof of the mouth, behind the maxillary bone.

Paleocene The first epoch in the Cenozoic.

Palmate Flattened and radiating from a central point, as the palm of a hand or a flattened antler of a deer.

Parietal eye A light-sensitive structure in the skull of some early vertebrates and some modern reptiles.

Parietal opening An orifice in the skull of some early vertebrates and some modern reptiles and present in embryonic mammals.

Parturition Birth.

Pelage Fur.

Periodic arousal The brief (several hours) return to activity from a dormant state (such as during hibernation).

Phallus Penis.

Pheromone A chemical emitted by one individual that affects the behavior of another of the same species.

Photocycle The complete light/dark period or the annual change in light/dark periods; thus either a daily or an annual photocycle.

Photoperiod The amount of light and darkness within a 24-hour period.

Phytoestrogen A plant compound that behaves as a mammalian hormone.

Pineal gland A small gland at the upper part of the brain that secretes melatonin.

Pinna The external ear, or earflap. The plural is *pinnae*.

Pituitary gland An endocrine gland lying at the base of the brain, ventral from the hypothalamus. It consists of three parts, or lobes, which are quite different in function.

Placenta The structure that exchanges nutrients and oxygen as well as waste products and carbon dioxide between the embryo and the mother; also known as the "afterbirth."

Placental lactogen A hormone, produced by the placenta, that stimulates the growth and development of mammary tissue during pregnancy.

Plague A bacterial disease of mammals, usually transmitted by fleas.

Plantar The flat underside of a foot.

Plantigrade A type of foot, or manner of walking, in which the entire plantar surface is placed on the ground, as in a bear or Raccoon (*Procyon lotor*) (or a human).

Pleistocene The geological epoch, characterized by alternating ice ages and mild interglacial periods.

Pliocene The epoch prior to the Pleistocene.

Polygynandry A mating system in which a group of males mate randomly with a group of females.

Polygyny A mating arrangement involving one male and several or many females, as in fur seals and sea lions.

Portal vein A blood vessel that carries blood from one structure to another, without passing through the heart.

Postauricular Located behind the ear.

Posterolateral Located on the side, toward the rear.

Postmandibular foramen A hole on the inner surface of the lower jaw (see fig. 10).

Postorbital breadth The greatest width of the skull behind the eyes.

Postorbital process A lateral projection from the top of the skull, just behind the eyes (see fig. 4), characteristic of squirrels.

Precocial The condition at birth in which the newborn is fully furred (or feathered, in birds), has its eyes open, and is usually capable of walking and feeding on solid food within a few minutes or few hours after birth. See *altricial*.

Prehensile Capable of holding, such as the somewhat prehensile tail of an opossum.

Premaxillary The bone at the front of the skull, in front of the maxillary, and bearing the incisors.

Premolar A tooth (or teeth) between the canine and molar; preceded (in time) by a milk tooth.

Preputial gland A gland that releases a fluid into urine, giving it a distinctive odor.

Primer A scent that affects sexual development of another individual of the same or the opposite sex.

Progesterone A hormone produced by the ovary, especially after ovulation, and also produced by the placenta in most species of mammals.

Prolactin A hormone from the anterior pituitary. It has many functions but is especially critical in the production of milk.

Promiscuous A mating system in which a female mates randomly with one or more males.

Pseudopregnancy The appearance of pregnancy without the presence of an embryo.

Pubic symphysis The juncture of the pelvic bones.

Pubis The anterior margin of the pelvis.

Pulmonary circulation The pathway of blood between the lungs and the heart.

Quadrate bone The bone in the skull to which the dentary articulates.

Releaser A stimulus, such as a scent that elicits an immediate behavioral response in another individual.

Reservoir An animal in which a parasite lives and serves as a source of the parasite for other species.

Rugose Rough or wrinkled.

Ruminant A mammal with a four-chambered stomach, each chamber having a different function, holding food in different stages of digestion and physical breakdown.

Sagittal crest The ridge or ridges at the top of the skull in many mammals, at the juncture of the parietal bones (see fig. 4).

Setpoint In a hibernating mammal, the temperature that triggers dormancy.

Speciation The evolution of one or more species from a single species.

Sperm Mature sex cells of the male.

Sphenoidal fissure An opening in the skull at the rear of the eye cavity (see fig. 96).

Suborbital Located below the eye.

Supraorbital Located above the eye.

Sympatric Occupying the same area.

Systemic circulation The pathway of blood from the heart throughout the body and back to the heart.

T Abbreviation meaning "tail."

Target organ The structure or organ that is sensitive to a given hormone.

Tarsals The bones in the ankle.

Territory A part of a home range that the occupant defends against intruders of the same species.

Testis The male gonad. The plural is *testes*.

Thermoneutral An environmental temperature at which a mammal does not need to alter its metabolism in order to preserve its normal body temperature.

Thermoregulation Control of body temperature.

TL Abbreviation meaning "total length."

Torpidity A condition in which breathing rate, heartbeat, and body activity decline with a simultaneous drop in body temperature. See *dormancy*.

Tuberculate A condition of premolar and molar teeth in which the surface is provided with rounded projections (tubercles).

Tularemia A highly infectious bacterial disease of many mammals.

Turbinate bone A delicate thin bone covered by arterioles and venules, in the nasal passages.

Tympanic bulla See *auditory bulla*.

Ultrasonic Sound waves with a frequency above 20,000 cycles per second (or hertz) and that are above the range of hearing of most humans.

Ungulate A hoofed mammal, such as a pig, deer, or horse.

Unicuspid A tooth with a single projecting surface.

Uterus The part of the reproductive tract of the female in which the embryo develops.

Vector An animal, usually a mite or an insect, which transmits a disease-causing parasite from a reservoir to an uninfected animal.

Venter The underparts or belly and chest of an animal.

Ventral On the underside.

Venule A minute vein leading to a larger vein, usually part of a system of branching venules.

Vestigial A nonfunctional structure that does not fully develop, in contrast to its development in species in which it is functional.

Vibrissa A stiff facial hair, sensitive to touch. The plural is *vibrissae.*

Vomeronasal organ An odor-sensitive recess in the roof of the mouth of most mammals that is absent in whales and humans.

White adipose tissue (WAT) Body fat, exclusive of brown adipose tissue.

Zygomatic arch A bony bridge on the side of the skull, over the outside of the eye (see fig. 4).

Zygomatic plate The flat, more or less vertical bone at the anterior side of the eye cavity.

REFERENCES

Mammal Ecology, Origins of Mammals

Anderson, S., and J.K. Jones Jr. 1984. *Orders and families of recent mammals of the world.* New York: John Wiley & Sons.

Banks, R.C., R.W. McDiarmid, and A.L. Gardner. 1987. *Checklist of vertebrates of the United States, the U.S. Territories, and Canada.* Washington, D.C.: U.S. Fish and Wildlife Service Resource Publication 166.

Chapman, J.A., and G.A. Feldhamer, eds. 1982. *Wild mammals of North America: Biology, management and economics.* Baltimore: Johns Hopkins University Press.

Cheeseman, C.L., and R.B. Mitson, eds. 1982. *Telemetric studies of vertebrates.* Zoological Society of London, series 49. New York: Academic Press.

Cockrum, E.L. 1962. *Introduction to mammalogy.* New York: Ronald Press.

De Blasé, A.F., and R.E. Martin. 1981. *A manual of mammology with keys to families of the world.* Dubuque, Iowa: William C. Brown.

Fagen, B. 2000. *The Little Ice Age.* New York: Basic Books.

Grove, J.M. 1988. *The Little Ice Age.* London: Methuen & Co.

Gunderson, H.L. 1976. *Mammalogy.* New York: McGraw-Hill Book Company.

Hall, E.R. 1981. *The mammals of North America.* 2d ed. 2 vols. New York: John Wiley & Sons.

Hamilton, W.J., Jr. 1939. *American mammals: Their lives, habits and economic relations.* New York: McGraw-Hill Book Company.

Ingles, L.G. 1965. *Mammals of the Pacific states: California, Oregon and Washington.* Stanford: Stanford University Press.

Ladurie, L.E. 1971. *Times of feast, times of famine: A history of climate since the year 1000.* London: George Allen and Unwin.

McKenna, M.C., and S.K. Bell. 1997. *Classification of mammals above the species level*. New York: Columbia University.

Matthews, L.H. 1969. *The life of mammals*. 2 vols. London: Weidenfeld & Nicolson.

Murie, O.J. 1954. *A field guide to animal tracks*. Boston: Houghton Mifflin Company.

Palmer, R.S. 1954. *The mammal guide*. Garden City: Doubleday and Co.

Savage, A., and C. Savage. 1981. *Wild mammals of northwest America*. Baltimore: Johns Hopkins University Press.

Stoddard, D.M., ed. 1979. *Ecology of small mammals*. New York: Chapman & Hall.

Van Gelder, R.G. 1982. *Mammals of the national parks*. Baltimore: Johns Hopkins University Press.

Vaughn, T.A., J.M. Ryan, and N.J. Czaplewski. 1999. *Mammalogy*. 4th ed. Philadelphia: W.B. Saunders Co.

Wilson, D.E., and D.M. Reeder. 1992. *Mammal species of the world*. Washington, D.C.: Smithsonian Institution Press.

Wilson, D.E., and F.R. Cole. 2000. *Common names of mammals of the world*. Washington, D.C.: Smithsonian Institution Press.

Reproduction, Social Groups

Bronson, F.H. 1989. *Mammalian reproductive biology*. Chicago: The University of Chicago Press.

Cowie, A.T., I.A. Forsyth, and I.C. Hart. 1980. *Hormonal control of lactation*. Berlin: Springer-Verlag.

Crews, D., ed. 1987. *Psychobiology of reproductive behavior*. Englewood Cliffs, N.J.: Prentice Hall.

Dewsbury, D.A., ed. 1981. *Mammalian sexual behavior: Foundations for contemporary research*. Stroudsburg, Penn.: Hutchinson Ross Publishing Co.

Flowerdew, J.R. 1987. *Mammals: Their reproductive biology and population ecology*. London: Edward Arnold.

Hayssen, V., A. van Tienhoven, and An. van Tienhoven. 1993. *Asdell's patterns of mammalian reproduction*. 2d ed. Ithaca: Cornell University Press.

Jameson, E.W., Jr. 1988. *Vertebrate reproduction*. New York: John Wiley & Sons.

Mead, R.A. 1981. The physiology and evolution of delayed implantation in carnivores. In *Carnivore behavior, ecology and evolution*, ed. J.L. Gittleman, 437–64. Ithaca: Cornell University Press.

Mepham, T.B. 1987. *Physiology of lactation*. Philadelphia: Milton Keynes.

Neville, M.C., and C.W. Daniel. 1987. *The mammary gland*. New York: Plenum Publishing Co.

Sadleir, R.M.F.S. 1969. *The ecology of reproduction in wild and domestic mammals*. London: Methuen & Co., Ltd.

Soloman, N.G., and A.J. French. eds. 1997. *Cooperative breeding in mammals*. Cambridge, U.K.: Cambridge University Press.

Stoddart, D. 1980. *The ecology of vertebrate olfaction*. London : Chapman and Hall.

van Teinhoven, A. 1983. *Reproductive physiology of vertebrates*. 2d ed. Ithaca: Cornell University Press.

Vandenbergh, J.G., ed. 1983. *Pheromones and reproduction in mammals*. New York: Academic Press.

Population Fluctuations, Thermoregulation

Cossins, A.R., and K. Bowler. 1987. *Temperature biology of animals*. London: Chapman and Hall.

Hales, J.R.S. 1984. *Thermal physiology*. New York: Raven Press.

Merritt, J.F. 1984. *Winter ecology of small mammals*. Pittsburgh: Special Publication of the Carnegie Museum of Natural History 10.

Schmidt-Nielsen, K. 1964. *Desert animals: Physiological problems of heat and water*. Oxford, U.K.: Clarendon Press.

Wunder, B.A. 1985. Energetics and thermoregulation. In *Biology of New World* Microtus, Special Publication, ed. R.H. Tamarin, 812–44. Lawrence, Kans.: American Society of Mammalogists.

Seasonal Dormancy

Carey, C., G.L. Florant, B.A. Wunder, and B. Horwitz. 1993. *Life in the cold*. Boulder, Colo.: Westview Press.

Kayser, C. 1961. *The physiology of natural hibernation*. Oxford, U.K.: Pergamon Press.

Lyman, C.P., J.S. Willis, A. Malan, and L.C.H. Wang. 1982. *Hibernation and torpor in mammals and birds*. New York: Academic Press.

Malan, A., and B. Canguilhen, eds. 1989. *Living in the cold*. Montrouge, France: John Libbey Eurotex, Ltd.

Musacchia, X.J., and L. Jansky. 1981. *Survival in the cold: Hibernation and other adaptations*. Amsterdam: Elsevier/North Holland.

Senses

Brown, R.E., and D.W. Macdonald, eds. 1985. *Social odours in mammals*. 2 vols. Oxford, U.K.: Clarendon Press.

Doty, R. L., ed. 1976. *Mammalian olfaction, reproductive processes and behavior.* New York: Academic Press.

Fenton, M. B. 1985. *Communication in the Chiroptera.* Bloomington: Indiana University Press.

Griffin, D. R. 1958. *Listening in the dark.* New Haven: Yale University Press.

Kellogg, W. N. 1961. *Porpoises and sonar.* Chicago: University of Chicago Press.

Lewis, B., ed. 1983. *Bioacoustics: A comparative approach.* London: Academic Press.

Popper, A. N., and R. R. Fay, eds.. 1995. *Hearing by bats.* New York: Springer-Verlag.

Purves, P. R., and G. Pilleri, eds. 1983. *Echolocation in whales and dolphins.* New York: Academic Press.

Sales, G. 1974. *Ultrasonic communication by animals.* London: Chapman and Hall.

Stoddart, D. M. 1980. *The ecology of vertebrate olfaction.* London: Chapman and Hall.

Vandenbergh, J. G., ed. 1983. *Pheromones and reproduction in mammals.* New York: Academic Press.

Walther, F. R. 1983. *Communication and expression in hoofed mammals.* Bloomington: Indiana University Press.

Migration and Movements

Aidley, D. J., ed. 1981. *Animal migration.* Cambridge, U.K.: Cambridge University Press.

Gauthreaux, S. A., Jr. 1980. *Animal migration, orientation and navigation.* New York: Academic Press.

Lidicker, W. Z., and R. L. Caldwell. 1982. *Dispersal and migration.* Stroudsburg, Penn.: Hutchinson Ross Publishing Co.

Lott, D. F. 1991. *Intraspecific variation in the social systems of wild vertebrates.* Cambridge, U.K.: Cambridge University Press.

Schmidt-Koenig, K., and W. T. Keeton. 1978. *Animal migration, navigation and homing.* New York: Springer-Verlag.

Stenseth, N. C., and W. Z. Lidicker, Jr., eds. 1992. *Animal dispersal: Small mammals as a model.* London: Chapman and Hall.

Mammals and California Society

Allen, G. M. 1942. *Extinct and vanishing mammals of the Western Hemisphere, with the marine species of all the oceans.* Special Publication No. 11. Washington, D.C.: American Committee for International Wild Life Protection.

Beran, G.W. and J.H. Steele. 1994. *Handbook of zoonoses*. 2d ed. 2 vols. Boca Raton: CRC Press.

Bruff, J.G. 1949. *The journals, drawings and other papers of J. Goldsborough Bruff, April 2, 1849–July 20, 1851*. Edited by Georgia Willis Read and Ruth Gaines. New York: Columbia University Press.

California Department of Fish and Game. 1980. *At the crossroads*. Sacramento: California Department of Fish and Game.

Craighead, F.C., Jr. 1979. *Track of the grizzly*. San Francisco: Sierra Club Books.

Davis, J.W., L.H. Karstad, and O. Trainer, eds. 1970. *Infectious diseases of wild mammals* Ames: Iowa State University Press.

Grenfell, B., and A. Dobson. 1995. *Ecology of infectious diseases in natural populations*. Cambridge, U.K.: Cambridge University Press.

Klinghammer, E., ed. 1979. *The behavior and ecology of wolves*. New York: Garland STPM Press.

Maloney, A.B., ed. 1945. *Fur brigade to the Bonaventura: John Work's California Expedition 1832–1833 for the Hudson's Bay Company*. San Francisco: California Historical Society.

McCullough, D.R., and R.H. Barrett, eds. 1992. *Wildlife 2001: Populations*. New York: Elsevier Applied Science.

Meyer, K.F. 1955. *The zoonoses in their relation to rural health*. Berkeley: University of California Press.

Rich, E.E. 1967. *The fur trade and the Northwest to 1857*. Toronto: McClelland & Stewart.

Roe, F.G. 1970. *The North American buffalo*. 2d ed. Toronto: The University of Toronto Press.

Rorabacher, J.A. 1970. *The American buffalo in transition: A historical and economic survey of the bison in America*. Saint Cloud, Minn.: North Star Press.

Saum, L.O. 1965. *The fur trader and the Indians*. Seattle: University of Washington Press.

Steinhart, P. 1990. *California's wild heritage*. Sacramento: California Department of Fish and Game.

Storer, T.I., and L.P. Tevis, Jr. 1955. *California grizzly*. Berkeley: University of California Press.

Wishart, D.J. 1979. *The fur trade of the American west, 1807–40: A geographical synthesis*. Lincoln: University of Nebraska Press.

Young, S.P., and E.A. Goldman. 1944. *The wolves of North America.* Washington, D.C.: American Wildlife Institute.

Food of Mammals

Barker, L.M., M. Best, and M. Domjan, eds. 1977. *Learning mechanisms in food selection.* Waco, Tex.: Baylor University Press.

Chivers, D.J., and P. Langer. 1944. *The digestive system in mammals: Food, form and function.* Cambridge, U.K.: Cambridge University Press.

Hudson, R.J., and R.G. White. 1985. *Bioenergetics of wild herbivores.* Boca Raton, Fla.: CRC Press.

Schulkin, J. 1992. *Sodium hunger: The search for a salty taste.* Cambridge, U.K.: Cambridge University Press.

Identification of Mammals

Jones, C., R.S. Hoffmann, D.W. Rice, R.J. Baker, M.D. Engstrom, R.D. Bradley, D.J. Schmidly, and C.A. Jones. 1997. *Revised checklist of North American mammals north of Mexico, 1997.* Lubbock, Tex.: Occasional Papers of the Museum of Texas Technological University, No. 173.

McKenna, M.C., and S.K. Bell. 1997. *Classification of Mammals above the Species Level.* New York: Columbia University Press.

Laudenslayer, W.F., Jr.; and W.E. Grenfell Jr. 1983. A list of amphibians, reptiles, birds and mammals of California. *Outdoor California* (January–February): 1–14.

Wilson, D.E., and D.M. Reeder. 1993. *Mammal species of the world, a taxonomic and geographic reference.* Washington, D.C.: Smithsonian Institution Press.

Marsupialia

Hartman, C.G. 1952. *Possums.* Austin: University of Texas Press.

Hunsaker, D., ed. 1977. *The biology of marsupials.* New York: Academic Press.

Lee, A.K., and A. Cockburn. 1985. *Evolutionary ecology of marsupials.* Cambridge, U.K.: Cambridge University Press.

Stonehouse, B., and M. Gilmore, eds. 1977. *The biology of marsupials.* London: Macmillan Press, Ltd.

Szalay, F.S., M.J. Novacek, and M. McKenna, eds. 1993. *Mammal phylogeny: Mesozoic differentiation, multituberculates, monotremes, early therians and marsupials.* New York: Springer-Verlag.

Tyndale-Biscoe, H., and M. Renfree. 1987. *Reproductive physiology of marsupials*. Cambridge, U.K.: Cambridge University Press.

Insectivora

Carraway, L. 1990. *A morphologic and morphometric analysis of the genus* Sorex vagrans *species complex of the Pacific coast region*. Lubbock, Tex.: The Museum of Texas Tech University, Special Publications, No. 32.

Gorman, M.L., and R.D. Stone. 1990. *The natural history of moles*. Ithaca: Cornell University Press.

Junge, J.A., and R.S. Hoffmann. 1981. *An annotated key to the long-tailed shrews (Genus* Sorex) *of the United States and Canada, with notes on Middle American* Sorex. Occasional Papers of the Museum of Natural History, The University of Kansas, No. 94.

Merritt, J.F. Kirkland, Jr., and R.K. Rose, eds. 1994. *Advances in the biology of shrews*. Pittsburgh: Carnegie Museum of Natural History.

Chiroptera

Altringham, J.D. 1996. *Bats: Biology and behavior*. Oxford, U.K.: Oxford University Press.

Barbour, R.W., and W.H. Davis. 1969. *Bats of America*. Lexington: University Press of Kentucky.

Fenton, M.B. 1985. *Communication in the Chiroptera*. Bloomington: Indiana University Press.

Fenton, M.B., P.A. Racey, and J.M. Raynor, eds. 1987. *Recent advances in the study of bats*. Cambridge, U.K.: Cambridge University Press.

Findley, J.S. 1992. *Bats: A community perspective*. Cambridge, U.K.: Cambridge University Press.

Griffin, D.R. 1958. *Listening in the dark*. New Haven: Yale University Press.

Hill, J.E., and J.D. Smith. 1984. *Bats: A natural history*. London: British Museum (Natural History).

Kunz, T.H., ed. 1982. *Ecology of bats*. 3 vols. New York: Academic Press.

Nactigall, W., ed. 1986. *Bat flight—Fledermausflug*. Biona Report 5. Stuttgart: Gustav Fischer.

Popper, A.N., and R.R. Fay, eds. 1995. *Hearing by bats*. New York: Springer-Verlag.

Racey, P.A., and F.M. Swift, eds. 1995. *Ecology, evolution and behav-*

iour of bats. Symposia of the Zoological Society of London, No. 67. Oxford, U.K.: Oxford University Press.

Ransome, R. 1990. *The natural history of hibernating bats.* London: Christopher Helm.

Schober, W. 1984. *The lives of bats.* London: Croom Helm/Arco.

Slaughter, B. H., and D. W. Walton. 1970. *About bats: A chiropteran biology symposium.* Dallas: Southern Methodist University Press.

Wimsatt, W. A., ed. 1970–1977. *Biology of bats.* 3 vols. New York: Academic Press.

Yalden, D. W., and P. A. Morris. 1975. *The lives of bats.* Newton Abbot: David & Charles.

Carnivora

Andersen, H. T., ed. 1969. *The biology of marine mammals.* New York: Academic Press.

Barnes, C. T. 1960. *The cougar or mountain lion.* Salt Lake City: Ralton Co.

Bekoff, M. 1978. *Coyotes: Biology, behavior and management.* New York: Academic Press.

Buskirk, S. W., A. S. Harestad, M. G. Raphael, and R. A. Powell. 1994. *Martens, sables, and fishers.* Ithaca: Cornell University Press.

Dobie, J. F. 1950. *The voice of the coyote.* Boston: Little, Brown & Co.

Ewer, R. F. 1973. *The carnivores.* Ithaca: Cornell University Press.

Gittleman, J., ed. 1989. *Carnivore behavior, ecology, and evolution.* Ithaca: Cornell University Press.

Gittleman, J., ed. 1996. *Carnivore behavior, ecology, and evolution.* Ithaca: Cornell University Press.

Grinnell, J., J. S. Dixon, and J. M. Linsdale. 1937. *Fur-bearing mammals of California: Their natural history, systematic status and relation to man.* Berkeley: University of California Press.

Harrison, R. J., R. C. Hubbard, R. S. Peterson, C. E. Rice, and R. J. Schusterman. 1968. *The behavior and physiology of pinnipeds.* New York: Appleton, Century, Crofts.

Kenyon, K. W. 1969. *The sea otter in the eastern Pacific Ocean.* North American Fauna 68. Washington, D.C.: U.S. Government Printing Office.

King, C. 1989. *The natural history of weasels and stoats.* Ithaca: Cornell University Press.

King, J. E. 1983. *Seals of the world.* Ithaca: Cornell University Press.

Leydet, F. 1977. *The coyote: Defiant songdog of the west.* San Francisco: Chronicle Books.

Powell, R.A. 1982. *The Fisher: Life history, ecology and behavior.* Minneapolis: University of Minnesota Press.

Prater, S.H. 1971. *The book of Indian animals.* Oxford, U.K. Oxford University Press.

Rice, D.W. 1998. *Marine mammals of the world: Systematics and distribution.* Lawrence: Allen Press.

Ridgway, S.H., and R.J. Harrison. 1981. *Handbook of marine mammals.* 2 vols. London: Academic Press.

Scheffer, V.B. 1958. *Seals, sea lions and walruses.* Stanford: Stanford University Press.

Verts, B.J. 1967. *The biology of the striped skunk.* Urbana: University of Illinois Press.

Young, S.P., and E.A. Goldman. 1946. *The puma: Mysterious American cat.* New York: Dover Publications Inc.

Cetacea

Boyd, I.L., ed. 1993. *Marine mammals: Advances in behavioural and population biology.* Zoological Society of London Symposium 66. Oxford, U.K.: Oxford Science Publications, Clarendon Press.

Coffee, D.J. 1977. *Dolphins, whales and porpoises: An encyclopedia of sea mammals.* New York: Macmillan.

Gaskin, D.E. 1982. *The ecology of whales and dolphins.* Exeter, N.H.: Heinemann Educational Books, Inc.

Jones, M.L., S.L. Swartz, S. Leatherwood, and P.A. Folkens. 1984. *The gray whale.* New York: Academic Press.

Klinowska, M. 1991. *Dolphins, porpoises and whales of the world.* The IUCN Red Book Data. Gland, Switzerland: The World Conservation Union.

Leatherwood, S., and R.R. Reeves. 1983. *The Sierra Club handbook of whales and dolphins.* San Francisco: Sierra Club Books.

Leatherwood, S., R.R. Reeves, W.F. Perrin, and W.E. Evans. 1982. *Whales, dolphins and porpoises of the eastern North Pacific and adjacent arctic waters: A guide to their identification.* Washington, D.C.: U.S. Department of Commerce.

Matthews, L.H. 1975. *The whale.* New York: Crescent Books.

Matthews, L.H. 1978. *The natural history of the whale.* London: Weidenfeld & Nicolson.

Mörzer Bruyns, W.F.J. 1972. *Fieldguide of whales and dolphins.* Amsterdam: Tor.

Payne, R. 1983. *Communication and behavior of whales.* AAAS Selected Symposium 76. Lawrence: Allen Press.

Rice, D.W. 1977. *A list of marine mammals of the world.* 3d ed. National Oceanic and Atmospheric Administration Technical Report NMFS SSRF-711. Washington, D.C.: U.S. Government Printing Office.

Rice, D.W. 1998. *Marine mammals of the world: Systematics and distribution.* Society for Marine Mammalogy, Special Publication Number 4. Lawrence: Allen Press.

Slijper, E.J. 1976. *Whales.* London: Hutchinson & Company.

Slijper, E.J. 1976. *Whales and dolphins.* Ann Arbor: University of Michigan Press.

Thewissen, J.G.M., ed. *The emergence of whales: Evolutionary patterns in the origin of Cetacea.* New York: Plenum Press.

Thomas, J.A., R.A. Kastelein, and A.Y. Supin. 1992. *Marine mammal sensory systems.* New York: Plenum Press.

Watson, I. 1981. *Sea guide to whales of the world.* New York: E.P. Dutton.

Winn, L.K., and H.E. Winn. 1985. *Wings in the sea: The humpback whale.* Hanover, N.H.: University Press of New England.

Perissodactyla

Berger, J. 1986. *Wild horses of the Great Basin: Social competition and population size.* Chicago: University of Chicago Press.

Groves, C.P. 1974. *Horses, asses, and zebras in the wild.* Hollywood, Fla.: Ralph Curtis Books.

MacFadden, B.J. 1992. *Fossil horses: Systematics and evolution of the family Equidae.* Cambridge, U.K.: Cambridge University Press.

Prothero, D.R., and R.M. Schoch, eds. 1989. *The evolution of perissodactyls.* Oxford Monographs on Geology and Geophysics, No. 15. Oxford, U.K.: Oxford University Press.

Artiodactyla

Barrett, R.H. 1978. The feral hog on the Dye Creek Ranch, California. *Hilgardia* 46: 283–355.

Bubenik, G.A., and A.B. Bubenik. 1990. *Horns, pronghorns and antlers.* New York: Springer-Verlag.

Caton, J.D. 1977. *The deer and antelope of America.* Boston: Hurd & Houghton.

Chapman, D., and N. Chapman. 1975. *Fallow deer: Their history, distribution and biology.* Lavenham, Suffolk: Terence Dalton, Ltd.

Geist, V. 1971. *Mountain sheep: A study in behavior and evolution.* Chicago: University of Chicago Press.

Goss, R.J. 1983. *Deer antlers: Regeneration, function and evolution.* New York: Academic Press.

Jones, F.L. 1950. A survey of the Sierra Nevada Bighorn. *Sierra Club Bulletin* (June): 29–76.

McQuivey, R.P. 1978. *The desert Bighorn Sheep of Nevada.* Biological Bulletin No. 6. Reno: Nevada Department of Wildlife.

Monson, G., and L. Sumner, eds. 1980. *The desert Bighorn.* Tucson: University of Arizona Press.

Murie, O.J. 1951. *The elk of North America.* Washington, D.C.: Wildlife Management Institute.

Putman, R. 1988. *The natural history of deer.* Ithaca: Cornell University Press.

Schaller, G.R. 1967. *The deer and the tiger: A study of wildlife in India.* Chicago: University of Chicago Press.

Taylor, W.P., ed. 1965. *The deer of North America.* Harrisburg: Stackpole Books; Washington, D.C.: Wildlife Management Institute.

Thomas, J.W., and D.E. Toweill, eds. 1982. *Elk of North America: Ecology and management.* Harrisburg: Stackpole Books.

Wallmo, O.C. (Ed.) 1981. *Mule and Black-tail Deer of North America.* Lincoln: University of Nebraska Press.

Yoakum, J. 1967. *Literature of the American Pronghorn Antelope.* Reno: U.S. Department of the Interior, Bureau of Land Management.

Rodentia

Genoways, H.H., and J.H. Brown, eds. 1993. *Biology of the Heteromyidae.* Lawrence, Kans.: American Society of Mammalogists.

Graur, D., W.A. Hide, and W.-H. Li. 1991. Is the guinea pig a rodent? *Nature* 351: 649–651.

Hall, E.R. 1981. *The mammals of North America.* 2d ed. 2 vols. New York: John Wiley & Sons.

Jameson, Everett W. Jr. 1999. Host-ectoparasite relationships among North American chipmunks. *Acta Theriologica* 44 (3): 225–231.

King, J.A. 1968. *Biology of* Peromyscus *(Rodentia).* Lawrence, Kans.: American Society of Mammalogists.

Korth, W.W. 1994. *The Tertiary record of rodents of North America.* New York: Plenum Press.

Larson, E.A. 1981. *Merriam's Chipmunk on Palo Escrito.* Big Pine, Calif.: Wacoba Press.

Linsdale, J.M. 1946. *The California Ground Squirrel: A record of observations made on the Hastings Natural History Reservation.* Berkeley: University of California Press.

Linsdale, J.M., and L.P. Tevis Jr. 1951. *The Dusky-footed Wood Rat: A* record of observations made on the Hastings Natural History Reservation. Berkeley: University of California Press.

Luckett, W. P., and J. L. Hartenberger, eds. 1985. *Evolutionary relationships among rodents: A multidisciplinary analysis.* New York: Plenum Press.

MacClintock, D. 1970. *Squirrels of North America.* New York: Van Nostrand Reinhold Co.

McCabe, T. T., and B. D. Blanchard. 1950. *Three species of* Peromyscus. Santa Barbara: Rood Associates.

Prakash, I., and P. K. Ghosh, eds. 1975. *Rodents in desert environments.* The Hague: Dr. W. Junk b.v. Publishers.

Rue, L. L., III. 1964. *The world of the Beaver.* New York: J. B. Lippincott Co.

Steele, M. A., J. F. Merritt, and D. A. Zegers. 1995. *Ecology and evolutionary biology of tree squirrels.* Martinsville, Va.: Virginia Museum of Natural History.

Tamarin, R. H., ed. 1985. *Biology of the New World.* Microtus. Lawrence, Kans.: American Society of Mammalogists.

Warren, E. R. 1927. *The Beaver.* Monographs of the American Society of Mammalogists, No. 2. Baltimore: Williams & Wilkins Co.

Lagomorpha

Myers, K., and MacInnes, C. D., eds. 1981. *Proceedings of the World Lagomorph Conference (1979).* Guelph, Ontario: University of Guelph.

Orr, R. T. 1940. *The rabbits of California.* Occasional Papers of the California Academy of Sciences, No. 19. San Francisco: California Academy of Sciences.

INDEX

Note: page numbers in **bold** refer to main discussion of the species; "pl." refers to plates.

Series Design:	Barbara Jellow
Design Enhancements:	Beth Hansen
Design Development:	Jane Tenenbaum
Cartographer:	Bill Nelson
Composition:	Impressions Book and Journal Services, Inc.
Text:	9/10.5 Minion
Display:	ITC Franklin Gothic Book and Demi

ABOUT THE AUTHORS

E.W. Jameson, Jr., is professor emeritus of zoology at the University of California, Davis, and the author of *Vertebrate Reproduction* (1988) and *Patterns of Vertebrate Biology* (1981).

Hans J. Peeters is professor emeritus of biology and zoology at Chabot College. His illustrations have appeared in *Birds of North America* (1997) as well as in several other bird guides.